Mountains and Plains

DENNIS H. KNIGHT

Mountains

and Plains

The Ecology of Wyoming Landscapes

YALE UNIVERSITY PRESS
NEW HAVEN AND LONDON

Published with assistance from the Louis Stern
Memorial Fund.

Designed by Nancy Ovedovitz and set in Minion type
by The Composing Room of Michigan, Inc. Printed
in the United States of America by Thomson-Shore,
Dexter, Michigan.

Library of Congress Cataloging-in-Publication

Knight, Dennis H.
    Mountains and plains : the ecology of Wyoming
landscapes / Dennis H. Knight.
        p.   cm.
    Includes bibliographical references (p.   ) and
index.
    ISBN 0-300-05545-5
    1. Ecology—Wyoming.   2. Landscape ecology—
Wyoming.   I. Title.
QH105.W8k58   1994
574.5′264′09787—dc20   93-37673
                                          CIP

A catalogue record for this book is available from the
British Library.

The paper in this book meets the guidelines for
permanence and durability of the Committee on
Production Guidelines for Book Longevity of the
Council on Library Resources.

10 9 8 7 6 5 4 3 2 1

For Judy, Christy, and Charley

Nature is an open book. . . . Each grass-covered hillside is a page on which is written the history of the past, conditions of the present, and predictions of the future. . . . Let us look closely and understandingly, and act wisely, and in time bring our methods of land use and conservation activities into close harmony with the dictates of nature.

—*John E. Weaver, 1954*

Contents

Preface   ix

Acknowledgments   x

I / Wyoming and the Rocky Mountain West

1   Introduction   3

2   Landscape History   10

3   Modern Environments   23

II / Along Creeks and Rivers

4   Riparian Landscapes   43

III / Plains and Intermountain Basins

5   Grasslands   67

6   Sagebrush Steppe   90

7   Desert Shrublands and Playas   108

8   Sand Dunes, Badlands, Mud Volcanoes,
    and Mima Mounds   120

IV / Foothills and Mountains

9   Escarpments and the Foothill
    Transition   133

10 Mountain Forests   153

11 The Forest Ecosystem   174

12 Mountain Meadows and Snowglades   193

13 Upper Treeline and Alpine Tundra   201

V / Landscapes of Special Interest

14 The Yellowstone Plateau   215

15 Jackson Hole and the Tetons   233

16 The Black Hills, Bear Lodge Mountains,
    and Devil's Tower   242

VI / Sustainable Land Management

17 Using Wyoming Landscapes   257

Epilogue   265

Appendix A: Latin Names for Plants
Referred to in the Text by Common Name   267

Appendix B: Latin Names for Birds,
Mammals, Reptiles, Amphibians, and
Invertebrates Referred to in the Text by
Common Name   273

Notes   277

Glossary   283

References   291

Index   333

Preface

In this book I have tried to summarize the abundant information now available on the ecology of the Rocky Mountains, intermountain basins, and western Great Plains in Wyoming and adjacent parts of Colorado, Utah, Idaho, Montana, South Dakota, and Nebraska. This information is essential for the proper management of grasslands, shrublands, woodlands, forests, and meadows. As important, it can add enjoyment to living and traveling in the region.

The book has six parts. After an introduction to the historical development of the diverse environments found in the region and a section on the riparian landscapes along streams and rivers, there are sections on plains and basins, foothills and mountains, and three landscapes of special interest—the Yellowstone Plateau, Jackson Hole and Grand Teton National Park, and the Black Hills. The book ends with a discussion of sustainable land management. The structure and depth of each chapter varies according to the information available and my perception of what most readers will find interesting.

Plant ecology is emphasized because vegetation, more than any other biotic feature, gives character to the landscape. Plant-animal interactions, nutrient cycling, current issues in land management, and the ecology of disturbances are discussed as well. Time and space have not permitted a synthesis of the literature on wetlands or the management of croplands and livestock, but there are discussions of weed ecology, irrigation, and the influences of herbivores. The bibliography should be helpful to readers desiring information on topics that I have not addressed.

Recognizing the diverse group of people interested in western landscapes, I have used scientific terms sparingly. Most of the text should be easy to understand for individuals curious about ecology and natural resource management. A glossary and detailed index are included. Common names for plants and animals are used throughout the text, but for those who need them, the Latin names are listed in appendixes A and B.

Acknowledgments

I have benefited greatly from the encouragement, knowledge, and talents of many friends and colleagues during the development of this book. William A. Reiners was especially encouraging and helped to arrange a sabbatical leave when I needed time to work in the libraries of neighboring universities. Harry Harju, Collin Fallat, and Martin Raphael helped in obtaining travel grants from the Wyoming Game and Fish Department, the Wyoming Department of Agriculture, and the U.S. Forest Service. Other agencies contributed indirectly through grants that have enabled me and my students to conduct research on the ecology of Wyoming landscapes over the past twenty-seven years. Our primary supporters have been the University of Wyoming–National Park Service Research Center, the Wyoming Water Resources Research Center, the U.S. Forest Service, the National Science Foundation, the Bureau of Reclamation, and the U.S. Department of Energy. As important, we have benefited from the resources and stimulating educational environment provided by the people of Wyoming through their university.

I am especially indebted to seven individuals who spent an extraordinary amount of time in reviewing the early drafts of this book. William Reiners and George Jones read every chapter, offering many helpful suggestions, and Kendall Johnson, Larry Munn, Jason Lillegraven, Richard Reider, and J. D. Love helped on several occasions with sections that I was hardly qualified to write.

Their comments led to significant improvements for which I am very grateful.

Many others helped as well in providing information or in reviewing one or several chapters. In keeping with the multidisciplinary emphasis of the book, they represent various professions, including land management, geology, geography, history, and economics as well as botany, zoology, and ecology. These individuals, listed by the section to which they contributed, are: Landscape History and Modern Environments: Jane Beiswenger, Donald L. Blackstone, Jr., Martha Christensen, Robert Dorn, Samuel J. Hundley, T. A. Larson, Barry Lawrence, Billie Lundberg, Richard Marston, Brainerd Mears, Jr., Glen Mitchell, Patty Moe, Brett Moline, Charles Nations, Michael Patrick, Richard Reider, Paige Smith, William K. Smith, Patrick Thrasher, and Danny Walker; Riparian Landscapes: William L. Baker, Thomas Ball, Donald J. Brosz, James Cagney, Robert Dorn, Steven Kiracofe, Richard Marston, Quentin Skinner, Bruce H. Smith, Kenneth Stinson, and William Wilson; Grasslands, Shrublands, and Foothills: Thomas Ball, Alan Beetle, Don Despain, James Detling, Jerrold Dodd, Ken Driese, S. G. Froiland, Harry Harju, Kimball T. Harper, Steven Kiracofe, William A. Laycock, Jeffrey A. Lockwood, Glen Mitchell, Richard Reider, William Romme, Leonard Ruggiero, Nancy Stanton, Kenneth Stinson, Dan Uresk, Thomas Veblen, Neil West, and William Wilson; Mountain Landscapes: William L. Baker, Dwight Billings, Don Despain,

Kathleen Doyle, Timothy J. Fahey, W. F. J. Parsons, Richard Reider, William Romme, Leonard Ruggiero, William K. Smith, Dan Uresk, Thomas Veblen, Tad Weaver, and Joseph B. Yavitt; Yellowstone, Jackson Hole, and the Black Hills: Ronald Beiswenger, Mark Boyce, Steven Cain, Don Despain, Kathleen Doyle, S. G. Froiland, Harry Harju, J. D. Love, Jim Peaco, William Romme, Paul Schullery, Terri Schulz, and Dan Uresk; Land Management: Martha Christensen, Linda Joyce, Judy Knight, and Charles Nations. In addition, Robert Dorn, Ronald Hartman, and E. B. Nelson have helped greatly over the years with plant identifications and proper nomenclature.

In addition to the reviewers, I have benefited from the talents of Linda Marston in drafting the maps, Allory Deis and Thomas Lund in drawing most of the figures, and Marilee Doyle in printing most of the photographs. The computer-generated image of Wyoming was produced by Lawrence M. Ostresh, and several people provided or helped to locate photographs that I did not have in my collection. Jean Thomson Black and Lorraine Alexson at Yale University Press provided encouragement and many helpful suggestions. It was a pleasure to work with them. For attending to many other details, I thank Michele Barlow, Rebecca Christensen, Marilee Doyle, Chris Garber, Jimmie Joe Honaker, Jane Struttman, and Ramona Wilson.

Finally, I gratefully acknowledge the support and patience of my wife, Judy, and our two children, Christy and Charley. They helped in many ways, and I was fortunate to enjoy their company on some of my trips around Wyoming—a place with high mountains and extensive plains that are interesting, beautiful, and challenging.

I

Wyoming and the Rocky Mountain West

C H A P T E R 1

Introduction

Straddling the Continental Divide and with a mean elevation of 2,030 m (6,700 ft), the landscapes of Wyoming are today very similar to those of presettlement times. Wildlife is abundant, and most of the area is still dominated by native vegetation. From the lowest point (939 m, 3,100 ft), where the Belle Fourche River flows across the border into South Dakota, to the summit of Gannett Peak in the Wind River Mountains (4,207 m, 13,804 ft), there is a rich diversity of plant and animal life that has long captured the interest of scientists, tourists, outdoor enthusiasts, and others—all of whom share an appreciation for the rigors and beauty of the western Great Plains and the Rocky Mountains.

Change in western landscapes has occurred slowly for various reasons. Early explorers concluded that certain areas should be preserved in their natural condition because of their unusual characteristics. This philosophy set the stage for establishing Yellowstone National Park in 1872, the first national park in the world. The first national forest—the Shoshone—was established nearby in 1891, and Devil's Tower became the nation's first national monument in 1906. Grand Teton National Park was established in 1929, and fifteen wilderness areas have been set aside by Congress within Wyoming. National parks and wilderness areas account for about 9 percent of the state's land area (figs. 1.1–1.5).

Another reason for the persistence of natural landscapes is the climate. Growing seasons are too short and cool for crops over most of the area above 7,000 ft—an area encompassing 37 percent of Wyoming. Water availability is a problem at lower elevations, but only about 3 percent of the state is irrigated. Many fields cultivated in the late 1800s were abandoned because of low production, and now, through the natural process of succession, they are often difficult to distinguish from the adjacent unplowed rangeland. Some of these lands, originally allocated to individuals through such laws as the Homestead Act of 1862, were subsequently purchased by the federal government in the late 1930s and eventually became known as National Grasslands (for example, the Thunder Basin National Grassland in northeastern Wyoming and the Pawnee National Grassland in northern Colorado).

Despite climatic limitations, agriculture is still an important industry, with hay (including alfalfa) occupying about two-thirds of the cropland. Other major crops include sugar beets, barley, wheat, corn, and beans. Most cultivation is concentrated in the Bighorn Basin, Green River Basin, Wind River Basin, Powder River Basin, and in parts of the western Great Plains where a combination of low elevation, good soil, and water availability provide conditions adequate for competitive agriculture. Still, less than 5 percent of the land in Wyoming is under cultivation (Wyoming Agricultural Statistics 1991). In contrast, livestock grazing is widespread over the unplowed rangelands, and during some years the production

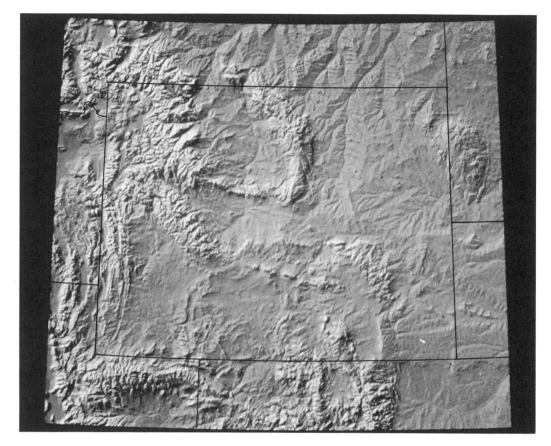

F I G. 1.1   A computer-generated image showing the topography of Wyoming and neighboring parts of Montana, South Dakota, Nebraska, Colorado, Utah, and Idaho. See fig. 1.2 for the names of mountains and basins, and fig. 4.1 for the names of major rivers. The total area of Wyoming, including Yellowstone National Park, is 253,325 km^2 (97,809 mi^2). The state is located between 104°3' and 111°3' west longitude and 41° and 45° north latitude. Courtesy of Lawrence M. Ostresh.

value of meat and wool is double the value of crops (Moline et al. 1991). Timber harvesting in mountain forests is a much smaller industry, contributing only about 10 percent the value of livestock and crops combined in the late 1980s (B. Moline, pers. comm.).

The various land uses in Wyoming can be ranked according to their contribution to the gross state product (GSP). From 1984 to 1988, about 25 percent of the GSP was generated from mining, compared to about 3 percent from agriculture (Wyoming Data Handbook 1991). Economists emphasize that the GSP is not a measure of total economic impact and that agriculture is more important than its contribution to the GSP

would suggest, but more precise numbers on total economic impact are not available. Everyone seems to agree that mining is the most important industry economically and that agriculture and tourism are about equal—though ranked a distant second and third.

In order of economic return, Wyoming's leading industries are oil, natural gas, coal, trona, and bentonite. The extraction of these resources can be disruptive to landscapes, especially in some counties. Yet for Wyoming as a whole, less than 0.2 percent has been mined thus far and land reclamation has been quite effective in some areas. Paved roads cover about two times more land than do mines.

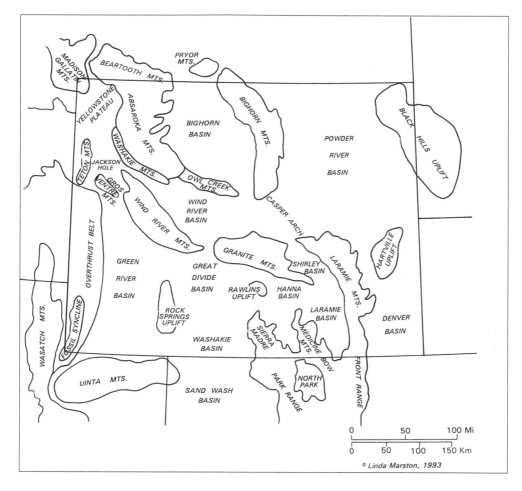

F I G. 1.2   Mountains and basins illustrated in fig. 1.1 (adapted from Blackstone 1988). See Love and Christensen (1985) for a detailed geology map of Wyoming.

Significant changes have occurred in some areas, especially along rivers and where mining and agriculture have been feasible. Wyoming and adjacent states, however, still provide many opportunities for learning about the natural history of presettlement times. Bison have been replaced by cattle, sheep, and horses in most areas, but pronghorn antelope, elk, and deer still abound. Some rangelands have been grazed heavily by livestock, but bison may have done the same. Grizzly bears, elk, and wolves no longer wander across the basins as they once did, but, except for wolves, they are still found in some mountain ranges.[1] Mountain forests have been fragmented by timber harvesting, but extensive tracts of virgin forest remain.

ECOLOGY

Most ecologists would argue that it is important to protect air and water quality, preserve biological diversity, and practice sustainable land management. Yet advocacy for environmental protection is not ecology per se. Strictly speaking, ecology is a science dedicated to the understanding of interactions between plants, animals (including humans), microorganisms, and their environment. The interactions are studied to be understood, not to be judged as good or bad. Ecologists strive to be objective through systematic measurements and experimentation. Many also work to ensure that decision-makers are provided with the best ecological information available.

Like other sciences, the goal of modern ecological research is to understand patterns in nature sufficiently well to make predictions about the effects of environmental change on organisms. Such changes can occur through time or across landscapes. Some ecologists work at a theoretical level, searching for generalizations that apply everywhere, while others tend to be more problem-oriented and address interactions associated with, for example, the management of specific forests and rangelands.

The subject matter of modern ecology can be divided into six interrelated subdisciplines: physiological ecology, population ecology, community ecology, ecosystem ecology, landscape ecology, and global ecology. Physiological ecologists attempt to understand how living organisms are adapted to their environment, while population ecologists draw on such information to understand better why the population sizes of organisms fluctuate through time and vary from place to place. Community ecologists emphasize the in-

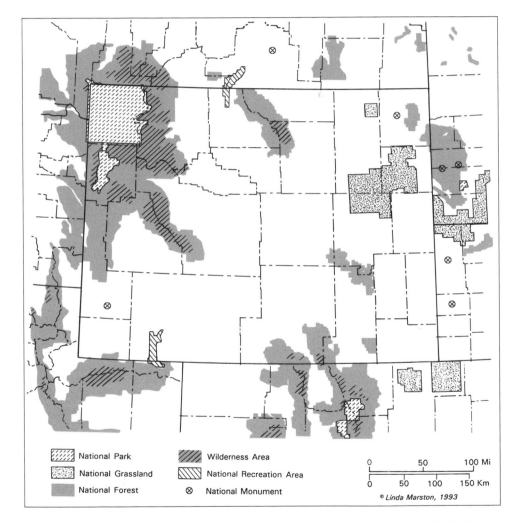

F I G. 1.3   Land jurisdiction by the U.S. Forest Service (national forests and national grasslands) and the National Park Service (parks, monuments, and recreation areas). The white area is mostly private in the eastern third of the region, but the Bureau of Land Management administers much of the land at lower elevations in the western two-thirds. Forty-seven percent of Wyoming is federal land; approximately 9 percent has been designated by Congress as wilderness and national parks.

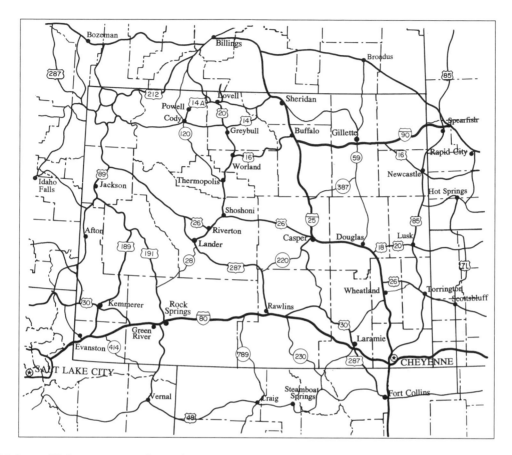

F I G. 1.4   Highways, towns, and counties. Wyoming is approximately 563 km (350 mi) from east to west and 450 km (280 mi) from north to south.

teractions between coexisting species, and ecosystem ecologists study how energy, water, and nutrients move between the soil, atmosphere, living organisms, and detritus (fig. 1.6). As the populations of a community or ecosystem change, other ecosystem attributes change as well (for example, rates of photosynthesis, herbivory, and nutrient cycling).

The term *landscape* is often used interchangeably with *ecosystem*. For example, the Yellowstone landscapes made famous by the artist Thomas Moran are part of an area now commonly referred to as the Greater Yellowstone Ecosystem. Landscape ecologists working at the scale of human activity typically focus on such large heterogeneous areas and are curious about how the landscape mosaic has changed in recent decades or centuries, and how it might change in the future. Another primary goal is to determine both the causes and effects of different landscape patterns.[2] Aerial photographs, satellite imagery, and maps are especially important tools. At a much larger scale, global ecologists focus on continents or even the whole earth, addressing such topics as the effects of human activities on the concentration of carbon dioxide in the atmosphere.

The exact boundaries of an ecosystem or landscape are established according to the objectives of a study or discussion. The area included may be as small as a pond or as large as several counties or states, but in all cases, ecologists recognize that energy, water, nutrients, and organisms move back and forth across whatever boundaries are established. For this reason, ecosystems are viewed as "open systems." Consequently, they are not easily defined. Ecosystem research focuses on inter-

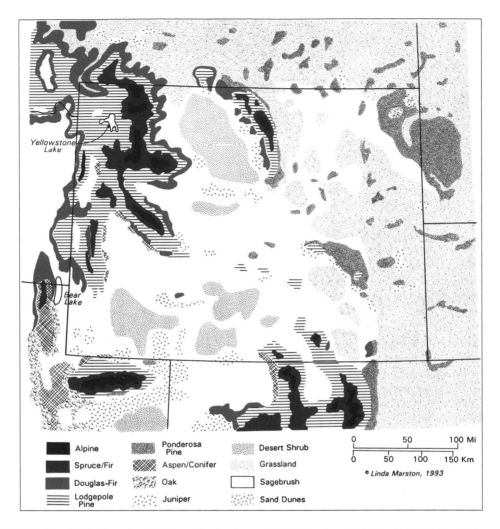

FIG. 1.5   The distribution of major types of natural vegetation. Because of map scale, two vegetation types are not shown, namely, the foothill grasslands and shrublands that fringe the mountains, and the riparian woodlands, shrublands, and meadows that occur along rivers and streams (see fig. 4.1). For a map of cultivated land, see Fallat et al. (1987). Vegetation boundaries are based partially on maps by Choate 1963; Küchler 1966; Garrison et al. 1977; Fallat et al. 1987; Despain 1973; and the Greater Yellowstone Coordinating Committee 1987.

acting species or on groups of species having similar functions—in particular, the "producers" (or green plants), herbivores, carnivores, omnivores, detritivores, and decomposers. Of interest to many ecologists is how the abundance of different species and the functions they perform are affected by varying environmental conditions and how they change with time following disturbances (secondary succession). Some ecologists specialize in botany, while others concentrate on bacteria, fungi, birds, insects, and other groups of organisms.

Ecology is a diverse discipline that fosters an improved understanding of organisms, ecosystems, landscapes, and the biosphere as well as natural resource management. At each level of organization, the systems are complex, with components that often are not yet well known and that

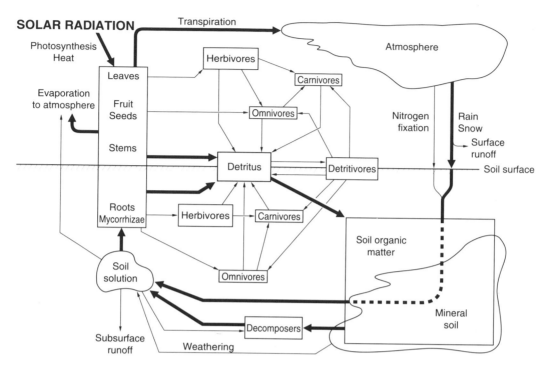

F I G. 1.6   A diagram illustrating the major components (boxes) and interactions (arrows) of a terrestrial ecosystem. Arrow width indicates the amount of energy or water moving along a particular pathway. Temperature, growing season length, water and nutrient availability, and other factors determine the rates of transfer that are possible in specific ecosystems. The irregular shapes indicate sources of water and nutrients. Complex food webs exist above and below the soil surface, with detritus providing an important link between both. Simple diagrams such as this one do not convey the complexity caused by the diverse group of organisms represented by each box. Changes in one component lead directly or indirectly to changes in others.

number in the thousands or millions. The energy and nutrient flow diagrams that ecologists draw for ecosystems or landscapes sometimes appear simple compared to those drawn by engineers for computers and other electronic devices, but there is a significant difference: natural systems were not constructed by humans, and much remains to be learned about the components and processes required for their long-term maintenance.

CHAPTER 2

Landscape History

Millions of Years Ago

The exposed bedrock of canyons, escarpments, and mountains has provided the data that geologists need to interpret the early development of Rocky Mountain landscapes (Love 1960, 1989; Knight 1974, 1990; Blackstone 1988; Snoke et al. 1993). They have learned that throughout the Paleozoic era (fig. 2.1) the area now known as Wyoming was subjected to mountain building, volcanism, erosion, uplifting, subsidence, and periodic inundations by sea water. Igneous rocks were covered by sedimentary strata, uplifted, and then exposed again by erosion.

By the middle of the Paleozoic era, about 350 million years ago, Wyoming was covered by salt water. The climate was tropical and marine life flourished. As time passed, the seas receded and advanced many times, depositing sand in some places and finer material elsewhere. Thick beds of limestone were formed during one widespread submergence—beds that now are known throughout the region as Madison limestone. About 300 million years ago, the shifting seas created enormous swamps, lagoons, and tidal flats that were frequented by dragonflies with 75-cm wingspans. Primitive vascular plants were abundant, including tree-sized clubmosses, horsetails, and ferns in the swamps, and pine, spruce, and fir on the uplands (Dott and Batten 1976). More than 280 million years passed before bison, antelope, and other familiar mammals wandered across the plains.[1] The Ancestral Rocky Mountains developed in Colorado and southeastern Wyoming about 300 million years ago, but in less than 100 million years they had been leveled by erosion and most of Wyoming was again at or below sea level—and still near the equator (fig. 2.2).

THE MESOZOIC ERA

Various episodes of volcanism occurred during the Mesozoic era (66–245 million years ago), leaving a deposit of fine ash that later was altered to form bentonite, which is now ecologically and economically important in the region. The colorful "redbeds" of the Chugwater and Spearfish formations, caused by iron oxides, were also formed at this time. The landscape remained near sea level and had low relief until near the end of the era (the Late Cretaceous period) when low-angle faulting in the earth's crust near the Wyoming-Idaho border created the Hoback, Wyoming, Salt River, and Sublette mountain ranges (Blackstone 1988; Lillegraven and Ostresh 1988). Known collectively as the Overthrust Belt and composed of sedimentary rock, the area is now an important source of oil and natural gas.

The landscapes of the Mesozoic era would have been especially exciting for biologists. By then, dinosaurs had become abundant in the still-tropical climate. Forty genera are now recognized in the fossil record from Wyoming (B. Breithaupt, pers. comm.). Some dinosaurs were large herbivores, suggesting that the vegetation must have been plentiful. Carnivorous species preyed on the

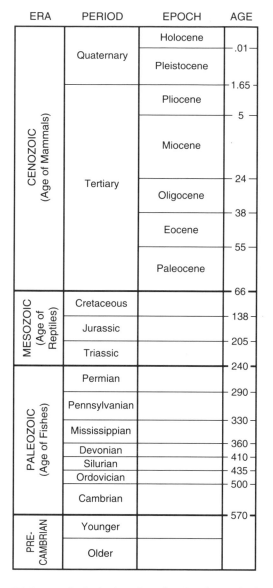

ERA	PERIOD	EPOCH	AGE
CENOZOIC (Age of Mammals)	Quaternary	Holocene	.01
		Pleistocene	1.65
	Tertiary	Pliocene	5
		Miocene	24
		Oligocene	38
		Eocene	55
		Paleocene	66
MESOZOIC (Age of Reptiles)	Cretaceous		138
	Jurassic		205
	Triassic		240
PALEOZOIC (Age of Fishes)	Permian		290
	Pennsylvanian		330
	Mississippian		360
	Devonian		410
	Silurian		435
	Ordovician		500
	Cambrian		570
PRE-CAMBRIAN	Younger		
	Older		

F I G. 2.1 Geologic time chart showing the age (millions of years ago) of different eras, periods, and epochs. Adapted from Lageson and Spearing 1988.

herbivores. Large and small, these reptiles were an important feature of the landscape for more than 100 million years. Curiously, the dinosaur fossils of the Rocky Mountain region are similar to those in Africa (Dott and Batten 1976)—a result of North America and Africa having been connected when dinosaurs were the dominant group of animals (see fig. 2.2).

Also during the Mesozoic era, flowering trees such as magnolia, palm, fig, breadfruit, sassafras, cinnamon, sweetgum, and willow became more common and coexisted in the wetlands with large ferns, horsetails, and clubmosses (Tidwell 1975). Wyoming coal was formed from the remains of these and other species during the Cretaceous and Tertiary periods (mostly about 90 to 55 million years ago). On the uplands, forests dominated by conifers, ginkgo, and cycads were common (Tidwell 1975).

THE CENOZOIC ERA

By about 66 million years ago, at the end of the Mesozoic era and at the beginning of the Cenozoic, dinosaurs had disappeared, and a rich diversity of mammals and birds had developed. The Rocky Mountains began to form during the Laramide Orogeny,[2] and the continents continued to drift apart after the breakup of Pangaea (in the Jurassic). All but the midcontinental Cannonball Sea had disappeared from the North American landmass (Lillegraven et al. 1979; Lillegraven and Ostresh 1988). Flowering plants (angiosperms) became more common, with broad-leaved trees joining the needle-leaf conifers as the dominant forest plants (Dorf 1942, 1964; Axelrod 1968; Leopold and MacGinitie 1972; MacGinitie et al. 1974; Wolfe 1978; Wing 1981). Still, Wyoming's climate was tropical; fossils of crocodiles and tar-

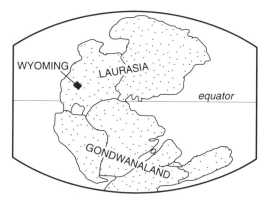

F I G. 2.2 Approximate location near the equator, about 250 million years ago, of the land area now known as Wyoming. The regions labeled Laurasia and Gondwanaland constituted the supercontinent Pangaea. Adapted from Lageson and Spearing 1988.

sirlike mammals date back to this time (Stuckey 1990; Beard et al. 1991).

Some of the trees and shrubs found in the region during the Tertiary period (2–66 million years ago) are still found in temperate climates today (for example, alder, beech, birch, black locust, chestnut, cottonwood, cypress, dogwood, elm, fir, ginkgo, hickory, maple, oak, pine, redwood, spruce, sweet gum, sycamore, walnut, and willow). Others, such as palms and breadfruit, are now found in subtropical regions (Brown 1962; Tidwell 1975). Some of the trees were huge, as indicated by the fossil logs in the Shirley Basin south of Casper that reach 25 m in length and 1.5 m in diameter (Riedl 1959). These fossil logs have not yet been identified, but they are probably the same species as those found in Yellowstone National Park (chestnut, dogwood, magnolia, maple, oak, redwood, sycamore, walnut, and others; Dorf 1964).

One of the most detailed studies of Cenozoic botany in Wyoming was done by Wing (1981). He examined plant fossils in the early Eocene Willwood formation of the Bighorn Basin, concluding that 38–54 million years ago the area had a subtropical climate with a dry season and that the characteristic vegetation was an evergreen broadleaved forest with some conifers. Coniferous forests must have been characteristic of the higher mountains, but snowfall at that time was probably a rare event even there. Many of the fossils suggest species that are common in the region today, but others are extinct or known only from Southeast Asia (for example, ginkgo and dawn redwood). Wing emphasized that the various species undoubtedly were segregated into different communities along gradients of temperature, water availability, and salinity. He estimated that the annual precipitation in the basin was 150–250 cm, much higher than the 10–60 cm that falls in the same area today. Greasewood was also present, suggesting that localized saline environments existed then as they do now.

The first mountains to form in Wyoming, as we know them today, were the Wind River, Gros Ventre, and Medicine Bow ranges (see fig. 1.2). Somewhat later the Bighorns, Black Hills, and Uintas were formed (Lillegraven and Ostresh 1988). Initially these mountains were covered with a mantle of sedimentary rocks. As weathering and erosion occurred, mountain peaks were lowered and basin floors raised. Basin filling was hastened by volcanic activity in northwestern Wyoming and westward that produced colossal volumes of ash in the late Eocene and Oligocene epochs (about 50–36 million years ago; Lillegraven and Ostresh 1988). Trees were buried and fossilized, and continued erosion, sedimentation, and ash deposition led to the eventual burial of most mountains.

Through these processes, topographic relief was gradually reduced until most of Wyoming was again comparatively flat, broken here and there by meandering rivers and hills of Precambrian granite—the tops of the mostly buried mountains. Huge freshwater lakes covered 10–25 percent of Wyoming at various times, first by Lake Lebo in the Powder River Basin (early Paleocene epoch, 63 million years ago) and later by Lake Tatman in the Bighorn Basin, Fossil Lake on the eastern parts of the Overthrust Mountain Ranges, and Lake Gosiute in the Green River Basin, just east of Fossil Lake (early Eocene epoch, 52 million years ago; Lillegraven and Ostresh 1988). Fish that were fossilized in the sediments of Fossil Lake are now the primary attraction of Fossil Butte National Monument near Kemmerer (fig. 2.3).

Regional uplifting began again about 10 million years ago in the late Miocene epoch, eventually bringing the landscape to near its present elevation. This uplifting was accompanied by localized faulting and folding that created the Tetons—the youngest mountain range in the Rocky Mountain region (Love and Reed 1971). A new cycle of erosion was initiated that eventually removed the sedimentary cover of the buried mountains. Fossilized tree trunks, some still standing, were exposed in the Absaroka Mountains (Dorf 1964).

Uplift and erosion occurred slowly, allowing some rivers to cut canyons through the older rocks (fig. 2.4). Examples include the Bighorn, Wind River, Snake River, Laramie River, Devil's Gate, Sweetwater, and Platte River canyons. Some badlands were also created during the uplifting process, through the erosion of claystones, siltstones, and other substrata. The diversity of exposed rock types is fundamental in explaining the

F I G. 2.3   The fossil of a 40-cm fish (*Diplomystus dentatus*) that lived about 50 million years ago in freshwater Fossil Lake, located near Kemmerer in present-day southwestern Wyoming. Some of the fossil beds in this area are protected in Fossil Butte National Monument.

variable soils and topography that play such an important role in determining modern landscape patterns.

Mountain building, combined with the now more isolated position of the North American continent, led to further climatic changes. Aridity was increased by rainshadows on the leeward, eastern sides of mountains, which caused the demise of some forests and favored the establishment or expansion of shrublands and grasslands (beginning about 7–5 million years ago; Axelrod 1985). Drought-tolerant plant species persisted or immigrated from nearby (Axelrod 1985; Leopold and Denton 1987). Some species—such as sagebrush, needle-and-thread grass, and bluegrass—immigrated from the Old World across the Bering Land Bridge that connected Alaska and Siberia at the time. Halophytic plants found in modern inland salt marshes evolved from plants that are characteristic of marine coastal environments, such as saltgrass, cordgrass, and alkaligrass. The grasslands created new niches for mammals, and the xerophytic nature of many of the plants led to the accumulation of fuels that could carry fires during periodic dry seasons. Such fires would kill many of the trees that still grew on the upland, thereby hastening the spread of grassland and restricting woodlands to ravines, valley bottoms, or ridges that burned less often (Axelrod 1985).

Thousands of Years Ago

THE QUATERNARY PERIOD

Volcanism and glaciation characterized the Quaternary period (the past 2 million years). Unimaginable volcanic eruptions led to the creation of the Yellowstone Plateau about 660,000 years ago

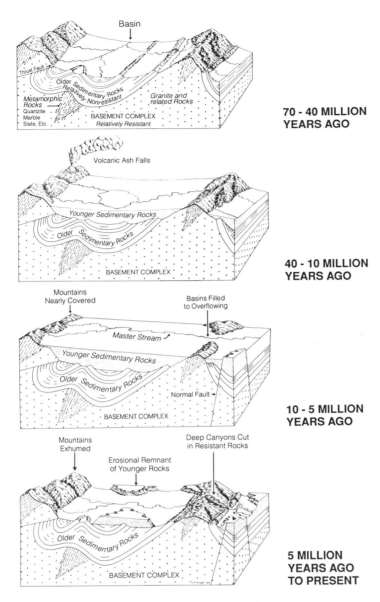

FIG. 2.4 Four diagrams illustrating the development of Wyoming landscapes during the past 70 million years. Note how the basins were filled with sediments 5–10 million years ago, and how rivers have cut canyons through mountain ranges since that time because of regional uplifting. Drawings by S. H. Knight.

(see chapter 14), and the climate of North America continued to cool. Snow fell heavily at higher latitudes and in the Rocky Mountains, often forming deep banks that persisted through the summer. Increased snow cover caused higher reflectivity of solar radiation, cooling the climate further, and many of the snow masses were converted to ice. Most of the glaciers began in this way during the Pleistocene epoch. Huge continental glaciers moved southward, forcing the Missouri River to flow southward into the the Gulf of Mex-

ico rather than into Hudson Bay. Continental glaciers never reached Wyoming, but features such as terminal and lateral moraines, U-shaped valleys, cirques, glacial outwash plains, and kettle topography (figs. 2.5, 2.6) indicate that glaciation occurred in most Wyoming mountain ranges (see table 10.2).

Tremendous amounts of water accumulated in the ice for thousands of years before the glaciers began to recede. Glacier formation and thawing occurred at least four to six times during the

F I G. 2.5 Cirques carved by glaciers in the alpine tundra of the Wind River Mountains. Elevation on the summit is approximately 3,344 m (11,000 ft).

Photograph by Austin S. Post, University of Washington.

FIG. 2.6   Jenny Lake in Grand Teton National Park, formed by terminal and lateral glacial moraines at the base of U-shaped Cascade Canyon in the Teton Range. Forests dominated by lodgepole pine, subalpine fir, and Engelmann spruce occur on the moraines around the lake. Douglas-fir and limber pine are common on the lower slopes of the Tetons, whereas subalpine fir and Engelmann spruce dominate the more mesic north (left) slopes of the canyon. Whitebark pine is common near the alpine zone. Mountain big sagebrush steppe occurs on the glacial outwash plains in the foreground. The highest peak is Grand Teton (4,197 m; 13,766 ft); the elevation of Jenny Lake is 2,061 m (6,779 ft). National Park Service photo by Bryan Harry.

Pleistocene (Mears 1974), and with each episode of melting, floods would shape broad riparian terraces and create glacial outwash plains.[3] Huge lakes also formed, including Lake Bonneville in Utah, Lake Missoula in southwestern Montana, and Lake Wamsutter in central Wyoming (Grasso 1990)—along with the smaller Fremont, Jackson, and Yellowstone lakes in northwestern Wyoming. The grinding of rock by ice led to the formation of a fine glacial dust, which, like the fine volcanic dust of the Tertiary period, was washed and blown onto the basins and plains to the east. Deposition of this material, known as loess (pronounced "lūss"), contributed significantly to the development of fertile soils in areas that now support grasslands and shrublands. Mountain soils were subjected to flushing by large volumes of water each spring during snowmelt, which eroded soil particles and slowed the accumulation of nutrients. Soil development in the mountains was most rapid where soft shales and schists were exposed instead of such harder rocks as granites, rhyolites, limestones, and some sandstones.

At the beginning of the Quaternary period (Pleistocene epoch) about 2 million years ago, the flora of the region was very similar to what it is today (Tidwell 1975). Coniferous forests, shrublands, and grasslands were common. Redwoods, cypress, and ginkgo had become extinct in the area, as had many species of broad-leaved trees. Cottonwoods and aspen are still widespread, and a few other species persist locally (birch, elm, oak, maple). Notably, the mammalian fauna still includes the mammoth, mastodont, rhinoceros, peccary, camel, sabertooth cat, giant beaver, short-faced bear, and horse, as well as wolves, pronghorn antelope, prairie dogs, and, more recently, the ancient bison (Walker 1987). Many of these mammals became extinct in North America during the Holocene,[4] apparently because of hunting by humans, climatic warming following the glacial period, increased competition, overspecialization, and reduced food availability caused by more frequent droughts (Anderson 1974; Frison 1975; Frison et al. 1978).

The interpretation of ecological changes during the Quaternary has been facilitated by the analysis of fossil pollen and seeds as well as leaf and animal fossils. Wind-blown pollen grains and seeds often are deposited in lakes and wetlands. Gradually they settle to the bottom, where they are preserved in anaerobic sediments. Paleoecologists now take vertical cores from the bottom sediments, and they are often able to identify the plants that produced the small fossils. Changes with depth suggest how the vegetation has changed over hundreds, thousands, or even millions of years.

Only a few pollen studies have been done in Wyoming or nearby, but they provide evidence of continued landscape changes through the Pleistocene to the present (fig. 2.7). For example, data collected from the Yellowstone Plateau show that forests dominated by Douglas-fir and limber pine were common about 127,000 years ago during an interglacial period (Baker 1976, 1983). Beginning about 15,000 years ago, when the Yellowstone Plateau was again covered with glaciers (the Wisconsin glacial advance), tree growth was restricted to a narrow band in the foothills (Baker 1983). A tundralike environment developed over much of the surrounding lowlands (Burkhardt 1976; Mears 1981, 1987; Beiswenger 1991; Whitlock 1993). The mean annual temperature probably was 10–13°C lower than it is today (Mears 1981). Permafrost was widespread, and several of the prevalent mammals are now characteristic of the

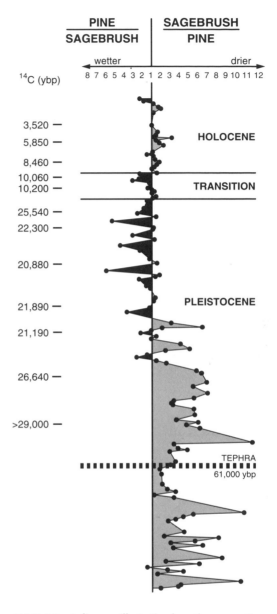

FIG. 2.7 A diagram illustrating how the proportion of pine and sagebrush pollen has changed during the past 60 million years. A higher proportion of pine pollen suggests wetter periods. A larger number on the left indicates that pine pollen was more abundant than sagebrush; a larger number on the right indicates that sagebrush pollen was more abundant. Based on data from Grays Lake, southeastern Idaho. Modified from Beiswenger 1991.

Arctic—caribou, collared lemming, and barren-ground muskox (Walker 1987). The mammoth roamed the intermountain basins as well. About 11,500 years ago, Engelmann spruce formed a spruce parkland in some areas, and several centuries later whitebark pine and subalpine fir became more common (11,000–9,500 years ago). Continued warming about 9,500 years ago led to the expansion of lodgepole pine at higher elevations and Douglas-fir in the foothills (Whitlock 1993).

A comparatively warm, dry period (the Altithermal) occurred for 3,000 years in the early Holocene, from about 7,000 to 4,000 years ago, with a concomitant expansion of big sagebrush, greasewood, juniper, and grasses (Beiswenger 1991). Reider et al. (1988) concluded from soil profiles that even some north slopes up to 2,400 m elevation in the mountains were covered by grasses during this time—indicating a warmer, drier climate. Subsequent cooling marked the beginning of the Neoglacial period, which continues to the present and has led to the expansion of forests (Beiswenger 1991; Whitlock 1993). Careful observations indicate that grassland soils sometimes are overlain by forest soils at lower elevations in the mountains (Reider 1983, 1990; Reider et al. 1988).

Paleoecologists have also used woodrat middens to interpret Holocene history. Woodrats (also known as packrats) accumulate seeds, cones, bones, and other material in their nests over long periods. Wells (1965, 1970a, 1970b) studied two woodrat middens in the Laramie Basin that were dated at 1,860 and 4,060 years ago. The remains of ponderosa pine and Rocky Mountain juniper were found in the middens, from which Wells concluded that much of the basin was once covered by juniper-pine woodland. Today, the basin is grassland and sagebrush steppe. Wells's conclusion has been questioned by Mears (1981) and Elliott-Fisk et al. (1983), who argued that the middens occurred near escarpments, where the pine and juniper are still found today. Trees in the basin typically are restricted to escarpments, a point that Wells recognized. He concluded, however, that the climate was more mesic during and just after the Altithermal and that the grasslands that characterize the Laramie Basin today may "constitute the first climatically induced episode of treelessness at this locality in post-Wisconsin time." Wells's critics, however, are convinced that the ba-

sin has been essentially treeless for the past 10,000 years (Mears 1981; Elliott-Fisk et al. 1983). In either case, the vegetation of today is strikingly different from the luxuriant subtropical forests that were so common in Wyoming 45–50 million years ago.

Most glaciers in Wyoming eventually disappeared. Those that remain are small and confined to high mountain valleys. Warming during the Altithermal probably caused the retreat of Engelmann spruce and subalpine fir to higher mountain slopes, but now these species characterize the mountains along with lodgepole pine, Douglas-fir, and aspen. Grasslands and shrublands dominate the lowlands, but there is still evidence of the once-colder tundra environment—soil fungi that are also known from the Arctic (Beiswenger and Christensen 1989) and curious soil patterns caused by permafrost (fig. 2.8; Mears 1987).

FIG. 2.8   Fossil ice wedge exposed at the Rawlins sanitary landfill, which suggests that a tundralike environment existed in that area during the Pleistocene (Mears 1981, 1987). Elevation 2,098 m (6,900 ft). Photo courtesy of Brainerd Mears, Jr.

Centuries and Decades Ago

Wyoming landscapes have changed little during the past few centuries. Landslides and flash floods have produced dramatic local changes, but at the time scale of centuries and decades, changes brought about by short-term droughts, fire, and human activity are of greater interest than geologic changes. In addition to examining fossil pollen and seeds, scientists from various disciplines have used the historical record contained in tree rings, soil profiles, anthropological data, and old journals and photographs to learn about change. What do such records reveal?

First, climatic change continues to occur even at the time scale of centuries. In a study of soil profiles in southeastern Wyoming, Reider et al. (1988) concluded that changes in rainfall and temperature caused the boundary between big sagebrush steppe and grasslands to shift back and forth. These climatic fluctuations occurred from about 200 to 3,000+ years ago. Soil strata that suggest drier conditions also had fewer artifacts of human activity (Reider et al. 1988), perhaps because the number of bison was lower during those periods.

A second type of change pertains to burning, an ecological process strongly affected by humans. Fires have been common wherever the environment has permitted the development of sufficient fuel or where grazing by bison, elk, pronghorn antelope, and other herbivores has been sufficiently infrequent to allow fuel accumulation. With fuel buildup, fires are inevitable. If they are not started by humans, they may be started by lightning. Semiarid deserts and plains probably burned infrequently in prehistoric times, as today, but fire was surely a regular occurrence in riparian zones, montane forests, some grasslands, and the more dense shrublands and woodlands. Because fires would kill many young trees and some shrubs, but usually not the grasses and forbs, the effect of fire in the foothills was to create savannas, especially at lower elevations in the eastern half of Wyoming, where thick-barked, fire-resistant ponderosa pine has been common for many centuries. Fires also restricted woodland development to rocky escarpments, where fire could not spread so easily. European settlers suppressed many fires, frequently causing tree and shrub density to increase (see chapter 9).

The first significant influence of Europeans on Wyoming landscapes was through the trapping of beaver in the early 1800s (Blair 1987; Crowe 1990). Many trappers, including Wilson Price Hunt, Jedediah Smith, and Jim Bridger, were lured to the region because beaver pelts were highly valued for the fabrication of felt used in the manufacture of clothing, primarily men's hats. Overtrapping caused a sharp decline in beaver populations, and without maintenance, many beaver dams must have failed—causing soil erosion and the development of deep gullies in some areas where ponds, meadows, and willow-dominated shrublands had existed previously (see chapter 4). Fortunately, the market for beaver pelts diminished rapidly in the late 1830s. Beaver now are quite abundant.

Large herbivores were also affected by humans. Bison and other animals had been subjected to hunting by Indians for thousands of years, but hunting pressure increased with the advent of the railroad and the tendency of immigrants to kill large numbers of animals just for fun if not for eastern markets (Blair 1987). Bison were essentially extinct as a free-roaming animal by the 1880s, about the same time that domestic livestock herds were becoming very large. By 1900, the population of all large mammals (mule deer, whitetail deer, elk, pronghorn antelope, bighorn sheep, bear, and bison) had been reduced to as few as 60,000 (Blair 1987). Fortunately, the conservation movement fostered by President Theodore Roosevelt at the turn of the twentieth century led to improved management, and over a period of ninety years or so, big game populations in Wyoming have been brought back to an estimated population of 1.1 million animals—more than twice the human population of the state. Compared to 454,000 humans, there are now about 543,000 mule deer, 413,000 antelope, 81,000 elk, 51,000 whitetail deer, 13,000 moose, 7,000 bighorn sheep, and 3,000 bison[5] (data from the Wyoming Game and Fish Department). Of course, at one time the bison in the western Great Plains numbered in the millions (Blair 1987).

A useful approach for determining landscape changes during the past 100–200 years is the comparison of present-day conditions with those described in early journals. The observations of trappers, explorers, and settlers in Wyoming

F I G. 2.9   A 1906 photograph of the forests in the Medicine Bow Mountains above Keystone after a period of logging. U.S. Forest Service photograph provided by the American Heritage Center, University of Wyoming.

during the period 1805–78 were summarized by Dorn (1986), who came to the following conclusions:

1. Bison were present in all parts of Wyoming, including subalpine meadows, but were especially common on the eastern plains. Evidence of heavy bison grazing was common.
2. Range fires occurred, but they were not mentioned often; grazing by native animals may have minimized fuel accumulation.
3. Forest fires were common.
4. Big sagebrush was very common.
5. Grizzly bears, black bears, wolves, and elk were common on the lowland grasslands and shrublands but declined with increased hunting pressure (or, in the case of predators, lower prey populations).
6. Some streams were ephemeral and had steep-walled gullies.
7. Grasshopper and Mormon cricket plagues occurred before settlement, as they do now.

8. Vegetation patterns have changed very little during the past century.

Generally, human influences on western landscapes were minor in the mid-1800s, when the early explorers were writing their journals (see also Schullery and Whittlesey 1992). Probably fewer than ten thousand people resided in Wyoming Territory at the time (Larson 1977). Notably, however, more than 350,000 immigrants traveled through Wyoming along the Oregon Trail between 1841 and 1868. Crossing the Continental Divide at South Pass was a much easier route than the alternatives in Colorado and Montana. Along with the wagon trains came thousands of livestock. These animals grazed along the way and sometimes were concentrated in small areas. For example, one group worked unsuccessfully for five days to drive its herd across the Platte River near Casper, and during this time the grass was devoured for miles around (Larson 1977, 49). The environmental impacts of people and livestock

F I G. 2.10.   A 1943 photograph of the same area shown in fig. 2.9. New forests dominated by lodgepole pine have developed in a very short time. The elevation of Keystone is 2,706 m (8,900 ft). A similarly rapid recovery of ponderosa pine forest in the Black Hills has been documented by Amundson (1991, 86–87). U.S. Forest Service photo provided by the American Heritage Center, University of Wyoming.

were probably much greater along the Oregon Trail than in the region as a whole.[6]

The effect of livestock became more widespread after the completion of the Union Pacific Railroad in 1869 (Mitchell and Hart 1987), which made much of the nation a market for the beef, lamb, and wool raised on western rangelands. Large herds of cattle were driven northward from Texas and many were shipped westward by rail. Sheep herds numbering ten thousand animals were not uncommon (Wentworth 1948). Great profits could be earned by big city investors and western stockgrowers. Livestock were moved to marginal rangelands until, as Mitchell and Hart (1987) suggested, "the land was filled entirely with cattle." Fortunately for rangeland condition, the livestock boom ended a few years later, in 1887, following a crash in the beef market that coincided with the dry summer of 1886 and the devastating winter of 1886–87, which killed 40–60 percent of the cattle in most of Wyoming (Abbott and Smith 1955; Larson 1978). Arguably, never again would there be so many cattle on western rangelands.[7] Vegetation changes caused by livestock in the late 1800s and early 1900s are difficult to assess, but they could have been substantial in some areas, especially in the intermountain basins to the west (R. F. Miller et al. 1993; see chapters 5 and 6).

Mountain landscapes also were subjected to human influences in the 1800s. Railroad ties were required by the nation's first transcontinental railroad (completed in 1869), along with lumber for new railroad towns and timbers for gold, silver, and copper mines. Some forests were essentially clearcut (figs. 2.9, 2.10).[8] Elsewhere, forests were thinned as the tie-hackers selected only the trees they wanted. Slab piles are still commonly found, indicating the locations of sawmills. The ties would typically be floated down creeks and rivers to railheads, especially during the spring when water levels were high. Such activity, along with placer mining for gold, must have caused significant changes in the riparian zone of some watersheds.

Another kind of historical record became available when the first photographers traveled west. Several studies have compared early and modern photographs of the same area. Kendall Johnson (1987) retraced the route of William Henry Jackson, the famous landscape photographer who had accompanied the 1870 Hayden Expedition across southern Wyoming—prior to the cattle boom. Johnson was able to locate and rephotograph fifty-six scenes approximately one century later. The photo comparisons led him to conclude:

1. Except where plowing has occurred, the plains grasslands of today are very similar to those of a hundred years ago. Overall, the effects of livestock grazing appear not to have been much different than the effects of native ungulates, although shifts in species abundance may have occurred.
2. Big sagebrush has increased, decreased, or remained about the same depending on land use and site characteristics, but it was abundant in the intermountain basins long before livestock grazing began.
3. Woodlands and forests on the uplands have become more dense, probably due to fire suppression, but trees have not invaded adjacent vegetation types over large areas.
4. River bottoms have changed dramatically due to impoundments, irrigation, cultivation, livestock grazing, and settlements; and some river margins now have more trees than in 1870.
5. Areas of low plant production, such as desert shrublands, retain a nearly identical appearance to those of 1870.

Photographs must be used cautiously in drawing conclusions, but they provide additional insights on landscape change during the past century.[9]

Since Wyoming became a state in 1890, changes have continued to occur in some areas because of human influences. Forest landscapes have been fragmented through timber harvesting, and some rangelands have been plowed. Erosion has been accelerated. Livestock have replaced bison almost everywhere, and many plant species have been introduced from other continents, some becoming weeds. Towns, cities, highways, power lines, and reservoirs have been built. Riparian landscapes and some grassland and forested landscapes have been altered most dramatically. Still, Wyoming and nearby states are remarkably similar to what they were when native Americans first encountered white-skinned explorers from the East.

CHAPTER 3

Modern Environments

Wyoming and other Rocky Mountain states are characterized by abrupt topographic relief and numerous types of exposed bedrock. The vegetation includes sagebrush, greasewood, and saltbush shrublands in the intermountain basins, grasslands on the Great Plains, juniper and mountain-mahogany shrublands in the foothills, and forests and alpine meadows in the mountains. The plants that characterize each vegetation type have distinctly different environmental tolerances for temperature extremes, flooding and drought, length of growing season, soil conditions, and disturbances —whether fire, floods, logging, landslides, avalanches, insect epidemics, or plowing.

All environmental factors are affected by topographic position and elevation. For example, temperature typically decreases with elevation, and the accumulation of blowing snow on the lee side of ridges or on north-facing slopes greatly augments the amount of water that percolates into the soil. Animal distribution patterns also are influenced by topography and elevation. Many animals migrate between contrasting environments; for example, deer and elk move from mountains to foothills to survive the winter.

The sunny days, short summers, and long winters that characterize much of Wyoming can be attributed to (1) high elevation, (2) northerly latitudes (between 41° and 45° north) with frequent invasions of polar air masses, (3) a continental location that minimizes the moderating influence of the oceans on seasonal temperature variation, and (4) a location on the leeward side of several mountain ranges that increases aridity. Together with topography, these factors influence airflow patterns and govern the extremes and variability in precipitation, temperature, evaporation, and solar radiation.

Understanding the environment is one of the greatest challenges that ecologists face. Environmental characteristics that seem especially important to plants and animals in the region are topography, elevation, rainfall, temperature, wind, soils, and disturbances. To simplify the discussion of such factors, convenient numbers such as averages often are calculated, but organisms respond to the full complex of interacting environmental factors—not to one individual factor. Moreover, extremes usually are more important than averages. The presence of a species implies that it is tolerant to the major hazards that characterize that environment, and because young individuals (plant or animal) are more susceptible than adults to such hazards, the distribution of a species depends more on the environmental conditions prevailing at the time of seedling establishment— conditions that usually are not known at the time of observation. Little wonder, then, that good correlations are hard to find between species distribution and individual environmental factors expressed as averages. The challenge is to integrate, in a convincing manner, the diverse factors that determine whether an organism will survive.

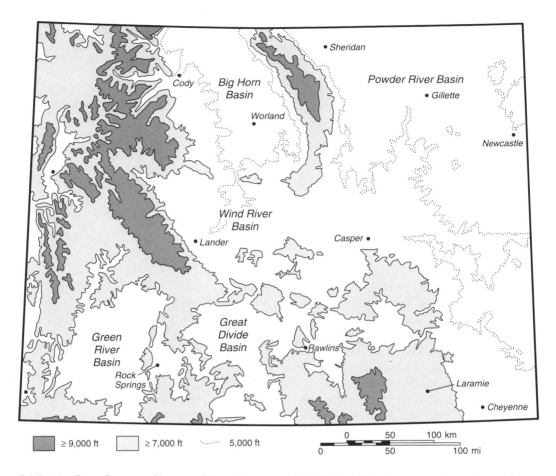

F I G. 3.1   General topographic map of Wyoming. Dark areas are above 9,000 ft (2,736 m) and gray areas are above 7,000 ft (2,128 m). The dotted line is the 5,000-ft (1,520 m) contour. Adapted from Martner 1986.

Topography

Wyoming landscapes span an elevational range of 3,232 m, from the lowest point on the plains (969 m, where the Belle Fourche River flows into South Dakota) to the highest mountaintop (Gannett Peak at 4,207 m). Thirty-seven percent of Wyoming is above 2,134 m elevation and 10 percent is above 2,743 m (7,000 ft and 9,000 ft, respectively; fig. 3.1). About fifty mountain peaks rise to 3,950 m (13,000 ft) or above (mostly in the Wind River Mountains). Only a few major rivers flow into the state before leaving in a different direction: the North Platte River and Laramie River from Colorado; the Bear River and the Henrys Fork and Blacks Fork of the Green River from Utah; and the Clarks Fork of the Yellowstone River from Montana. All others originate in Wyoming and flow outward into the Missouri, Colorado, or Columbia river drainages (see chapter 4).

Elevation affects a variety of environmental factors. Annual precipitation increases with elevation, at least to the upper mountain slopes if not to the summit, while temperature decreases (Martner 1986). These trends create a more mesic environment in the mountains. Yet higher rates of evaporation resulting from lower atmospheric pressure, combined with cooler soil temperatures that restrict the rate of water uptake by roots, create a less mesic environment than might be expected (Smith and Geller 1979). In fact, high south-facing slopes that receive direct solar radiation throughout the year may be as dry as deserts at much lower elevations. In contrast, cooler north slopes at low elevations may have vegetation normally found at much higher elevations. Such observations have led ecologists to conclude that topographic position and exposure are often more important than elevation in determining plant and animal distribution patterns. Nevertheless, it is intriguing to observe how some of the most common plant genera in Wyoming have different species adapted to the lowland, foothill, montane, and alpine environments (fig. 3.2).

Based on elevation, topography, and location with regard to the Continental Divide, several physiographic regions can be identified (see fig. 1.2). The Northern Great Plains characterize the eastern third of the state, with elevations ranging from 970 to 2,280 m. Summer precipitation is

more frequent on the plains than farther to the west, which apparently has a major influence on the flora and fauna. For example, Mack and Thompson (1982) discuss why bison might have been much more numerous on the Great Plains than in the Great Basin, attributing the abundance to more reliable summer rains and consequently more summer food for lactating cows. Over a period of more than a million years, the grasses of the Great Plains evolved adaptations to withstand grazing pressure from large herbivores, and hence the grasslands there appear to be more tolerant of livestock grazing than those in the Great Basin.

To the west of the Great Plains is a region of intermountain basins (see fig. 1.2). Most of the basins range in elevation from about 1,200 to 2,200 m. The Continental Divide passes through the middle of the intermountain basin region, splitting near Rawlins to form the Great Divide Basin with no drainage to either ocean. At this point the Continental Divide is only about 2,150 m (7,000 ft) above sea level and is characterized by sagebrush steppe rather than the coniferous forest or alpine tundra typical along the divide to the north and south.

Scattered through the Great Plains and intermountain basins are escarpments formed by resistant bedrock, usually sandstones and limestones (fig. 3.3). The vegetation of such escarpments is distinctive, partly because of the coarse, rocky substrate, but also because of topographic influences on snow accumulation. Such escarpments usually constitute an important component of the foothills, with the predominant vegetation being shrublands dominated by mountain big sagebrush, mountain-mahogany, or skunkbush sumac; or woodlands characterized by juniper, limber pine, ponderosa pine, and rarely, oak (see fig. 1.5). These vegetation types add scenic diversity to the Wyoming landscape and are important winter ranges for elk and deer.

Constituting about 30 percent of Wyoming, the mountains are characterized by lakes, streams, coniferous forest, aspen groves, subalpine meadows, abrupt treelines, and alpine tundra. Soils, precipitation patterns, and other environmental factors vary greatly throughout these mountain ranges, with freezing temperatures possible every month of the year at higher elevations. Geogra-

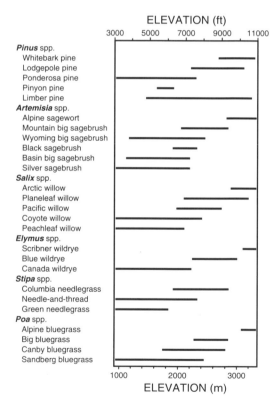

FIG. 3.2 Different species of the same plant genus are often found at higher or lower elevations, indicating different morphological and physiological adaptations to the varying climatic conditions along mountain slopes.

F I G. 3.3   Limber pine and Rocky Mountain juniper grow only along fracture planes of this sandstone escarpment in the Frontier formation near the Ferris Mountains north of Rawlins. Wyoming big sagebrush and greasewood shrublands are found on the adjacent fine-textured soils. Elevation 2,128 m (7,000 ft).

phers have classified the Laramie, Medicine Bow, and Sierra Madre mountains in the southeast as part of the Southern Rocky Mountains, and most of the other mountain ranges as part of the Middle Rocky Mountains (Fenneman 1931). The Northern Rocky Mountains begin with the Madison, Gallatin, and Beartooth ranges.

The plains, basins, foothills, and mountains can be subdivided further using such geomorphic features as dunes, arroyos, buttes, playas, deflation hollows, mima mounds, and braided or meandering stream channels. Such features add diversity to the landscape, creating more habitats for plants, animals, and microorganisms. Similarly, glacial moraines, hogback ridges, colluvial or alluvial fans (bajadas), canyons, and nivation hollows diversify the foothills. The ecological significance of

such geomorphic influences is a recurring theme in this book.

Precipitation

As in other semiarid environments, precipitation is a climatic factor with a major influence on plants and animals as well as human activity. In particular, the amount of plant growth in grasslands and shrublands (also known as net primary productivity [NPP] and expressed as grams/square meter/year) is typically highly correlated with rainfall or some other measure of water availability (Webb et al. 1978; Joyce 1981). Where precipitation is higher, such as in montane forests, other factors may be more limiting, for example, temperature or the length of the growing season.

Mean Annual Precipitation (inches), 1951-80

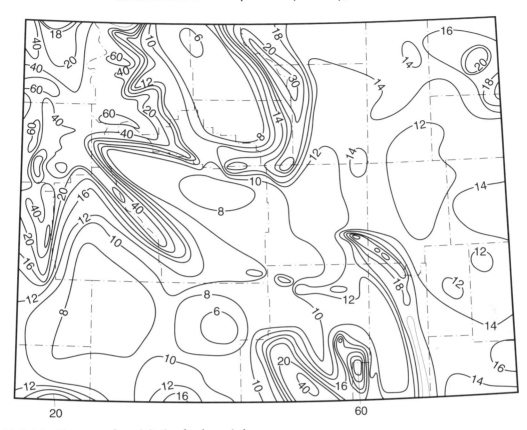

F I G. 3.4   Mean annual precipitation for the period 1951–80. (The values shown are inches; 1 in. = 2.54 cm.) Adapted from Martner 1986.

In all cases, water availability to plants is affected as much by the infiltration and water-holding characteristics of the soil as by the amount and timing of precipitation.

Across Wyoming, mean precipitation varies by a factor of 10, from 15 to 150 cm each year (fig. 3.4). The two driest areas are in the Great Divide Basin (near Wamsutter on Interstate 80) and in the northern part of the Bighorn Basin (between Lovell and Powell). In general, the intermountain basins in the western two-thirds of the state are drier, with averages of 15–30 cm, than the Great Plains region to the east, with averages of 30–40 cm a year. The foothills and mountains receive 40–150 cm.

With the prevailing westerly airflow, mountains receive more precipitation than the lowlands because air moving across the mountains is cooled as it rises, thereby causing more condensation, which often develops into rain or snow. On the leeward sides of mountains, however, the lower slopes and basins are drier than might be expected because descending air masses are subjected to warming, which increases the potential for evaporation instead of precipitation. This phenomenon is the well-known rainshadow effect.

Most Wyoming basins and plains are in a rainshadow, with the "shadow effect" on the Great Plains being ameliorated during years when more than average amounts of moist air from the Gulf of Mexico penetrate to the western Great Plains. The Arizona monsoon, a northerly flow of moist air from the Gulf of California, may cause added summer precipitation in southeastern Wyoming, a phenomenon that could account for the presence of Gambel oak on the west slopes of the Sierra Madre (see chapter 9). Some years have

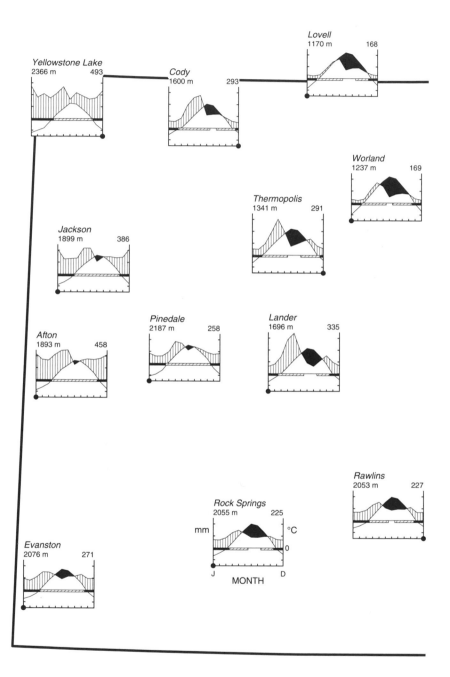

F I G. 3.5   Climate diagrams for twenty-five low-elevation sites in Wyoming, drawn using data from Martner (1986). The graphs show mean monthly precipitation and mean monthly temperatures from January to December. Precipitation is more evenly distributed during the year in the western half of the state. Black areas indicate periods of drought, when the precipitation line drops below the temperature line. Each unit on the vertical axis is 20 mm of precipitation (water equivalent) or 10°C. The number on the left is elevation, on the right, mean

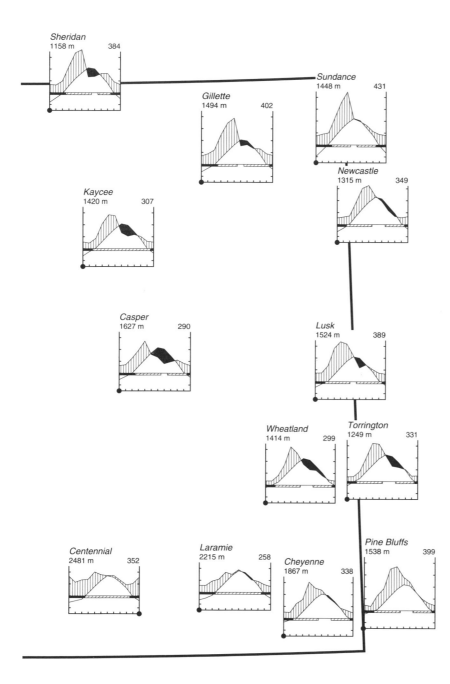

annual precipitation (mm). A break in the horizontal bar in the summer indicates the frost-free period; a black bar indicates that the mean daily minimum during a month is below freezing and a hatched bar indicates months when the lowest temperature is below freezing. See the Rock Springs diagram and Walter (1973) for detail. Walter et al. (1975) provide climate diagrams for many meteorological stations in North America and around the world.

high amounts of precipitation, apparently caused by flows of such air masses from the south. Other years, when these same air masses do not reach the state or are diverted to the east, are extremely dry. As with rain in the lowlands, snowfall in the mountains varies greatly from one year to the next.

Equally important as the amount of precipitation is its form, duration, and intensity. Rain that occurs during warm periods may evaporate quickly, sometimes even before it can infiltrate the soil. Heavier rainstorms can cause infiltration to 10–30 cm, but plant water stress soon develops if significant rainfall events are separated by more than a week or two, as they usually are.

In contrast, snow usually accumulates during the winter, when evaporation is minimal. In the spring, the accumulated snow melts during a short period when the potential for evapotranspiration (ET) is low. Consequently, the water is more likely to penetrate deeply into the soil. Moreover, the deep soil water is not as readily evaporated, and if the soils are deep enough and have a high water-holding capacity, plant growth is sustained without significant water stress for much of the growing season. Such deep percolation seems important for the survival of big sagebrush in most of the intermountain basins. Stoddart (1941) and Martner (1986) present data that show how spring and early summer precipitation is more characteristic of the eastern plains, while the western two-thirds of the state often has a more even distribution of precipitation during the year (and a larger proportion that occurs in the winter as snow [fig. 3.5]).

Snow accumulation in the mountains is especially noticeable, with drifts often more than 6 m deep. The soils near such drifts are wet throughout the growing season (which may be short because of the time required for snow melting). Late-lying snowdrifts account for the development of nivation hollows or snowglades—places where snow accumulation prevents the establishment of plants that are characteristic of the surrounding area (see chapter 12).

Periodic droughts characterize the mountains as well as the lowlands,[1] but long-term data on precipitation and temperature in Cheyenne, Lander, and Yellowstone National Park, dating as far back as 1870, suggest that no clear trend to a wetter or drier environment is apparent. Martner (1986) observed, however, that the mean annual precipitation from 1951 to 1980 was 2–12 percent less in some areas than from 1941 to 1970.

Temperature, Frost-free Period, and Evaporation

As commonly observed, there is a clear correlation between temperature and elevation, with a cooling rate (adiabatic lapse rate) of 9.8°C per 1,000 m elevation (5.6°F/1,000 ft) for northwestern Wyoming (Dirks and Martner 1982) and 6.5°C per 1,000 m (3.6°F/1,000 ft) for the state as a whole (Martner 1986). There is also a tendency toward cooler mean annual temperatures from south to north, but only by a few degrees on average because the effect of elevation clearly obscures the effect of latitude.

Because there is such great topographic relief in Wyoming, temperature varies greatly. For example, the mean daily high temperature in July ranges from 32°C (90°F) on the Great Plains and in the Bighorn Basin to less than 24°C (<75°F) in the higher mountains. The mean daily low temperature for July ranges from 13°C (55°F) to near freezing (fig. 3.6). Of course, nighttime low temperatures in the mountains can drop to below freezing on any day during the growing season. The mean frost-free period varies from 125 to fewer than 25 days, with the longest periods being on the Great Plains, in the Bighorn Basin, and in a small area near Riverton in the Wind River Basin (fig. 3.7), where, predictably, the largest amount of cropland occurs. Growing season length is an important limiting factor for plant growth. Many native species have evolved adaptations for preventing frost damage, but often this is not the case for crop plants. Significantly, although temperature has an important effect on the initiation of plant growth in the spring, the termination of growth in the summer is more often a result of summer drought stress, especially in the lowlands, where precipitation is comparatively low.

Topography also influences growing season length and mean temperature because the more dense cold air on the mountains drains to the

Little Brooklyn Lake
3158 m   1208

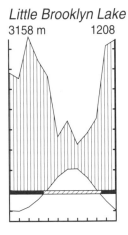

Foxpark
2763 m   402

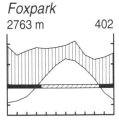

Laramie
2215 m   258

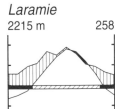

Cheyenne
1867 m   338

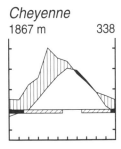

FIG. 3.6   Climate diagrams in southeastern Wyoming that illustrate climatic changes with elevation. See fig. 3.5 for explanation. Based on data from Martner 1986 and the Wyoming Water Resources Center in Laramie.

lower elevations, often along waterways, creating temperature inversions. Thus, Jackson and Cheyenne are at about the same elevation (1,892 m and 1,848 m) (6,209 ft and 6,062 ft, respectively), but Jackson is considerably cooler because of its location within the Jackson Hole basin, which traps cold air drained from the adjacent mountains (Martner 1986).

One of the most significant effects of temperature is on the potential rates of evapotranspiration. Higher temperature and lower humidity cause higher evapotranspiration, which can lead quickly to summer drought stress. Martner (1986) calculated potential evapotranspiration (PET) using available Wyoming weather station data, with the results showing how PET is on the average greater than annual precipitation in the plains, basins, and foothills, but usually less than annual precipitation in the mountains (fig. 3.8). Actual moisture availability depends on the form and timing of precipitation plus soil characteristics, as discussed previously, but significant water stress can be expected when PET exceeds annual precipitation in any given year. Conversely, nutrient leaching and different soil development processes occur when water is more abundant. Only 10 percent of Wyoming has a precipitation-evaporation (P-E) ratio greater than 1 (Ostresh et al. 1990).

Topography and wind also play an important role in moisture availability and water stress development. Topography is important because south slopes are warmer than north slopes, and wind is important because it accelerates evapotranspiration and leads to snow redistribution and sublimation. Those areas blown free of snow are drier during the growing season; those areas where snow accumulates are more moist. The resulting patterns of snow accumulation often cause the patchy distribution pattern of contiguous vegetation types (the vegetation mosaic).

Strong winds are common in Wyoming, especially in the south (Martner 1986). Wind velocity can be attributed in part to the prevailing westerly winds being funneled through the Rocky Mountains at a low point in the Continental Divide (about where Interstate 80 crosses the divide). This wind pattern has prevailed for millions of years, contributing to the active development of a narrow, one-hundred-mile-long field of sand

Mean Length of Frost-Free Season (days), 1951-80

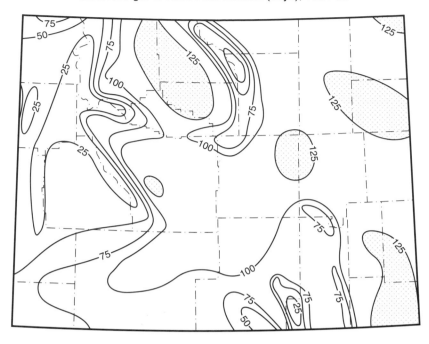

Mean Number of Days Annually with Maximum Temperature ≥90°F, 1951-80

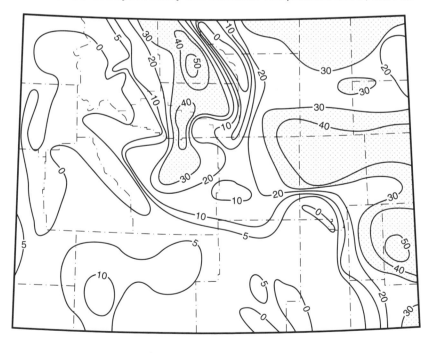

F I G. 3.7   The longest frost-free periods (top) and the longest periods of high temperatures (bottom) are found at low elevations on the eastern plains and in the Bighorn and Wind River basins. Adapted from Martner 1986.

Estimated Mean Annual Potential Evapotranspiration (inches), 1951-80

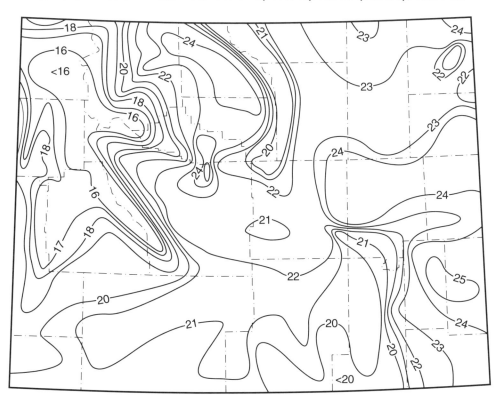

F I G. 3.8   Potential evapotranspiration (PET) in
Wyoming (1 in. = 2.54 cm). Adapted from Martner
1986.

dunes north of I-80 between Rock Springs and
Rawlins (see chapter 8).

Diurnal and Seasonal Climatic Changes

Fluctuations in environmental factors from night
to day and month to month are more significant
than annual averages. The extreme in diurnal fluctuations is experienced by plants and animals living in alpine tundra, where air temperatures near
the ground may be 27°C (80°F) or more during
the day and below freezing at night. Such rapid
temperature fluctuations occur because of the
high elevation, where heating during the day occurs rapidly owing to less atmosphere through
which sunlight must pass. For the same reason,
heat is lost rapidly to a cold, typically cloud-free
sky at night. Many plants cannot tolerate such

drastic changes in temperature over such a short
period. Temperature acclimation on a twenty-
four-hour time scale must be more difficult for
plants than acclimation to warming or cooling
over a period of days or weeks, as usually occurs at
lower elevations.

Soil Characteristics

Landscape mosaics in the Rocky Mountain region
are often caused by changes in the soils, especially
in nonforested areas. For plants, the soil provides
a reservoir of water and nutrients, whereas for animals and microorganisms the soil determines indirectly the amount of food that will be available.
The carrying capacities for wildlife and livestock,
as well as for agricultural production, depend on
soil characteristics. Often viewed as an inert substrate, the soil is better thought of as a matrix of

organic matter, clay, silt, sand, and stones that provides habitat for an abundance of roots, invertebrates, and microbial life. Indeed, the amount of energy consumed by belowground organisms in grassland ecosystems can exceed the amount used by animals aboveground (see chapter 5).

A multitude of different soils exist in Wyoming, each the result of interacting forces during thousands of years and the varying influences of climate, topography, geologic substrate, fauna, vegetation, and time. Sometimes the soils change gradually, forming a gradient (catena) from low, relatively moist environments to high, comparatively dry environments. In other places the soils change abruptly, usually a reflection of abrupt changes in topography or geologic substrate. The soils are often shallow and not well developed, largely because of recent glaciation in the mountains, an abundance of erodible slopes, and a comparatively dry, cool environment that slows soil development. Deep soils occur in lower topographic positions, the result of more rapid soil development or downslope sediment accumulation.

The physical and chemical structure of soils is heavily influenced by the nature of the parent material, whether bedrock or transported material. Igneous and metamorphic rocks, such as granites, basalts, and quartzite, are found primarily in the mountain ranges. They weather very slowly and consequently the soils are usually shallow and coarse. Sedimentary strata are exposed more frequently in the foothills and lowlands, though they may be found in the mountains as well. Sandstones and limestones are slower to weather, often forming escarpments and ridges with shallow soils. Where some weathering has occurred, sandstones usually contribute to the formation of a coarse-textured, sandy soil with a high infiltration capacity. Sometimes the resulting sand is sorted by wind into extensive dune fields (for example, the Killpecker Sand Dunes northeast of Rock Springs). Glacial moraines in much of the Rocky Mountains also lead to the development of coarse soils with high infiltration rates.

Fine-textured and deeper soils often develop from the more easily eroded and weathered shales, mudstones, and siltstones. The salinity, alkalinity, and sodicity of such soils may be high or low,

depending on bedrock origin. Infiltration rates may be slower in fine-textured soils, but the water-holding capacity is higher. Valley or arroyo soils between hogback ridges usually have developed on such bedrock and can be quite fertile. Elsewhere, fine-textured soils may be saline or high in toxic elements such as selenium. Saline playas develop where fine-textured sediments accumulate in closed basins with no outward drainage. Some shale-derived soils (for example, from the Mowry, Steele, Belle Fourche, Pierre, and Thermopolis formations) are high in bentonitic clays that swell and shrink during wetting and drying cycles, a feature that often prevents the establishment of young plants. Similarly, soils high in soluble salts, gypsum (hydrated calcium sulfate), and sodium may form crusts that restrict plant establishment. Some of the more erodible sedimentary bedrocks, such as shales and siltstones, lead to the formation of classic badland topography (for example, Hell's Half Acre west of Casper).

Volcanism has contributed to soil development throughout the state, sometimes improving soil fertility. Fine, windblown volcanic ash is common in sedimentary rocks, thereby influencing soil formation. Hard igneous rocks such as rhyolite, formed from the solidification of lava flows, have contributed to the development of coarse-textured soils in the Absaroka Mountains and Yellowstone Plateau.

Predictably, as climate changes along elevational gradients, so do soil characteristics (Thorp 1931; Munn 1977; Weaver 1978). Organic matter content increases from lowland up to montane grasslands, but then decreases farther up into the alpine zone. Weathering and plant production are highest at midelevations, where more moisture is available but where the climate is still not as cold as higher on the mountain. Severe climates tend to restrict soil development as well as vegetation development. The ratio of precipitation to evaporation also changes with elevation. With a *P-E* ratio of less than 1, as would be expected most years in the lowlands, salts and nutrients accumulate in the soil. A layer of lime (calcium carbonate) develops near the average depth of water infiltration, forming a hardpan or *caliche* layer that may restrict root development. In contrast, no hardpan develops in the mountains where the *P-E* ratio is greater than 1. There, the soils tend to be acidic, and the potential for nutrient leaching is higher. Table 3.1 lists and defines the major soil types that can be expected with some of the major vegetation types in region.

There are five soil features that seem especially critical in determining vegetation patterns in Wyoming, namely, infiltration rate, depth, water-holding capacity, salinity, and aeration. Infiltration is the rate at which water percolates into the soil and is especially important in semiarid regions, where the potential for evaporation is high. Water that percolates rapidly to a depth of at least 5–10 cm is less likely to be evaporated directly and can be used for plant maintenance and growth. Coarse-textured soils in semiarid regions (<37 cm precipitation annually; Sala et al. 1988) tend to have higher infiltration capacities and consequently higher rates of plant productivity—a principle referred to as the inverse texture effect because coarse-textured soils usually have lower productivity in more mesic environments (Noy-Meyer 1973; Sala et al. 1988).

The inverse texture principle can be extended to rock outcrops where the soil is thin or even nonexistent but where water is funneled into cracks. Plant roots capable of growing through such fissures, such as the roots of mountain-mahogany and limber and ponderosa pine, probably have more water available to them than would be the case in fine-textured, deeper soils. In fact, shrubs and trees throughout the region are often found on rock outcrops or coarse soils, not on fine-textured soils (see fig. 3.3).

Water-holding capacity is a function of soil texture, soil depth, and the amount of organic matter. Fine-textured soils, with a high percentage of silt and clay (fig. 3.9), hold more water per unit volume than do coarse-textured soils. They also enable more plant growth if the soil is saturated at least once during the year (such as often occurs near snowdrifts). Abrupt transitions between grasslands and some shrublands are often associated with changes in soil depth or soil texture, with the more drought-tolerant grasses being dominant on shallower soils with a low water-holding capacity (see fig. 6.8).

Soil aeration is determined by texture and depth to the water table. Aeration is important to plants because air spaces in the soil determine oxygen availability for root respiration. Oxygen is less readily available if the spaces are filled with water, which occurs more often on fine-textured soils. Greasewood apparently is well adapted to saturated soils with low levels of aeration, while big sagebrush cannot tolerate standing water within 10 cm of the surface (Ganskopp 1986). Well-drained, well-aerated soils are necessary for the survival of big sagebrush.

Other soil characteristics important in determining Wyoming landscape patterns are salinity, alkalinity, and sodicity—traits that are chemical in nature and that depend on parent material, the amount of water that has evaporated from the soil, and a water regime with little or no potential for leaching. Saline soils can be defined as any soil with concentrations of soluble salts that produce an electrical conductivity of 4 dS/m or more (Branson et al. 1967, 1970, 1976). Alkalinity refers to soils where the less soluble calcium carbonate is important (pH 7.8–8.2), and sodicity implies that sodium occupies 15 percent or more of the soil adsorption sites. Plant survival on such soils is dependent on an ability to maintain cells with a solute concentration higher than the solute concentration of the soil solution. If this is not the case, water flows out of the plant into the soil. Plants that tolerate saline, alkaline, and sodic soils are known as halophytes, for example, greasewood and Gardner saltbush. Other less salt-tolerant species, such as big sagebrush, are sometimes found growing with halophytes, but usually in areas where salt concentrations are lower, such as near snowdrifts or along drainages where the leaching of salts can occur (see chapter 7).

The weathering of some geologic strata leads directly to the formation of saline, alkaline, or sodic soils, especially the Cretaceous marine shales (Steele, Niobrara, Cody, Pierre). Such soils often develop in basins where there is considerable water accumulation on the surface during wet periods and where subsequent evaporation leads to the accumulation of salts that often form white crusts over the soil surface.[2] Soils derived from bentonite tend to be sodic because this mineral is composed largely of montmorillonitic clay, which develops in high-sodium, sedimentary environments.

TABLE 3.1 Soil orders and subgroups characteristic of vegetation types in Wyoming

| Vegetation type | Soil Order[a] | | | | |
	Entisols	Inceptisols	Aridisols	Mollisols	Alfisols
LOWLANDS					
Mixed-grass prairie	—	—	BoHAI[b]	ArABO[b]	—
	—	—	BoNAI	—	—
Sagebrush steppe					
Black sagebrush	—	—	TyCOI	—	—
	—	—	TyHAI	—	—
Big sagebrush	—	—	TyHAI	—	—
	—	—	BoHAI	—	—
Silver sagebrush	TyTFE[b]	—	—	—	—
	TyTPE	—	—	—	—
Riparian vegetation					
Lowland	—	—	—	TyHAO	—
Foothill	—	—	—	CuCBO	—
Montane	—	HiCAP[b]	—		—
Alpine	—	PeCAP	—		—
Desert shrublands					
Saltbush	—	—	TyHAI	—	—
	—	—	TyNAI	—	—
Mixed	—	—	TyNAI	—	—
Greasewood	—	—	BoNAI	—	—
	—	—	TySOI	—	—
	—	—	TyGOI	—	—
Basin grassland	—	—	BoHAI	—	—
	—	—	UsHAI	—	—
	—	—	BoCOI	—	—
Sand dunes (stabilized)	TyTPE	—	—	—	—
Badlands	TyTOE	—	—	—	—
Playas	—	—	TySOI	—	—
FOOTHILLS					
Shrublands					
Mixed	—	—	—	TyHBO	—
Mountain-mahogany	LiTOE	—	—	—	—
Threetip sagebrush	TyCOE	TyCCP	—	—	—
Mountain big sagebrush	—	—	—	TyCBO	—
Foothill grassland	—	—	—	ArHBO	—
	—	—	—	ArABO	—
Woodlands					
Oak woodland	TyUOE	—	—	—	—
Juniper woodland	TyTOE	—	—	—	—
Ponderosa pine	LiTOE	—	—	—	—
	LiUOE	—	—	—	—
Douglas-fir	—	—	—	—	TyHUA[b]
	—	—	—	—	TyHBA
Aspen	—	—	—	TyHAO	—
	—	—	—	TyHBO	—
MOUNTAINS					
Ponderosa pine	—	—	—	—	TyHBA

TABLE 3.1 (*Continued*)

Vegetation type	Soil Order[a]				
	Entisols	Inceptisols	Aridisols	Mollisols	Alfisols
Lodgepole pine	—	TyCCP	—	—	TyCBA
Engelmann spruce–subalpine fir	—	TyCCP	—	—	TyCBA
Douglas-fir	—	—	—	—	MoCBA
	—	—	—	—	TyHBA
Aspen	—	—	—	TyCBO	AqCBA
Subalpine meadows					
Dry	—	—	—	TyCBO	—
Moist	—	—	—	AqCBO	—
Wet	—	HiCAP	—	TyCAO	—
Alpine tundra					
Fellfield	—	TyCUP	—	—	—
Turf	—	TyCUP	—	—	—
	—	TyCCP	—	—	—
Wet	TyCAE	—	—	—	—

Note: — = absent or less common

[a] See glossary for definitions; table courtesy of Larry Munn.

[b] The characteristics of soil subgroups are provided in Soil Survey Staff (1992), *Keys to Soil Taxonomy* (5th ed.). Subgroup names are abbreviated as follows:

AqCBA	Aquic Cryoboralfs	MoCBA	Mollic Cryoboralfs	TyHAO	Typic Haplaquolls
AqCBO	Aquic Cryoborolls	PeCAP	Peregelic Cryaquepts	TyHBA	Typic Haploboralfs
ArABO	Aridic Argiborolls	TyCAE	Typic Cryaquents	TyHBO	Typic Haploborolls
ArHBO	Aridic Haploborolls	TyCAO	Typic Cryaquolls	TyHUA	Typic Haplustalfs
BoCOI	Borollic Calciorthids	TyCBA	Typic Cryoboralfs	TyNAI	Typic Natrargids
BoHAI	Borollic Haplargids	TyCBO	Typic Cryoborolls	TySOI	Typic Salorthids
BoNAI	Borollic Natrargids	TyCCP	Typic Cryochrepts	TyTFE	Typic Torrifluvents
CuCBO	Cummulic Cryoborolls	TyCOE	Typic Cryorthents	TyTOE	Typic Torriorthents
HiCAP	Histic Cryaquepts	TyCOI	Typic Camborthids	TyTPE	Typic Torripsamments
LiCOE	Lithic Cryorthents	TyCUP	Typic Cryumbrepts	TyUOE	Typic Ustorthents
LiTOE	Lithic Torriorthents	TyGOI	Typic Gypsorthids	UsHAI	Ustollic Haplargids
LiUOE	Lithic Ustorthents	TyHAI	Typic Haplargids		

Periodic Disturbances

The environment is usually perceived in terms of temperature, wind, water availability, soil characteristics, and other similar features, but just as important are the various disturbances that have been occurring in most ecosystems for millennia. A disturbance can be defined as any event, usually occurring over a comparatively short time, that kills some members of the ecosystem or causes a rapid destruction or redistribution of biomass within the ecosystem. Examples include fire, windstorms, burrowing by mammals, floods, and epidemics of certain insects.[3] Attempts to suppress such disturbances may create an environment in which some native species cannot survive.

Fire is a well-known natural disturbance. Wherever there is the potential for the accumulation of sufficient fuel—in the form of dried plant biomass—a fire is inevitable. Long before humans learned to use fires, lightning strikes provided the necessary ignition, even in grasslands. The number of lightning strikes to the ground is surprisingly large (often several hundred during a summer storm), and although not all lightning strikes caused a fire in presettlement times,

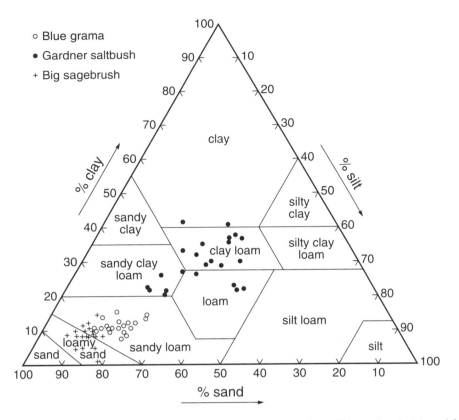

FIG. 3.9 Soil texture has a strong influence on plant distribution, as shown here for blue grama, Gardner saltbush, and big sagebrush. Adapted from Nichols 1964.

once started they could burn for weeks and months over a large area. Early explorers often noted smoke on the horizon. Native Americans frequently started fires themselves, probably to reduce tree or shrub density or to drive big game. Meriwether Lewis wrote about the scarcity of timber along the Missouri River in 1805: "This want of timber is by no means attributable to a deficiency in the soil to produce it, but owes its origin to the ravages of the fires, which the natives kindle in these plains at all seasons of the year. The country on both sides of the river, except some of its bottom lands . . . is one continued open plain, in which no timber is to be seen except a few . . . clumps of trees, which from their moist situation, or the steep declivities of hills, are sheltered from the effects of fire" (Thwaites 1905).

Fires could be annual events in some of the more productive grasslands, or they could be comparatively rare events in some high-elevation forests, where adequate fuels to sustain a fire require up to two hundred years or longer to develop, or where the climatic conditions required for fires are rare. In the foothills a fire might be expected every five to twenty-five years (Houston 1973; Fisher et al. 1987). With large concentrations of bison and other herbivores, there is less fuel and the chances of fire are lower. Grazing by domestic livestock reduces rangeland flammability as well, and combined with the suppression of most fires that do start, creates a favorable environment for the establishment and growth of trees and shrubs.

Although all fires are conspicuous disturbances, some are more severe than others. Crown fires, those that become extremely hot and burn through the forest canopy, may kill most of the trees and abruptly change the nature of the ecosystem. In contrast, surface fires often consume only detritus, killing few organisms. Perennial grasses and forbs, and some shrubs and trees, may simply sprout from surviving roots, leading to recovery within a year or two. Because plant communities are composed of some species that are tolerant of fire and others that are not, fire can have a significant effect on species composition. A good example is sagebrush steppe, where fire may kill most of the big sagebrush, creating a grassland until the big sagebrush becomes reestablished ten to twenty years later.

Windstorms constitute another periodic disturbance and are most important in forests, where trees laden with heavy snow are especially susceptible. Individual trees are often toppled, creating canopy gaps important to the survival of some understory species. Large numbers of trees may be blown down, as occurred in the Teton Wilderness Area in August 1987. Within a few minutes, the trees over an area of 6 km² were toppled, creating a landscape reminiscent of that after the eruption of Mount St. Helens (see fig. 10.2). Evidence of windstorms is easy to find in montane forests.

Fire and windstorms are physical disturbances, but biotic disturbances are just as important. Burrowing animals (including prairie dogs, pocket gophers, badgers, and harvester ants) create disturbances that often cause the death of some organisms but provide a suitable habitat for others. The gaps created in this way are individually small, but combined they modify the environment over large areas. Furthermore, burrowing animals move periodically to different locations, with the result that a large proportion of the land area is subjected to burrowing activity over long periods. Such disturbances are sometimes a nuisance for livestock owners, but there are beneficial influences as well—better infiltration of water, better aeration, a mixing of organic matter deeper into the soil, and generally higher plant productivity (see chapter 5).

Population explosions of certain insects also create disturbances. In Wyoming, the insects most commonly involved are grasshoppers and Mormon crickets in the lowlands and mountain pine beetles, spruce beetles, and western spruce budworms in coniferous forests. Grasshoppers and Mormon crickets sometimes reach densities of 50–250 per square meter in grasslands and shrublands, consuming most of the herbaceous biomass in a short time. Beetles may kill trees over large areas. Aside from influencing species composition, insect outbreaks accelerate nutrient cycling and may increase or reduce flammability.

Overall, the environment of any area involves the continuously varying interaction of many factors that changes either gradually or abruptly across the landscape and over time. No two areas have exactly the same environment; generalizations must be made with caution. The complexity caused by the interaction of so many variables makes the environment difficult to conceptualize, but usually a few factors are especially influential. In Wyoming these key factors include seasonal and elevational temperature differences; soil infiltration rates, water-holding capacity, and salinity; and moisture input that is heavily dependent in some areas on snow redistribution by the wind. Periodic disturbances by fire, wind, burrowing animals, and a few insect species further modify an already diverse landscape mosaic. Regionally, climatic characteristics have caused the formation of grasslands, shrublands, forests, and alpine tundra. Locally, the nature of the vegetation is dependent on soil characteristics, snowdrifting, topographic differences, and disturbances such as fire, flooding, and large windstorms. Detail to the landscape mosaic is added by burrowing animals in grasslands and meadows, by wind- or insect-caused patches in forests, and by frost-generated soil patterns in the alpine zone.

I I

Along Creeks
and Rivers

CHAPTER 4

Riparian Landscapes

Riparian landscapes occur along creeks and rivers, where water is readily available. They can be narrow or wide and are best defined by plants that are not found in drier environments. If free-flowing streamwater is impounded or diverted from the main channel, whether by beaver dams or irrigation systems, the nature of the riparian zone changes.

Riparian landscapes are valuable for several reasons. Though a mere 1 percent or less of the region is classified as riparian, an estimated 80 percent of native animals depend at some time during the year on this limited area for food, water, shelter, and migration routes (Olson and Gerhart 1982). Water flows only downward, but animals move along riparian corridors in both directions through an intricate network of waterways that sometimes pass through otherwise hostile environments (fig. 4.1). In addition, many riparian zones filter sediments and nutrients, thereby influencing water quality. Their capacity for storing water in alluvial sediments during the spring helps sustain streamflow later in the summer. Domesticated animals often congregate in riparian zones, as do people. Cities, towns, roads, and cultivated fields are usually located in these zones. Not surprisingly, most riparian zones at lower elevations are privately owned.

The riparian landscape can be viewed as a continuum from the smallest rivulet at high elevations to the largest river in the lowlands. Streams that begin with the melting of alpine snowbanks cross the full range of climatic conditions in the region. Elsewhere, streams are much shorter, such as those that originate as seeps or springs in the lowlands. Except at occasional waterfalls and rapids or at entrances to canyons, the climatic factors that influence riparian ecosystems change gradually, even imperceptibly. Incised streams with hard bottoms of gravel and boulders eventually become meandering streams that have mud or sand bottoms.

Typically, streamflow volume increases down the elevational gradient as small creeks become rivers. Streamflow changes seasonally as well, with the time of maximum flow usually occurring in the spring during snowmelt. At this time new channels are formed; old trees are toppled along eroding banks; new sediments are deposited as bars or on floodplains; and new terraces and levees or "debris jams" are formed. Inevitable changes in streamflow and channel location cause periodic changes in the riparian vegetation mosaic.

Physical factors that determine the nature of streamside vegetation include stream gradient, sinuosity, channel-width-to-depth ratios, topography, and soil type. Beaver dams, water development projects, and grazing by large mammals are key biological factors. The soils of riparian environments are important in determining plant species composition and are highly variable, having alluvial lenses of gravel, sand, silt, or clay.[1] The vegetation is also highly variable and may include marshes, meadows, shrublands, and tree-

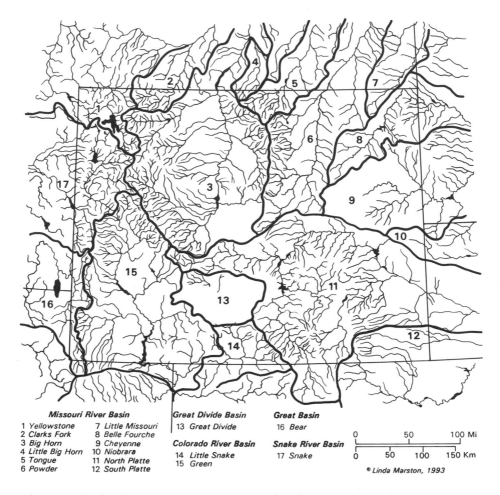

Missouri River Basin		
1 Yellowstone	7 Little Missouri	
2 Clarks Fork	8 Belle Fourche	
3 Big Horn	9 Cheyenne	
4 Little Big Horn	10 Niobrara	
5 Tongue	11 North Platte	
6 Powder	12 South Platte	

Great Divide Basin
13 Great Divide

Colorado River Basin
14 Little Snake
15 Green

Great Basin
16 Bear

Snake River Basin
17 Snake

0   50   100 Mi
0   50   100   150 Km

© Linda Marston, 1993

F I G. 4.1   A network of creeks and rivers connect the landscapes of the region. Many of the smaller creeks are ephemeral, lacking water during the driest part of the summer. The location of Yellowstone Lake and the larger reservoirs is shown. Drainages 14–17 lie west of the continental divide. Drainage 13 is the Great Divide Basin (see fig. 1.2), which is formed by a split in the continental divide and has no outlet.

dominated woodlands (table 4.1; see table 4.2 for a list of a characteristic animals). Some vegetation types are found on fine-textured soils; others occur more often on coarse soils.

From Rivulet to River

RIPARIAN LANDSCAPES IN THE MOUNTAINS

Riparian environments in the mountains are usually quite different from those in the lowlands because of steeper stream gradients. In particular, most stream banks have little soil accumulation because of continuous erosion, and narrow valleys limit channel shifting. Exceptions occur where the topography flattens, forming willow-dominated wetlands with finer textured soils and meandering channels. Mountain streams are also cooler because of higher elevations and cold-air drainage. Climatologists have discovered that watersheds are cold-air sheds as well.

At high elevations, streams begin as melting

TABLE 4.1 Characteristic plants of riparian vegetation types in Wyoming

	Broadleaf Woodland	Conifer–Broadleaf Woodland	Alder–Conifer Woodland	Tall Willow Shrubland	Short Willow Shrubland	Cinquefoil–silver Sagebrush Shrubland	Greasewood Shrubland	Riparian Meadow
TREES								
Blue spruce	—	X	X	—	—	—	—	—
Boxelder	X	—	—	—	—	—	—	—
Engelmann spruce	—	X	X	—	—	—	—	—
Green ash	X	—	—	—	—	—	—	—
Lodgepole pine	—	X	X	—	—	—	—	—
Narrowleaf cottonwood	X	X	—	—	—	—	—	—
Peachleaf willow	X	X	—	—	—	—	—	—
Plains cottonwood	X	—	—	—	—	—	—	—
Russian olive	X	—	—	—	—	—	—	—
SHRUBS								
Barrenground willow	—	—	—	—	X	—	—	—
Booth willow	—	—	—	X	—	X	—	—
Coyote willow	X	X	—	—	X	—	—	—
Drummond willow	—	—	—	X	X	—	—	—
Geyer willow	—	—	—	X	X	—	—	—
Grayleaf willow	—	—	—	—	X	—	—	—
Monticola willow	—	—	—	X	X	—	—	—
Planeleaf willow	—	—	—	—	X	—	—	—
Wolf willow	—	—	—	—	X	—	—	—
Basin big sagebrush	X	—	—	—	—	—	X	—
Common snowberry	X	—	—	—	—	—	—	X
Greasewood	—	—	—	—	—	—	X	X
Red-osier dogwood	X	X	X	X	—	—	X	—
Rubber rabbitbrush	X	X	—	—	—	—	—	—
Saltcedar	X	—	—	—	—	—	X	X
Shrubby cinquefoil	—	—	—	X	X	X	—	X
Silver buffaloberry	X	X	X	—	—	X	X	X
Silver sagebrush	X	X	X	X	X	X	X	X
Silverberry	X	X	—	—	—	—	—	—

45

TABLE 4.1 (continued)

	Broadleaf Woodland	Conifer–Broadleaf Woodland	Alder–Conifer Woodland	Tall Willow Shrubland	Short Willow Shrubland	Cinquefoil–silver Sagebrush Shrubland	Greasewood Shrubland	Riparian Meadow
Skunkbush sumac	X	X	—	—	—	—	—	—
Thinleaf alder	—	X	X	—	—	—	—	—
Water birch	X	X	X	—	—	—	—	—
Wood rose	X	X	X	—	—	—	—	X
GRASSES								
Alkali cordgrass	—	—	—	—	—	—	X	X
Alkali sacaton	—	—	—	—	—	—	X	X
Alkaligrass	—	—	—	—	—	—	X	X
Basin wildrye	X	—	—	—	—	—	—	X
Bluejoint reedgrass	—	X	X	X	X	X	X	X
Foxtail barley	X	X	—	—	—	—	—	X
Kentucky bluegrass	X	X	—	—	—	—	—	X
Meadow barley	X	—	—	—	—	—	—	X
Orchard grass	—	—	—	—	—	—	X	X
Saltgrass	—	—	—	—	—	—	X	X
Slender wheatgrass	X	—	X	X	X	X	—	X
Smooth brome	X	—	—	—	—	—	—	X
Thickspike wheatgrass	X	—	—	—	—	—	—	X
Timothy	X	—	—	X	X	X	—	—
Tufted hairgrass	—	X	X	X	X	X	—	X
Western wheatgrass	—	—	—	—	—	—	—	X
SEDGES								
Beaked sedge	—	—	X	X	X	—	—	—
Nebraska sedge	—	—	X	X	X	—	—	—
Water sedge	—	—	X	X	X	—	—	—
FORBS								
Horsetail	X	X	X	X	—	—	—	X
Iris	—	—	—	—	—	X	—	X
Mountain bluebells	—	X	X	X	X	—	—	—

Note: — = absent or less common

TABLE 4.2 Mammals, birds, reptiles, and amphibians found in riparian landscapes

MAMMALS

Beaver	Moose
Coyote	Mouse, Deer
Deer, White-tailed	Muskrat
Elk	Myotis, Little brown
Fox, Red	Raccoon
Mink	Skunk, Striped

BIRDS

Blackbird, Brewer's	Robin, American
Blackbird, Red-winged	Snipe, Common
Blackbird, Yellow-headed	Sparrow, Lincoln's
Dipper, American	Sparrow, Song
Eagle, Bald	Swallow, Bank
Flicker, Northern	Swallow, Northern rough-winged
Goldfinch, American	Swallow, Tree
Harrier, Northern	Swallow, Violet-green
Kestrel, American	Vulture, Turkey
Kingfisher, Belted	Woodpecker, Downy
Magpie, Black-billed	Woodpecker, Hairy
Nighthawk, Common	Wren, House
Owl, Great horned	

REPTILES

Garter snake, Red-sided
Garter snake, Valley
Garter snake, Wandering

AMPHIBIANS

Frog, Boreal chorus

snow or seeps. As the rivulets merge, a first-order stream develops with vegetation dominated by several species of sedges, grasses, and willows (table 4.1). Meadows and short willow shrublands (fig. 4.2) become more conspicuous as one first-order stream merges with another, forming a second-order stream, and as two second-order streams merge to form a third-order stream. Lower on the mountain, such shrubs as thinleaf alder and tall willows become more conspicuous (figs. 4.3, 4.4). In the foothills and high inter-mountain basins, trees predominate (fig. 4.5)—Engelmann spruce, narrowleaf cottonwood, lodgepole pine, aspen, and occasionally, blue spruce and balsam poplar. Some of the species occur on comparatively dry banks, others in very wet habitats. The adjacent uplands typically have coniferous forest, aspen woodlands, mountain meadows, or shrublands dominated by mountain big sagebrush. Wet meadows commonly have shrubby cinquefoil, mountain silver sagebrush, or barrenground willow.

Several studies have examined streamside vegetation in the mountains (Patten 1968; Starr 1974; Phillips 1977; Brunsfeld and Johnson 1985; Baker 1989; Kittel 1994). Collectively, the results suggest the following environmental factors as causes of the patchy, highly variable riparian mosaic: the duration of soil saturation, soil depth and texture, the frequency of flooding, depth to the water table, oxygen availability in the soil, the duration of snow cover, growing season length and temperature, big game browsing, and ice damage in the spring. All are affected by such terrain features as valley width and orientation, drainage basin area, and stream gradient and sinuosity. Abrupt changes in vegetation are typically associated with different riparian landforms (point bars, cutbanks, terraces, islands, and the channel itself). Beaver dams—and to a lesser extent, debris jams —can also have dramatic effects on the vegetation mosaic. Often within short distances, channel morphology changes between meandering, braided, and incised (fig. 4.6).

FIG. 4.2   Headwaters of the Wind River (2,797 m; 9,200 ft) near Togwotee Pass in the Absaroka Mountains. The riparian shrublands are dominated by Wolf willow; shrublands away from the creeks are domi- nated by mountain big sagebrush and Idaho fescue; and the forests are dominated by lodgepole pine, Engelmann spruce, and subalpine fir.

F I G. 4.3   Mid-elevation riparian shrublands on Jack Creek in the Sierra Madre, at about 2,584 m (8,500 ft) elevation. Common shrubs and trees include Geyer willow, Bebb willow, Booth willow, alder, Engelmann spruce, and lodgepole pine. The forests are dominated by lodgepole pine at lower elevations and Engelmann spruce and subalpine fir at higher elevations. Mountain big sagebrush, Idaho fescue, and numerous forbs dominate the dry slope in the foreground.

RIPARIAN LANDSCAPES IN THE LOWLANDS

Streams flow less rapidly in the flatter terrain of the lowlands, leading to the deposition of alluvial soils over sometimes broad floodplains. Channel meandering is typical except where rivers have cut canyons through once-buried mountain ranges or where steeper gradients lead to the formation of braided streams (fig. 4.6). Freshly deposited alluvium along the channel of meandering streams is usually composed of sand or gravel and is referred to as a point bar or depositional bar. These bars provide excellent sites for the establishment of cottonwoods and some species of willow. Across from the point bars are eroding cutbanks, where trees and other plants eventually fall into the water (fig. 4.7).

Away from the point bars and cutbanks is the active-channel shelf, where coarser materials are deposited during flood periods; beyond that is the floodplain, which has one or more terraces. The terraces are former floodplains that have become elevated because of stream downcutting. Terraces are rarely flooded, creating an environment that is sometimes transitional between the riparian and upland landscapes. Salt accumulation is often higher on terraces than on the floodplain, probably because leaching events, which can dissolve and remove salts, are less common. Consequently, halophytes (such as greasewood) typically occur

F I G. 4.4   Taller willows, such as Bebb willow, occur along creeks in the foothills, forming a riparian greenbelt in the otherwise arid landscape on Little Sage Creek (the north end of the Sierra Madre; elevation 2,280 m [7,500 ft]). Such areas often are subjected to heavy grazing and browsing by large herbivores, causing some of the shrubs to have a pruned appearance. Shrublands dominated by Wyoming big sagebrush occur on the upland.

on terraces, whereas the less salt-tolerant silver sagebrush and basin big sagebrush occur on the lower floodplain (Nichols 1964).

The soils and depth to water table are not uniform in riparian zones, causing considerable patchiness in the vegetation. In Jackson Hole, for example, new sandbars are often occupied by sandbar willow, flooded fine-textured soils have Booth willow, and unflooded terraces regularly have Geyer willow (Houston 1967; Miller 1979). Cottonwood regeneration commonly occurs along sand and gravel bars, frequently creating bands of trees that provide a living record of flooding patterns and channel migration (figs. 4.8, 4.9). The older trees typically occur some distance from the channel (Everitt 1968; Johnson et al. 1976; Irvine and West 1979; Akashi 1988).

The riparian zone at lower elevations is often characterized by tree-dominated woodlands. The greatest tree diversity is found in the eastern lowlands, where plains cottonwood, the most abundant riparian tree, occurs with ash, boxelder, lanceleaf cottonwood, peachleaf willow, and rarely, American elm (Allred 1941).[2] Shrubs that often occur in the understory include chokecherry, hawthorn, rubber rabbitbrush, silver buffaloberry, silver sagebrush, skunkbush sumac, wild rose, and various species of willow. At slightly higher elevations, narrowleaf cottonwood is the predominant tree in riparian woodlands. Occasionally, Colorado blue spruce can be found as well (for example, along the lower Hoback River, Snake River, and upper Wind River). Associated shrubs include red-osier dogwood, silver buffaloberry, silver sagebrush, thinleaf alder, water birch, and several species of willow. Willows frequently form dense thickets or shrublands. (See figs. 4.2, 4.3, and table 4.1.)

Portions of some lowland rivers are naturally devoid of trees (see fig. 4.10). In 1813, Robert Stuart wrote that an island in the North Platte River in Nebraska, "abounds with timber, but not a single tree decorates the main shore." The same was apparently true for some large rivers farther west, such as parts of the Sweetwater (Sun 1986; Johnson 1987) and Laramie rivers in Wyoming. Elsewhere, cottonwood trees were abundant. For example, in 1860, W. F. Raynolds wrote about the Wind River near Riverton: "The valley is a mile or more in width, and the immediate banks of the stream for 300 or 400 yards are covered with a thick growth of cottonwood" (Raynolds 1868). Similarly, Col. D. B. Sacket wrote in 1877, "All along the Bighorn River . . . much fine, large cottonwood timber grows" (Dorn 1986, 69).

The near lack of trees in some areas could have been caused by several factors—for example, frequent prairie fires, browsing by large mammals, a hydrologic regime that did not favor tree establishment, or stable channels dominated by grasses and sedges that reduced the chances of tree-seedling establishment (as discussed below). The specific cause in each area is still unknown and indeed may have varied at different times.[3] It

F I G. 4.5 Larger rivers at low elevations, and with more dependable streamflow, typically have woodlands dominated by narrowleaf or plains cottonwood. Wyoming big sagebrush is taller and more dense in ravines where snow accumulates, such as in this area along the Green River (1,885 m; 6,200 ft).

seems unlikely that seed dispersal is the explanation, as many riparian species (for example, cottonwood and willow) have seeds that are both buoyant and easily spread by wind.

Riparian vegetation devoid of trees on the eastern plains is commonly dominated by meadows of Baltic rush, Nebraska sedge, prairie cordgrass, redtop bent grass, silver sagebrush, and four widespread, introduced species—Kentucky bluegrass, smooth brome, sweetclover, and timothy. In areas where salts accumulate, such as on high terraces that are flooded less frequently, various halophytes are common (for example, greasewood, inland saltgrass, and alkali sacaton). Riparian marshes with cattails occur in wetter habitats.

Woodlands are now more common along some rivers than in presettlement times (Johnson 1987). Climatic conditions may be more favorable for cottonwood establishment, the frequency of tree-damaging floods may have been reduced by reservoirs, and return flows (that is, water that seeps through a field back into the groundwater or stream) from upstream irrigation may have facilitated tree establishment and growth (Baker 1990).[4]

The riparian mosaic at the confluence of the Bighorn and Shoshone rivers in northern Wyoming has been studied in some detail and illustrates landscape patterns that can be found elsewhere (fig. 4.11, and see fig. 4.9). Fertile soils and water availability in the Bighorn Basin sustain one of the most important agricultural regions in the state, but remnants of the presettlement riparian mosaic persist along those portions of the rivers located in the Bighorn Canyon National Recreation Area. Plains cottonwood woodland is

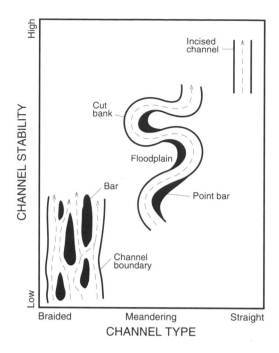

F I G. 4.6   Braided, meandering, and straight channels have different levels of resistance to changes in
channel location, providing different environments
for riparian plants. Bars are formed by the deposition
of sand, silt, and clay sediments.

the most conspicuous vegetation type along the
rivers. Associated tree species include peachleaf
willow, silver buffaloberry, and the introduced
Russian olive.[5] A variety of shrubs, forbs, and
grasses are found as well (table 4.1). Shrub and
herbaceous plant cover is low where tree density is
high because of inadequate light beneath the
woodland canopy and greater competition from
trees for water.

Interspersed in the floodplain woodlands are
shrublands dominated by basin big sagebrush,
greasewood, rubber rabbitbrush, saltcedar,[6] sandbar willow, skunkbush sumac, western snowberry, wild rose, and a variety of forbs and grasses.
Abrupt transitions from woodland to shrubland,
together with abundant evidence of fire-scarred
trees and logs, suggest that fire has played an important role in converting older woodlands to
shrublands. Cattail marshes occur in wet oxbows,
and meadows are found in drier areas, such as
recently abandoned channels.

Shifting Riparian Mosaics

Some riparian zones are relatively stable because
the creek or river is at the bottom of a V-shaped
valley incised in bedrock. In contrast, riparian
mosaics on the alluvial soils of broad floodplains
change rapidly (Maddock 1976). Both plant
growth and succession are more rapid. Fires are
frequent because of abundant fine fuels, and
spring floods tend to overflow banks, sometimes
shortening the channel and creating oxbows. A
complex and shifting mosaic develops.

The rapid change that can occur in lowland
riparian mosaics was documented by Akashi
(1988). Comparing a series of aerial photographs
taken over the Bighorn River, she showed how
some of the woodlands, shrublands, and mea-

F I G. 4.7   The Snake River in Grand Teton National
Park showing point bars on the right and a cut bank
on the left (elevation approximately 2,037 m; 6,700
ft). Douglas-fir is the dominant tree on the embank-
ment in the foreground, mountain big sagebrush
steppe occurs on the terrace and glacial outwash
plain, and lodgepole pine forests occur in the background on glacial moraines. See fig. 15.1 for a photograph taken approximately 90° to the left.

FIG. 4.8   A band of plains cottonwood seedlings along the Bighorn River east of Lovell (elevation 1,125 m; 3,700 ft). If the seedlings survive subsequent flooding events, a curvilinear band of trees will develop, such as those apparent in fig. 4.9.

dows are in different locations today than they were just fifty years ago. Fire was a major factor causing the change. Though moister than on the uplands, the dense, herbaceous biomass along the river is flammable in the fall. The trees are often killed by inevitable fires, leading to the formation of shrublands that are dominated by skunkbush sumac and other shrubs.

Another important factor causing rapid change along the Bighorn River is flood control. Akashi observed much less channel shifting today than before the enlargement of the Boysen Reservoir in 1952. Consequently, fewer and smaller point bars have been formed, resulting in less habitat for the establishment of cottonwood seedlings.[7] At the same time, existing woodlands continue to die from old age, fire, beaver cutting, and agricultural clearing. Old-growth cottonwood woodlands appear to be declining in the area Akashi studied, a trend observed in other areas as

well (Johnson et al. 1976; Weynand et al. 1979; Miller 1979). The major cause of cottonwood decline appears to be flood suppression. Just as fire suppression has led to changes in mountain forests, flood suppression has changed the riparian zone (Johnson et al. 1976, 1982; Fenner et al. 1985; Bradley and Smith 1986; Akashi 1988; Rood and Heinze-Milne 1989; Sedgwick and Knopf 1989; Baker 1990; Knopf and Scott 1990; Rood and Mahoney 1990; J. R. Miller et al. 1993).

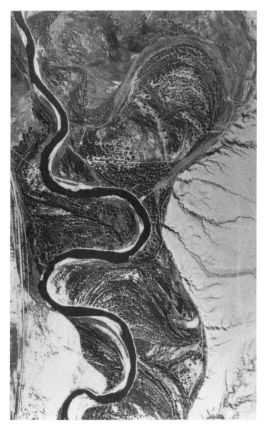

FIG. 4.9   Aerial view of the riparian woodlands along the Bighorn River east of Lovell. Former channel locations are apparent. Most of the area shown is dominated by plains cottonwood. The tree-ring record indicates that a new band of trees became established every ten to twenty years, with parallel bands becoming older and taller with distance from the river. Similar patterns are found along some other rivers, for example, the Powder River south of Arvada. Desert shrublands are found on the adjacent uplands.

F I G. 4.10   When this photo was taken by William Henry Jackson in 1870, from the summit of Independence Rock, southwest of Casper, trees and shrubs were rare or absent along the Sweetwater River (looking northeast toward Casper; river elevation approx-imately 1,763 m [5,800 ft]). The vegetation, dominated by alkali sacaton, fringed sagewort, American licorice, silver sagebrush, rubber rabbitbrush and other plants, is very similar today (Johnson 1987). Photo provided by the U.S. Geological Survey (USGS 284).

Interestingly, the situation along the Bighorn River is different from that on the nearby Shoshone River. The Shoshone is a more rapid, braided river, whereas the Bighorn is gentle and meandering. Cottonwood seedlings are observed more frequently along the Shoshone, perhaps because flooding there is more frequent, despite the Buffalo Bill reservoir 100 km upstream. A favorable habitat for seedling establishment is created more often. Similarly, Houston (1967) noted the absence of cottonwood reproduction along the lower Snake River, which has a regulated flow because of the Jackson Lake dam and extensive diking, but he commonly observed cottonwood regeneration along the nearby, free-flowing Gros Ventre River.

The riparian landscapes along meandering and braided streams change so rapidly that it is difficult to think about successional patterns in the traditional sense (Baker 1988). Before the biota has a chance to ameliorate the environment significantly, thereby creating a more stable climax community, the river channel typically shifts to a new location or a fire occurs because of rapid fuel accumulation. Campbell and Green (1968) concluded about the Rio Grande in New Mexico that "because of disturbances in the flood-prone channel, species form mosaics of seral stages of communities with different combinations of species dominating each stage; thus, the vegetation probably never reaches a 'climax.'" Youngblood et al. (1985) included a successional diagram for the

RIPARIAN SUCCESSION ALONG THE BIGHORN RIVER

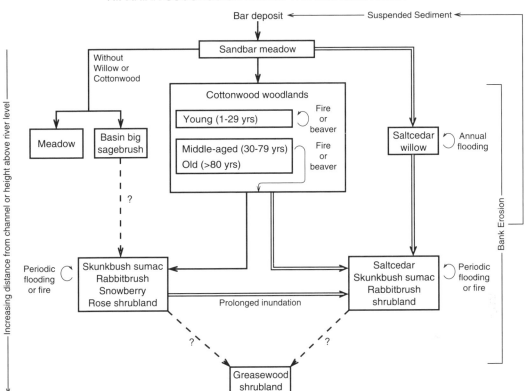

F I G. 4.11   A diagram illustrating probable succession on new point bars along the meandering channel of the Bighorn River east of Lovell (elevation 1,125 m; 3,700 ft). Succession has changed considerably in this area since the introduction of saltcedar, as indicated by the wider arrows. With flood control, many point bars are not adequate for the establishment of cottonwood seedlings. Therefore, the proportion of the landscape dominated by cottonwood decreases as older groves are converted to shrublands. Adapted from Akashi 1988.

riparian vegetation in western Wyoming but commented that "climax conditions within the riparian ecosystem may not be obtainable given the periodic fluctuations of environmental conditions."

Nevertheless, several successional trends have been observed. Cottonwoods and some willows are commonly considered to be pioneer species, with gravel bars usually colonized by narrowleaf cottonwood, sandbars by sandbar willow, and mudflats by Booth willow in western Wyoming and eastern Idaho (Reed 1952; Houston 1967; McBride and Strahan 1984; Youngblood et al. 1985; Baker 1990). Boggs (1984) noted that cottonwood reproduced only along the stream mar-

gin and that as the trees became older and more distant from the margin (sometimes after only sixty years), they died and gave way to shrubland or grassland. Akashi (1988) observed a similar pattern along the Bighorn River in Wyoming (see figs. 4.9 and 4.11). Thompson (1983) found that if the water table became deeper than 1.2 m, different species of willow and other shrubs or small trees would become more common than cottonwood (namely, red-osier dogwood, Rocky Mountain juniper, silver buffaloberry, and skunkbush sumac). Apparently, the shrublands and grasslands should be considered as the climax community along some lowland rivers.[8]

Bank Storage, Stream Hydrology, and Beaver

The plant growth on floodplains slows the movement of water and thereby greatly increases the potential for sedimentation when floods do occur. As sediments accumulate, the floodplain becomes broader and flatter. Consequently, the flooding waters are spread over an even larger area, which increases sedimentation still more. The accumulated sediments increase the amount of water that can be retained in the bank and floodplain soils, which increases the probability of streamflow throughout the year (Stabler 1985).

The potential for bank storage is difficult to measure, but a recent study west of Cheyenne provides some insights. As part of an effort to augment the water supply for Cheyenne, water was diverted from Douglas Creek on the west-facing slope of the Medicine Bow Mountains to the South Fork of Middle Crow Creek on the east side of the Laramie Mountains. The initial diversion was done in August 1985, after streamflow in the ephemeral South Fork had ceased for the summer. Three weeks passed before the stream was flowing 10 km downstream, probably because much of the initial water was retained in bank storage. Two years later, the dominant plants in the riparian zone had shifted to plants characteristic of wetter habitats (Wolff et al. 1989; Henszey et al. 1991).

Maintaining bank storage, however, has its costs in the form of water consumption by phreatophytes. Abundant riparian vegetation reflects not only stored water but also high levels of photosynthesis and transpiration. Every kilogram of plant tissue produced by photosynthesis may require as much as 400 kg of water because up to 99 percent of the water entering a plant through the root system is lost by transpiration from the leaves. This water is required to keep stomata open for the uptake of carbon dioxide and to maintain turgid leaves for better light interception. Water transpired by plants, however, is water lost for downstream uses—a conveyance loss along with seepage to groundwater aquifers and evaporation directly from the stream surface. Most managers view the water consumed by riparian vegetation as a small price to pay for the benefits of increased forage, shade for livestock, erosion control, sedimentation during floods, biological diversity, better wildlife habitat, and sustained late-summer flows.

Few data are available to estimate quantitatively the actual costs of maintaining bank storage or, in other words, the amount of water that is used by riparian vegetation and not available for streamflow. Two generalities, however, seem probable. First, streamflow depletion by phreatophyte transpiration is higher in small streams than in rivers; second, depletion is higher during late summer, when streamflow is low and transpiration is more rapid. Preliminary estimates suggest that reductions in streamflow by phreatophyte transpiration may range from 0.01 to 2.0 percent in June when streamflows are high, and from 0.2 to 48 percent in late August, when they are low (Jeffrey Foster, pers. comm.).

BEAVER

Beaver were, and continue to be, another important factor influencing riparian landscapes (fig. 4.12, 4.13). As the only mammal that builds dams, aside from humans, beaver have created many meadows and broadened many floodplains throughout the West—in both the lowlands and mountains. Their influence on riparian landscapes in the past cannot be fully comprehended without considering that the beaver population for thousands of years was much greater than it is now. In many western valleys, people today live on the income from the soil capital laid down by beavers centuries ago (J. Sedell, pers. comm.). Through dam construction on small and midsized streams, beaver backed up water over large areas, hastening the rate of sedimentation, meadow formation, and riparian vegetation development (Ives 1942). Although their dams occasionally failed, the network of dams up and down many streams served as a buffer.

Some have suggested that beaver are a "keystone species," affecting ecosystem structure and function far beyond their immediate requirements for food and space (Rutherford 1954; Call 1966; Naiman et al. 1986). Sedimentation behind the dams provides clearer water downstream, and greatly enhanced bank storage leads to sustained streamflow throughout the year. Neff (1957), working in Colorado, estimated that a valley with

F I G. 4.12   Centuries of beaver activity have created a relatively flat mountain meadow in what might have been a steep-sloping valley. Water sedge and beaked sedge are common in the wet meadows; lodgepole pine dominates the forested upland. Elevation 2,766 m (9,100 ft); Medicine Bow Mountains, southwest of Dry Park.

an active beaver colony had eighteen times more water storage in the spring and a higher streamflow in late summer than a drainage where the beaver had been removed. Higher late-summer streamflow benefits fish and wildlife as well as landowners. Beaver also thin woodlands by cutting trees, thereby stimulating the formation of new sprouts and the growth of understory plants.

In addition to modifying the structure of stream and riparian habitats, beaver also influence water chemistry (Skinner et al. 1984; Naiman et al. 1986; Parker 1986). Sediments and organic matter accumulate behind beaver dams, allowing for clearer, better oxygenated water below. Bottom sediments behind the dams are often anaerobic, but the pond water is still oxygenated by stream riffles. Moreover, the sediments provide a slow, steady source of such important nutrients as inorganic nitrogen and phosphorus, thereby modify-

ing water chemistry as well as improving conditions for riparian plant growth. Bank erosion affects streamwater chemistry as well (Parker 1986).

Somewhat similar to beaver dams are the debris jams that commonly occur along smaller wooded streams (fig. 4.14; Bilby and Ward 1991). Debris jams are most often initiated by a tree that falls across a stream, or by floating wood and other detritus that become lodged in the channel. Other floating material then accumulates behind such obstacles until a dam is created. Like beaver dams, the debris jams increase sedimentation and improve water quality. Gradual decomposition of the original log or logs eventually leads to dam failure and a sudden redistribution of sediments and accumulated debris, but by that time other jams are established nearby. Debris jams are a common and natural feature of many streams in

FIG. 4.13   This gully might have been formed by erosion when the beaver population was no longer able to maintain enough dams. Beaver are now active in this small creek south of Baggs (elevation 1,824 m; 6,000 ft). If the gully is filled again by sediment accu-mulation behind new dams, such as the one shown here, the water table would rise and the greasewood and Wyoming big sagebrush on the adjacent upland would be replaced by willow shrublands and riparian meadows.

the Rocky Mountains. Beaver dams may be larger and more permanent than debris jams because of the construction and maintenance activities of the beaver, but the effects of debris jams are nevertheless significant.

The human reaction to beaver dams and debris jams has often been drastic: remove them to "improve" streamflow or stream aesthetics. Debris jams sometimes appear messy, and beaver dams may inundate land that owners prefer not be flooded. Indeed, beaver can become a nuisance when their populations are large or when they interfere with irrigation projects.

What has been the effect of reduced beaver populations in the lowlands? Along some streams, the alluvial sediments that were originally deposited behind beaver dams have eroded. With fewer dams to dissipate the energy of spring floodwaters, new gullies can be created in a year or two, sometimes leaving fence posts hanging in midair.

As the channel cuts deeper, the water table is lowered, and surface sediments on either side of the gulley begin to dry out. The riparian vegetation changes from willows and moist meadows, which are relatively rare in the landscape and valuable for livestock and wildlife, to more xerophytic grasslands or greasewood and sagebrush shrublands. The riparian zone narrows, and gully erosion continues as long as the flows are rapid and unobstructed. Such flows provide less opportunity for bank storage and, in fact, through erosion are constantly reducing the potential for bank storage.

Ironically, erosion along lowland streams is sometimes of sufficient concern to call for building dams, which is far more costly than the problems caused by beaver. As one Wyoming rancher noted (Randall 1983), "Beavers can be a pain in the neck when you get too many in the wrong place. On the other hand, how are we going to get this erosion stopped if we don't use beaver? They work pretty cheap, and I've never seen a lazy one yet."

The role of beaver in preventing erosion and creating riparian habitat has long been recognized. It was therefore natural for land managers to consider transplanting beavers to severely eroded gullies on publicly owned desert shrublands in southwestern Wyoming. They observed, however, that the beaver had difficulty in making and maintaining dams out of greasewood and sagebrush.[9] One Bureau of Land Management (BLM) biologist, Bruce Smith, wondered whether the beaver would be more successful if construction material were provided. He cut several aspen trees in the distant foothills and placed them on the bank near an existing beaver lodge. Several days later a dam was being constructed, and

F I G. 4.14   Trees falling across streams create "debris jams" that provide a more diverse habitat for fish and other organisms. Here, Mullen Creek in the Medicine Bow Mountains is flanked by Engelmann spruce, subalpine fir and thinleaf alder; elevation 2,675 m (8,800 ft).

within two years much of the gulley behind the dam was filled. With the rising water level, sagebrush and greasewood became less common, and meadow grasses and willows more abundant.

The BLM experiment involved working with beaver rather than against them and has now been repeated successfully in several locations. Problem beaver are trapped where they are unwanted, moved to gullied desert streams, and provided with construction materials. Although it was natural to look for aspen initially, riparian ecologists have learned that even old tires wired together can serve as a stabilizing framework for new beaver dams. Once the tires are in place, the beaver finish and maintain the project as long as management allows for the growth of the required woody material (Skinner et al. 1984).

Significantly, gullies are not a new landscape feature created by human activities. Dorn (1986) reviewed the journals of explorers in the 1800s and found frequent mention of gullies across which early western travelers moved their wagons with difficulty. Some gullies could have resulted from excessive bison grazing or beaver trapping by the first fur traders. Yet even the best beaver construction sometimes fails, especially with exceptionally high spring floods. Beaver populations and dam maintenance can be diminished for other reasons as well, including disease. The evidence suggests that gullies were present in presettlement times, but they are probably more common now.

Livestock, Reservoirs, and Irrigation

Riparian habitats have long been important to humans living in the semiarid west. Beaver trapping and placer mining for gold took place there, and abundant forage and irrigation water facilitated agricultural developments. Most farms, towns, and cities were located nearby, as people were drawn by the resources and aesthetic qualities of the lowland riparian environment. Floods continued to occur, often causing great damage to these settlements, but usually the accepted solution for the flood problem was dam construction or channelization. Dams not only prevented most spring floods but provided opportunities for electric-power generation, agricultural development, and recreation. Floodplain amenities were too valuable to consider moving to higher ground.

Today, more people than ever before want access to the resources provided by riparian zones. The result is high land prices along streams and rivers and increased concern about riparian zone management (Swift 1984; Johnson et al. 1985). Three topics are especially interesting in the 1990s: livestock grazing in the riparian zone and its impact on fisheries and water quality, reservoir construction and management, and the effect of different irrigation systems on streamflow characteristics.

GRAZING

Ranchers now recognize that livestock concentrate where water, food, and shade are most abundant and nearby. Excessive grazing sometimes results. Large concentrations of bison could have had the same effect on streams as cattle or sheep. Indeed, early explorers observed bison herds of thousands of animals grazing along rivers (Dorn 1986; Skinner 1986). Osborne Russell wrote about southern Montana in 1835, "The bottoms along these rivers are heavily timbered with sweet cottonwood and our horses and mules are very fond of the bark which we strip from the limbs and give them every night as the buffaloe have entirely destroyed the grass throughout this part of the country" (Haines 1965).

The explorers, however, did not provide information on the amount of time that bison spent in a particular area. Bison would have wandered at will without fences, possibly grazing an area heavily but not returning to it for a year or more. Also, the presettlement upland forage might have been as attractive to the bison as that in the riparian zone. Today, water tanks and salt blocks are used to draw cattle to the upland, at least for a portion of the year, but in some areas it seems as though the only solution to preserving the riparian habitat is additional fencing or herding.

One of the undesirable consequences of heavy grazing is bank erosion, which leads to a decline in bank storage, water quality, and fish habitat (Platts 1981; Platts et al. 1983; Kauffman and Krueger 1984; Clary and Webster 1990; Schulz and Leininger 1990; Clary et al. 1992). Such erosion can occur at any elevation, especially on small

Good

Fair

Poor

F I G. 4.15 Excessive trampling by large animals or people can cause cool, narrow creeks with overhanging banks to become warm, shallow, and less suitable for fish. Also, the sediments (black) that contribute to bank storage are lost. With proper management, streams in poor condition can be restored.

meandering streams, reversing a natural tendency during stream development—sedimentation near the banks, where the flow is slower, and a gradual stream narrowing as sediments are stabilized by plants. The deeper, narrower streams often have overhanging banks that provide excellent fish cover (fig. 4.15).

With heavy grazing by hooved animals, whether livestock or big game, the stabilizing vegetation deteriorates, banks are eroded, water storage capacity declines, the streambed becomes wider, stream depth shallower, and water temperatures increase.[10] Fish cover is reduced and late-summer streamflows may be too shallow in the now-wider channel to maintain fish populations. Several studies have documented these trends and the fact that fish production is often improved on small streams by better livestock control (Gunderson 1968; Platts 1981, 1982; Robinson 1982; Kauffman et al. 1983; Platts et al. 1983; Kauffman and Krueger 1984; Hubert et al. 1985; Stabler 1985).

Of course, poor livestock management is not the only potential cause of stream degradation. Bank erosion often occurs with road construction and other developments, and nutrients can enter the water by seepage from feedlots and fertilized cropland. One of the largest sources of additional nutrients is from municipal sewage treatment plants with outflows into the river. Usually, nutrients are not removed by sewage treatment facilities. The effects of stream degradation may be as serious for the lakes and reservoirs below as they are for the streams. A case in point is the Flaming Gorge Reservoir near the Wyoming-Utah border, where the upper part of the reservoir has algal blooms each summer that interfere with recreational activities—possibly because of nutrient additions to the Green River from agricultural runoff (DeLong 1977; Parker 1986).

Drought and the browsing of shrubs can also influence riparian vegetation. During the droughts of the 1930s, Ellison and Woolfolk (1937) observed a loss of vigor and even the death of plains cottonwood in eastern Montana. Cottonwoods apparently do not tolerate moisture stress, which may account for their demise as groundwater becomes less available at increasing distances from the channel. Reduced flows due to impoundments or water diversions may also cause the death of cottonwoods (and probably other riparian species), essentially narrowing the riparian zone. Elsewhere, along the Gallatin River, willow mortality has been attributed to elk feeding during severe winters (Patten 1968). Kay (1993) concluded that willows were once more common in some parts of Yellowstone National Park, when, he suggests, the concentrations of elk on winter ranges were much lower (see chapter 14). Browse lines on willows, juniper, and aspen are now commonly observed throughout the region, some-

times the result of browsing by livestock as well as native ungulates.[11]

RESERVOIR CONSTRUCTION AND MANAGEMENT

Because of its high elevation, Wyoming is sometimes referred to as the "headwaters of the west." Snow-fed streams flow into all major drainage systems of the region (see fig. 4.1). One hundred and sixteen reservoirs with storage of 1,000 acre-feet or more retain some of this water for irrigation and other uses later in the summer (State Engineer's Office, pers. comm.), but about 85 percent of the streamflow generated in Wyoming still flows into neighboring states (Marston and Brosz 1990).[12] Further reservoir construction has been stalled by environmental concerns, economic constraints, litigation over downstream water rights, and currently, slow industrial development caused by distant markets.

The most direct negative effects of reservoir construction are the loss through inundation of riparian habitat and the alteration of the habitat downstream. Whether the loss of prime farmland, prime wildlife habitat, or one of the few remaining free-flowing rivers in the region, sacrifices are always necessary when new reservoirs are constructed.

One of the reasons that reservoirs are controversial is that their utility can be short-lived because of sedimentation, which continues regardless of location.[13] Frequently, sedimentation is accelerated by human activities that increase erosion. Dealing with large, sediment-laden reservoirs will be a problem for future generations. Beaver also cause sedimentation, but their activities are restricted to much smaller areas. The breaking of a beaver dam, or its abandonment due to sedimentation, is not a big issue.

The various purposes of reservoirs include flood control, irrigation, power generation, and recreation. Typically, water levels fluctuate considerably—high at the beginning of the growing season and after the flood season to maximize the water available for agriculture, but low by the end of the growing season because of a greater outflow than inflow. Periods of high demand for electricity may also cause water levels to drop. Fluctuating water levels create nearly barren shores around reservoirs when the water level is down, shores that sometimes become mudflats or sources of dust clouds. Most plants cannot tolerate the drastic water-level fluctuations that occur, although such sites are sometimes occupied by weedy species capable of rapid growth (such as dock, foxtail barley, goosefoot, knotweed, saltcedar, sowthistle, summercypress, sumpweed, sweetclover, and others; Akashi 1988).

In general, patterns of inundation, sedimentation, water-level fluctuations, and streamflow inevitably change following reservoir construction in ways that affect riparian landscapes. The seriousness of these changes must be judged after considering what is lost, what is gained, and how long the benefits are likely to last.

IRRIGATION AND STREAMFLOW

By whatever means, whether from reservoirs, pumps, or streamflow diversions, irrigation is attempted when resources are available to construct a conveyance system and the soils and growing season are adequate for the crops desired. In Wyoming, this land area is small, only about 2.4 percent of the state in 1987. The effects, however, are more far-reaching than the amount of land involved. One effect is that some formerly perennial streams now flow intermittently late in the summer because of water withdrawal upstream. Intermittent or ephemeral streams are less valuable for fish, wildlife, and livestock. In 1986, the Wyoming state legislature recognized fishing and recreation as beneficial uses of streamwater. Now the flows of designated stream segments are managed to maintain or improve existing fisheries.

Significantly, some irrigation leads to sustained streamflow later in the summer. A trade-off is involved: irrigation reduces streamflow, but by spreading spring floodwater over an area larger than the natural riparian zone, more land is used for the slow release of return flows. Consequently, the chances of a late-summer flow increase (Hasfurther 1992). With flood irrigation, much of the irrigation water simply percolates through the soil and back into the groundwater or stream later in the summer. Some major rivers, such as the North Platte near Saratoga, are reported to have higher late-summer flows now than before settlement because of widespread irrigation. Many landowners believe that late-summer flows will be higher if farmers upstream irrigate their land. In this way, irrigation essentially enlarges the riparian zone. Indeed, widespread cultivation and irrigation have greatly complicated the task of defining riparian-zone boundaries.[14]

Of course, some of the irrigation water is transpired or evaporated from the cropland, thereby reducing the amount of water returned to groundwater or streams. As in most western states, agriculture and evaporation from reservoirs account for more than 90 percent of the total water consumption in Wyoming (Gribb and Brosz 1990). Losses through evapotranspiration would constitute a high proportion of the total irrigation water in drainages where the amount of flood irrigation is low, where irrigation takes place on deep upland soils, or where the amount of water required for percolation back to the stream or groundwater is greater than the amount of water applied. Many factors are involved, including the hydraulic conductivity and water-holding capacity of the soil, climatic characteristics, and water requirements for crops.

Irrigation by flooding has disadvantages as well as the potential advantage of sustained streamflows. Nutrient leaching, for example, occurs more frequently with flooding, which leads to fertilizer losses and the cultural eutrophication of streams and groundwater. Salts that may have accumulated in the soil over long periods are transported to the stream or groundwater, sometimes creating water supplies downstream (or in wells) that are too saline for livestock, agricultural, or human uses (Wetstein 1992). Removing salts from the soil is a benefit, but adding salts to other water supplies is a detriment. Problems like these can assume staggering proportions, such as in the Colorado River drainage that originates partially in Wyoming's Green River. Formerly important freshwater supplies have sometimes been converted to supplies of salt water of little or no value without expensive desalinization.

Recently, some farmers have converted to sprinkler and drip irrigation systems that are more efficient because the water is spread more uniformly and a smaller portion of the available

F I G. 4.16   Sprinkler irrigation systems use water more efficiently than flood irrigation, but they can cause lower late-summer streamflow because of lower return flows. Photo taken northeast of Farson by Joseph G. Hiller (elevation 2,034 m; 6,672 ft).

irrigation water is transported to a particular field (fig. 4.16). Available water can thus be extended over a larger amount of land. There is also less potential for the loss of fertilizer by leaching, and consequently stream eutrophication occurs more slowly. Increased crop production and lower fertilizer requirements help to cover required equipment and pumping costs.

There is considerable appeal in using sprinkler irrigation systems, but they too have disadvantages. With less leaching there is a greater potential for salts to accumulate on the surface of the soil. Eventually, this salinization may make the land unsuitable for cultivation. One solution is to flood the soil periodically, but this alternative simply washes the salts into other bodies of water. The losses of one ecosystem are the gains of another.

Also, with sprinkler irrigation there is much less return flow. Less water may be taken out, but

less water is returned. Consequently, streamflows can be lower in the fall, as, for example, in the Star Valley in western Wyoming (Brosz 1986). As significant, spring flood peaks in the Salt River are considerably higher now than when irrigation was done by flooding, causing bank erosion in some areas. The stream may be returning to a more natural condition, but not everyone would agree that these new developments are desirable.

There is one clear conclusion for riparian landscapes: streamflow regulation, fire suppression, agriculture, irrigation, livestock grazing, and other human activities are creating a riparian habitat that is quite different from that of presettlement times. To varying degrees, alterations of the riparian zone are occurring in both the lowlands and mountains, especially where roads and summer homes have been constructed in valley bottoms, where large herds of livestock or big game congregate, and where the land is cultivated. Such uses may be sustainable if management is done correctly, but the riparian zone has been altered more extensively than any other landscape. Because riparian landscapes connect and are affected by all other landscape types, it will not be possible to attribute changes to a single factor. Moreover, although there is general agreement that many human influences have been detrimental, there is also a great diversity of streams in the region. Not all will be affected by development in the same way or to the same extent. Sweeping generalizations for all riparian ecosystems are not possible, but it seems clear that the proper use of riparian resources will require a coordinated effort by riparian scientists, fisheries biologists, engineers, and land managers.

III

Plains and
Intermountain Basins

CHAPTER 5

Grasslands

The most extensive grasslands in Wyoming occur east of the Rocky Mountains (on the western Great Plains) and in several intermountain basins (see fig. 1.5). Elevation ranges from a low of 964 m in the northeast to about 2,190 m in the Shirley and Laramie basins. Parts of the lowlands have been plowed for crop production (mostly winter wheat), but large areas of grass- and shrub-dominated vegetation remain (fig. 5.1). Pronghorn antelope are still common. Scattered woodlands of ponderosa pine, limber pine, and juniper occur on outcrops of resistant bedrock, such as sandstone, limestone, scoria, and some shales. Foothill grasslands and shrublands occur on upper slopes, and sand dunes and badlands are found in a few areas (see chapter 8). Aridisols are most common, with Mollisols occurring in some moist depressions where the NPP has been comparatively high for a longer period (see table 3.1).

Most Wyoming grasslands are known as either shortgrass prairie or mixed-grass prairie (Weaver and Albertson 1956; Singh et al., *Vegetation*, 1983; Singh et al. *Geography*, 1983). Shortgrass prairie, which occurs primarily in the southeastern corner of the state and southward into Colorado, is characterized by the dominance of two common short grasses—blue grama and buffalograss.[1] Mixed-grass prairie occurs over most of eastern Wyoming. Sometimes the grasses grow in small clumps and are known as bunch grasses, while rhizomatous grasses are sod-formers. Sod-formers seem to be more typical where the environment is more mesic, such as alluvial meadows or on the lower part of slopes.

Mixed-grass prairie can be divided into several types, but all are characterized by needle-and-thread grass, western wheatgrass, blue grama, Sandberg bluegrass, threadleaf sedge, needleleaf sedge, junegrass, Indian ricegrass, pricklypear cactus, scarlet globemallow, fringed sagewort, Hood phlox, and various species of milkvetch and locoweed. There may be fifty or more plant species per hectare.[2] Mixed-grass prairie in the foothills is typically dominated by bluebunch wheatgrass, little bluestem, and sideoats grama. On sandy soils, Indian ricegrass, prairie sandreed, sand dropseed, sand sagebrush, and yucca may be common. Saline soils lead to an increased abundance of such halophytes as alkali sacaton, four-wing saltbush, greasewood, and inland saltgrass. In some areas, the grasslands change (gradually or abruptly) into sagebrush steppe or woodlands dominated by ponderosa pine, limber pine, or juniper (for example, in the Powder River basin east of the Bighorn Mountains). Silver sagebrush is common on sandy soils and in riparian zones, whereas Wyoming big sagebrush is common in the intermountain basins throughout the state. Sand sagebrush is less common in Wyoming than in eastern Colorado, but it occurs in a few areas near the Nebraska border (north of Torrington, for example). In addition to shortgrass and mixed-grass prairie, there are also small tracts of tallgrass prairie on sandy soils or along streams

F I G. 5.1   Mixed-grass prairie south of Sundance. Ponderosa pine forests occur on the coarse-textured soils of Inyan Kara Mountain, which reaches an elevation of 1,936 m (6,368 ft).

(Livingston 1952; Branson et al. 1965; Limbach 1974; Singh et al., *Geography*, 1983). Common species include big bluestem, Canada wildrye, little bluestem, Indiangrass, prairie dropseed, side-oats grama, and switchgrass (Limbach 1974).

Wherever grasslands occur, plant species composition varies with changes in topographic position, such as from hilltops to valley bottoms (along soil catenas). Valley bottom soils are often deeper, finer textured, more mesic, and more fertile than soils on hilltops (Schimel et al. 1985). Topography also affects snowdrifting, with more snow accumulating in ravines or on leeward slopes, causing them to be more mesic than at other topographic positions. Depressions or playas with little or no drainage are also common, often (though not always) having saline or alkaline soils (Holpp 1977). Differences in soil depth, salinity, and texture cause considerable variation in what sometimes appears as a uniform vegetation cover.

Other patterns occur on a smaller scale. For example, cloning from root crowns, rhizomes, and stolons sometimes causes the formation of patches that are one to several meters across (such as blue grama, prairie sandreed, and yucca). Burrowing animals, such as badgers, ground squirrels, harvester ants, kangaroo rats, pocket gophers, and prairie dogs, also create small-scale disturbances that favor some species over others. Also, "fairy rings" may develop—an interesting but poorly understood phenomenon caused by certain species of saprophytic fungi (fig. 5.2).

Significantly, most of the grassland ecosystem is hidden; 75 percent or more of the plant biomass and most herbivores are in the soil (fig. 5.3; Stanton 1988). Reported ratios of belowground to aboveground biomass in mixed-grass prairie range from 3 to 6 (and up to 13 in shortgrass prairie; Sims and Coupland 1979). The extensive root system of many grassland plants provides a means of obtaining limited water and nutrients. Some species have deep root systems for extracting moisture from throughout the soil profile, while other species have shallow roots to take ad-

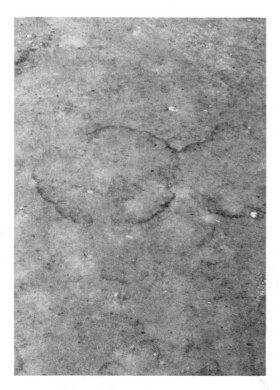

F I G. 5.2   An aerial view of grasslands in the Laramie Basin, showing "fairy rings," about 10 m in diameter, created by fungi that grow outward. A flush of nutrients is made available as the fungi die and decompose, just inward from the zone of major fungal activity, thereby stimulating plant growth (Hudson 1986). Elevation 2,213 m (7,280 ft).

vantage of light showers that wet only the surface soil.

Surviving in the Grassland Environment

Throughout the western Great Plains, the grassland environment is characterized by fire, extended periods of drought, the presence of large herbivores, and a sometimes short growing season. These factors have led to a vegetation composed largely of perennial grasses but with a substantial number of sedges and herbaceous forbs. Small shrubs are also common. Raunkiaer (1934) observed that most of the plants of temperate grasslands are perennials and have their perennating buds at or just below the soil surface.[3]

Why should grassland plants keep their perennial tissue close to the soil surface? First, the semiarid nature of mixed-grass prairie limits the production of perennial woody tissue above ground. For many grassland species, herbaceous tissue is apparently less expensive to produce and maintain than woody tissue.

Second, and perhaps more important, the preponderance of herbaceous plants in grasslands is a result of the frequency of fires, which usually kill the aboveground buds of woody plants. Buds near or below the soil surface are typically not damaged because the soil temperature during a prairie fire is comparatively low.[4] Moreover, the growth of perennial herbaceous plants from the root crown (sprouting) usually occurs rapidly.

Periodic drought and fire are commonly viewed as the primary reasons for the prevalence of perennial herbaceous plants in grassland environments (Axelrod 1985). Such plants, however, are also well adapted to tolerate grazing by large herbivores (which at one time included the camel and horse, as well as bison, elk, and pronghorn antelope in North America; see Milchunas et al. 1988). Because the perennating buds are in the soil, they are less accessible to herbivores. Only the easily regenerated herbaceous stems and leaves are eaten.

The ability to replace leaves and stems that have been eaten or burned depends on two other adaptive characteristics: the presence of special meristems and the capacity for energy storage in the undamaged perennial tissue below ground. Special meristems exist in the form of subterranean, dormant buds on rhizomes or root crowns that begin to grow when hormonal production above ground ceases with the loss of leaves and stems. This removal of apical dominance stimulates the dormant buds to begin growing, thereby replacing the eaten or burned plant structures. Another way that grass leaves can be replaced is through intercalary meristems, which are tissues capable of cell division that occur near the bases of the leaves and some stems. As grass leaves are eaten, new leaf tissue is produced by the leaf itself; no buds are involved (Hyder 1971; Langer 1979).

Regrowth, whether from subterranean buds or intercalary meristems, requires an energy source, and in this way grassland plants are also well adapted. Carbohydrates produced by photosyn-

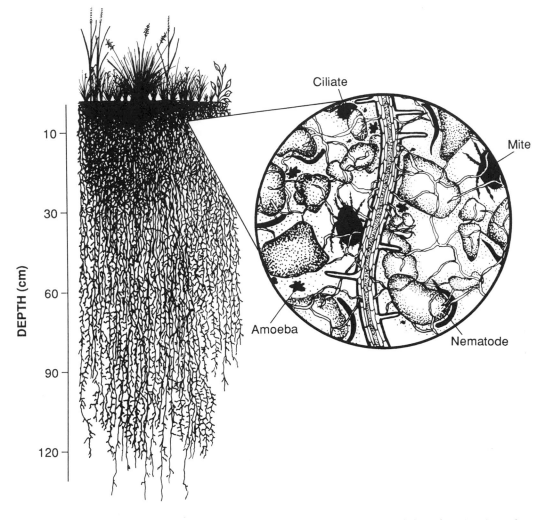

FIG. 5.3   Most grassland biomass is in the soil (below ground). The enlargement in this drawing shows
a single root with root hairs and the hyphae of mycorrhizal fungi. A film of water (stippled area) coats
each soil particle and provides habitat for nematodes
and numerous protozoans including amoebae and cil-
iates. Air spaces provide habitat for mites, insect larvae, and other invertebrates. Bacteria are extremely
abundant but are too small to illustrate. Magnification about 15×. Based on Weaver 1968 and Hunt et
al. 1987.

thesis are translocated to protected roots, rhizomes, corms, and bulbs for storage. Considering
that usually more than 75 percent of grassland
plant biomass is below ground, there is a large
amount of energy available to keep the plant alive
while new stems and leaves are produced.[5] Of
course, plants die from a lack of energy if fire,
herbivory, or drought are too frequent to allow for
the replacement of the stored energy spent in previous episodes of regrowth.

Most grassland plants have the characteristics
of xerophytes, that is, plants adapted to dry environments. Typically, they have small or narrow
leaves that sometimes have fine, light-colored
hairs (pubescence). Their small size reduces the
area exposed for transpiration and heat absorption on warm days, and pubescence slows the rate
of transpiration and reflects some solar radiation.
Usually, the leaves of grassland plants are facul-
tatively deciduous, falling or becoming senescent

when they use more carbon for maintenance than can be fixed by photosynthesis, whether caused by drought or low temperatures.

Although grassland plants are adapted to tolerate low water availability, some avoid drought stress by growing rapidly in the spring and completing their annual growth while water is readily available. Examples of drought-evading plants are early spring wildflowers, such as the sand lily and pasque flower. Other grassland plants, the cacti in particular, store water in tissues, which remain succulent throughout the growing season. Succulent plants would be grazed heavily by thirsty herbivores were it not for such special defense mechanisms as sharp spines.

Animals, too, must have special adaptations for surviving in the grassland environment (table 5.1). With most of the plant biomass below ground, it is logical that there would be many subterranean herbivores, such as nematodes, mites, and insect larvae, and various burrowing mammals, such as pocket gophers (Stanton 1988).[6] Above ground, the primary herbivores include grasshoppers, rabbits, prairie dogs, ground squirrels, and the larger mammals already mentioned. One of the major problems for large herbivores is ob-

TABLE 5.1   Mammals, birds, and reptiles found in grasslands, shrublands, and escarpments at lower elevations

MAMMALS

Badger	Ground squirrel, Wyoming
Bobcat	Jackrabbit, Black-tailed[b]
Chipmunk, Least	Jackrabbit, White-tailed[c]
Cottontail, Desert	Mouse, Deer
Cottontail, Nuttall's	Mouse, Northern grasshopper
Coyote	Pocket gopher, Northern
Deer, Mule	Prairie dog, Black-tailed[b]
Fox, Red	Prairie dog, White-tailed[c]
Fox, Swift	Pronghorn antelope
Ferret, Black-footed[a]	Woodrat, Bushy-tailed
Ground squirrel, Thirteen-lined	

BIRDS

Bluebird, Mountain	Meadowlark, Western
Eagle, Golden	Nighthawk, Common
Falcon, Prairie	Owl, Burrowing
Grouse, Sage	Sparrow, Brewer's
Harrier, Northern	Sparrow, Lark
Hawk, Red-tailed	Sparrow, Sage
Jay, Pinyon	Sparrow, Vesper
Kestrel, American	Thrasher, Sage
Lark, Horned	Towhee, Green-tailed
Lark bunting	Vulture, Turkey
Longspur, McCown's	

REPTILES

Bullsnake
Lizard, Northern sagebrush
Lizard, Spiny
Rattlesnake, Prairie

[a] Now rare
[b] Eastern grasslands of Wyoming
[c] Western two-thirds of Wyoming

taining enough protein from coarse grassland plants with tough cell walls and high concentrations of lignin and silica. Chewing this food source causes rapid tooth abrasion. Interestingly, hypsodont dentition arose in mammals at about the same time that grasslands became widespread, some 20,000–25,000 years ago (probably during the Miocene epoch). This dentition is characterized by continually erupting tooth crowns (Webb 1977).

Another adaptation for animal survival on grassland plants is the ruminant digestive system found in antelope, bison, deer, and elk (as well as in cattle and sheep). The four chambers of ruminant stomachs contain diverse bacteria and protozoans that facilitate digestion and the extraction of protein from the coarse food supply. Regurgitation and additional chewing is an important feature of this special digestive system, enabling ruminants to be more efficient in extracting food from coarse grassland plants. Nonruminants (such as horses) can also survive on the grassland, but they must consume much larger amounts of food because it passes more quickly through their gastrointestinal system and is less completely digested. To facilitate the digestion of plant tissues, all herbivores have longer intestines per unit of body weight than do carnivores.

Partitioning Grassland Resources

Despite the rigorous environment, many species of plants and animals coexist in grasslands. In fact, they often seem to occupy nearly the same space and tap the same resources. Closer study has demonstrated, however, that there are subtle differences in the ways that different species use resources. Some plant species, for example, complete their growth at a different time of year than do their neighbors. Understanding how the resources of an ecosystem are partitioned between neighboring or coexisting species is a topic that has intrigued ecologists for many years. Plants and animals that use different resources within the same community or at different times of the year are viewed as occupying different ecological niches. There is a tendency during evolution for the niches of coexisting species to become less

similar over time, thereby reducing the level of competition.

One of the first studies to suggest resource partitioning between grassland plants was done on root systems. Weaver (1954, 1958, 1968) observed that some plants have mostly deep roots, while other species have mostly shallow roots. In this way, neighboring plants might be tapping different parts of the soil for water and nutrient resources. Scarlet globemallow and slimflower scurfpea, for example, have roots down to 3 m or more (Weaver and Albertson 1956). In studying the root system of skeletonplant, Weaver dug by hand to about 6.5 m, at which point, "It seemed expedient to abandon the trench because of caving soil" (Weaver 1954). In contrast, 85 percent of blue grama roots are in the top 20 cm of the soil (Singh and Coleman 1977). Shallow-rooted plants (such as blue grama) apparently depend more on summer rainshowers; deeper-rooted plants depend on water that percolates to greater depths after snowmelt or heavy spring rains.

These differences mean that neighboring plants are not necessarily in direct competition for water and nutrients (Risser 1985). In a sense, the grassland is the reverse of the forest, where vertical stratification exists above ground (trees above, shrubs and herbaceous plants in the understory). Forest ecologists are concerned with falling branches and trees, whereas grassland ecologists think more about being buried in their trenches.

Grassland plants also are stratified seasonally, with cool-season plants completing their growth by early summer or in the fall (when moisture is available), and warm-season plants growing in the summer (until moisture becomes limiting). Junegrass, threadleaf sedge, and western wheatgrass are good examples of cool-season species, while blue grama, buffalograss, and little bluestem are examples of warm-season grasses (Moore 1977; Singh et al., Vegetation, 1983).

The differentiation of grassland plants is based largely on their physiological characteristics (Risser 1985). Cool-season species commonly have lower optimal temperatures for photosynthesis and have the C_3 metabolic pathway for carbon fixation. Such plants are usually less tolerant of high temperatures. Growing early in the season before water stress develops is therefore adaptive. In contrast, warm-season species have a higher optimal temperature for photosynthesis, higher light requirements, and higher water-use efficiencies (that is, more photosynthesis per unit of water uptake)—all physiological characteristics of the C_4 metabolic pathway (Limbach 1974; Kemp and Williams 1980; Risser 1985). The separation of grasses into C_3 and C_4 species has stimulated great interest among biologists and land managers, but as noted by Risser (1985), "Few unambiguous generalizations are possible about the ecological significance of the two pathways. . . . It is important to recognize that the photosynthetic pathway is only one of a myriad of adaptive strategies employed by rangeland plants."

The differences in C_3-C_4 physiology also appear to be influential in causing differences in geographic distribution and vegetation response to grazing. With regard to grazing, a critical time for ranchers and their livestock is in the spring, when hay supplies may be low and there is little new forage production on the rangeland. The first plants to become green are the cool-season C_3 species; they are the plants most likely to be grazed first. The warm-season species soon become important, but by that time there is more food available and less intense pressure on any one group. Consequently, early-spring grazing commonly leads to an increase in some warm-season C_4 species (Ode and Tieszen 1980; Risser 1985).

With regard to geographic distribution, Boutton et al. (1980) examined the relative abundance of C_3 and C_4 species along an elevational gradient from 1,825 m near Cheyenne up to 2,590 m in the Laramie Range. They observed that the amount of total biomass in warm-season (C_4) species was greater at the warmer, lower environments than in the cooler, mountain grasslands. Apparently, C_3 plants are better adapted to the cooler elevations than are C_4 plants.

Resource partitioning to minimize competition can also be observed in coexisting animals. For example, bison, jackrabbits, and some grasshoppers tend to eat more grasses, while other insects (including other grasshopper species) and antelope eat more forbs (Ellison 1960; Peden et al. 1974; Peden 1976; Vavra et al. 1977; Schwartz and Ellis 1981). Thus, competition between the rep-

resentatives of these two groups is minimized. When the food habits of bison, cattle, pronghorn antelope, and sheep are compared, bison and antelope are most different, cattle are similar to bison, and sheep are similar to antelope. From an evolutionary perspective, bison and antelope should have the most different food habits; they have coexisted for the longest time. Similarly, elk and deer have coexisted for millions of years, and their food habits are also quite different. Elk consume more grass, and deer more twigs and the leaves of broad-leaved plants (called browse). Of course, the diets of all coexisting animals are determined to some extent by what is available; considerable overlap may occur in some seasons.

Grasslands from an Ecosystem Perspective

ENERGY FLOW

As in all terrestrial ecosystems, green plants constitute more of the grassland biomass than any other group of organisms—96 percent according to Coupland and Van Dyne (1979). The next largest component is not the large herbivores, as is often assumed, but rather microorganisms (such as bacteria, fungi, nematodes, mites, and protozoans, which constitute most of the remaining 4 percent).[7] Of these organisms, the weight of fungi has been estimated at four times that of bacteria. Earthworms are rare in semiarid grasslands, but there may be several million nematodes and a hundred thousand mites in one cubic meter of soil (Stanton 1988). The total biomass of all mammals, birds, and insects combined is less than 1 percent of the ecosystem biomass, even in grasslands grazed by domestic livestock.

As in all ecosystems, the solar energy used for photosynthesis is only a small portion of the total impinging on the landscape (typically less than 2 percent). In southeastern Wyoming, the mean daily global radiation during the growing season is 27,000 kJ/m², with slightly less than half of that in wavelengths that can be used by chlorophyll for photosynthesis (photosynthetically active radiation, PAR). The energy made available to herbivores and detritivores through net primary production is only about 0.3–0.5 percent of the daily global radiation (0.85–1.4 percent of PAR; Sims and Coupland 1979), with half of the remainder going to the evaporation of water and the other half to heating the soil and other ecosystem components (French 1979). The small percentage of the solar radiation fixed by photosynthesis leads to a total NPP (above and below ground) of 500–900 g/m² in mixed-grass prairie[8]—enough to maintain a diverse fauna of herbivores and carnivores (Sims and Coupland 1979; Singh et al., *Vegetation*, 1983).

As expected in semiarid ecosystems, the net primary productivity of mixed-grass prairie is strongly influenced by water availability. Webb et al. (1978) reported a nearly linear correlation between aboveground NPP and evapotranspiration up to 500 mm, with about 1 g of new aboveground biomass resulting from each 1,000 g of water (fig. 5.4). Some investigators have hypothesized that nutrients may control NPP more often than is usually thought, with rain stimulating plant growth indi-

rectly by facilitating the mineralization of organic matter (Sala and Lauenroth 1982; McNaughton et al. 1983). Sala et al. (1988) found a linear correlation between NPP and precipitation up to about 370 mm in the western Great Plains, after which they surmised that nutrients become limiting.

Two other factors that influence grassland NPP are the rate of water infiltration into the soil and whether precipitation comes as snow or rain. Rain that does not infiltrate into the soil evaporates quickly and is generally unavailable to plants. In contrast, accumulated snow melts over a short time and is more effective in sustaining plant growth than a series of short showers, even though the total amount of precipitation is the same (see chapter 3). The amount of snow that accumulates during storms, or subsequently by drifting, is partially dependent on the height or roughness of the vegetation. Research has shown that taller vegetation can retain more snow and that increased

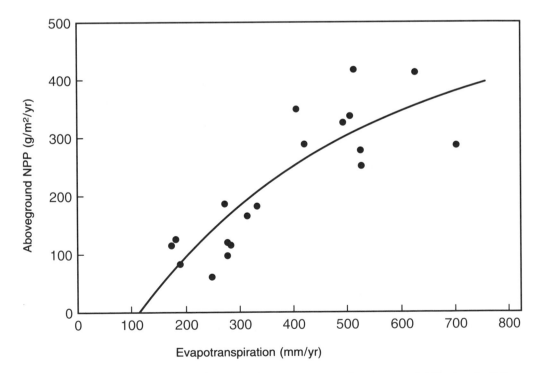

F I G. 5.4   The aboveground net primary production (ANPP) of grassland ecosystems is correlated with evapotranspiration. ET integrates the effects of tem-

perature and moisture availability (see fig. 3.8). Adapted from Webb et al. 1978.

snow retention leads to increased soil water re-charge and increased plant growth (Van Haveren 1974; Ries and Power 1981).

Primary productivity and other aspects of eco-system activity are often episodic, with pulses of plant growth, decomposition, and nutrient trans-fers occurring after rainfall events. Moreover, because the convectional rainfall of grasslands is often patchy, the amount of productivity and available biomass varies from place to place as well. In African grasslands and savannas, patchy rainfall patterns determine the movement of large herbivores and the nomadic people associated with them (Ellis and Swift 1988). Patchy rainfall could also have affected the movement of people, bison, and other animals in western North Amer-ica, thereby distributing the grazing pressure over a larger area (Coughenour 1991a, 1991b, 1993).

By what pathways does the energy fixed by pho-tosynthesis flow through mixed-grass and short-grass prairie ecosystems? For green plants, 65–85 percent is translocated below ground (Coleman et al. 1976; Singh and Coleman 1977), and about 30–40 percent is used by the plants for their own maintenance (Duckham and Masfield 1971; Det-ling et al. 1979; Painter and Belsky 1993). Of the energy left over after the needs of the plants are met, a large portion is converted by the end of the growing season to standing dead or detrital biomass—perhaps 60–80 percent (above and be-low ground) depending on the intensity of graz-ing. Detritus adds organic matter to the soil, leads to increased infiltration rates, provides the base for an extensive microbial food web (Hunt et al. 1987), and as it decomposes, provides the nutri-ents required by plants. Of course, fires can burn the aboveground living and detrital biomass as well, converting a substantial portion of the en-ergy to heat and having other significant effects in some grasslands (Knapp and Seastedt 1986).

On average, herbivores consume only about 10–30 percent of the total annual NPP (above and below ground), with more than twice as much energy going to nematodes and soil arthropods as to large herbivores, such as cattle (table 5.2; Cole-man et al. 1976). In a typical grassland, cattle eat less than 2 percent of the total energy available in plant biomass during the year. It is not surprising, then, that some ecologists have concluded that

light and moderate levels of grazing have little or no effect on plant species composition (Klipple and Costello 1960; Grant 1971; Milchunas et al. 1988).

Jones (1979) contrasted the amount of energy available for human use in a plant-cow-human ecosystem and a cropland ecosystem. More energy is available to people if plants are consumed in-stead of animals, because when animal food is desired, a large amount of the energy consumed by the animals is used for their maintenance and is converted into animal tissue or waste products that people do not eat. Further, the second law of thermodynamics dictates that every conversion of energy from one form to another—for example, from plant to animal tissue or from organic com-pounds to animal movement—leads to the pro-duction of heat. This principle accounts for the fact that energy flows rather than cycles through ecosystems; all solar energy fixed by photo-synthesis is eventually converted to heat, though it may persist as soil organic matter or fossil fuels for centuries or more.

Human food production is greater per unit area when animals eat plants rather than meat, but it is also true that, aside from gathering a few edible native plants and hunting native animals, the only way people obtain food from rangelands is from livestock raised for meat and milk. The alternatives are dry-land or irrigated agriculture, which typically lead to soil erosion and other envi-ronmental problems (Lauenroth et al., 1993).

NUTRIENT CYCLING IN GRASSLANDS

Organisms require nutrients for their survival as well as suitable temperatures and adequate energy and water. For example, calcium is used in plant cell walls; phosphorus is important for the storage and release of energy during metabolism; potas-sium is required for the regulation of cell water and manganese for the synthesis of chlorophyll; and nitrogen is an important component of amino acids, proteins, and chlorophyll. All nutrients or elements cycle through ecosystems, with losses oc-curring primarily during erosion, and in some cases, leaching. Losses are usually balanced by nu-trient inputs from rock weathering, rain and snow, aerosol deposition, and such nutrient-specific processes as nitrogen fixation. Studies on nutrient

TABLE 5.2   Estimated energy flow through a prairie ecosystem on the western Great Plains

Ecosystem Component	Energy Input	Lost by Respiration	Tissue Production	Production Consumption
SOLAR INPUT				
Global radiation	4,155,000			
Photosynthetically				
actively active radiation	1,966,000			
PRIMARY PRODUCTION				
Gross[a]	21,882	7,439		
Net[b]				
Aboveground			2,163	
Belowground			12,280	
Subtotal			14,443	
HETEROTROPHS				
Aboveground				
Herbivores				
Mammals[c]	105	92	13	0.13
Macroarthropods[d]	34	23	11	0.32
Carnivores[e]	8.3	7.5	0.8	0.10
Subtotal	147	123	25	
Underground				
Herbivores				
Macroarthropods[f]	127	66	61	0.48
Nematodes	50	42	7.9	0.16
Carnivores[g]	20	15	4.6	0.23
Detritivores				
Microorganisms[h]	12,560	9,632	2,929	0.23
Nematodes	72	61	12	0.16
Others	9.2	6.7	2.5	0.27
Subtotal	12,838	9,822	3,016	
TOTAL	12,986	9,945	3,041	

Source: Coupland and Van Dyne 1979

[a] Gross primary production (GPP), or total photosynthesis/m^2/yr, varies greatly from year to year, depending on water availability (see fig. 5.4).

[b] Net primary production is calculated as GPP less the amount of GPP used by plants for their own maintenance (respiration).

[c] Mammalian herbivores included cattle, antelope, rabbits, and ground squirrels; cattle grazing was light (one yearling steer or heifer per 10.8 ha for 180 days each year).

[d] Mostly grasshoppers and other insects

[e] Aboveground carnivores include the coyote, fox, birds of prey, and snakes.

[f] Mostly mites and insect larvae

[g] Belowground carnivores include mites.

[h] Bacteria, fungi, and actinomycetes

cycling must therefore consider the rate at which nutrients are being added and lost, as well as the rates and pathways by which they move through the ecosystem. All aspects of cycling in grasslands cannot be considered here, but a few processes seem particularly interesting and relevant when thinking about the western Great Plains.

As noted, grassland plants typically have tissues high in lignin and cellulose, which are resistant to decomposition. For this reason, and because the warm, moist conditions suitable for decomposition last for only a short period each year, organic matter tends to accumulate on the surface as detritus as well as in the soil. Among other effects, surface detritus improves infiltration rates (Knapp and Seastedt 1986), and soil organic matter increases the water and nutrient storage capacity of the soil while providing a more erosion-resistant soil structure. Soil organic matter is highly resistant to decomposition, but its mineralization provides inorganic nutrients for plant growth as well as a source of energy for bacteria and fungi—which in turn are food for nematodes, mites, and other microbes. Fire, soil disturbances, and grazing—above and below ground—can stimulate primary productivity because they increase the rate of mineralization and thereby improve nutrient availability (Ingham et al. 1985; Elliott and Coleman 1988; Stanton 1988; Holland and Detling 1990).

Nutrient loss through leaching is an unlikely phenomenon in most western grasslands because rarely is there enough precipitation to cause water percolation below the rooting zone. Wind erosion during droughts or episodes of heavy grazing is another potential cause of nutrient loss. Some nutrients, especially nitrogen, are lost through volatilization during the inevitable grassland fires, but only a small portion of the total nitrogen is lost in this way, because most of the nitrogen is in unburned soil organic matter (Woodmansee et al. 1981; Schimel et al. 1985). Woodmansee (1978) concluded that nitrogen accumulates in western grasslands (fig. 5.5), with losses induced almost entirely by such abiotically controlled events as fire, drought, and flash floods. Even then, nitrogen losses are probably small because so much of the nitrogen is bound in belowground biomass and soil organic matter, and because, unlike forests, there is little potential for nutrient leaching beyond the rooting zone. Woodmansee (1978) and Clark et al. (1980) analyzed the various potential mechanisms for nitrogen loss in Great Plains grasslands, concluding that denitrification

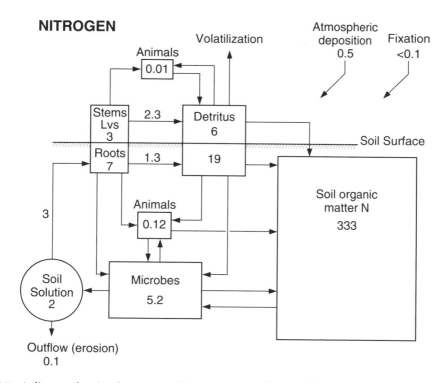

F I G. 5.5 A diagram showing the amount of nitrogen in major components (boxes; g/m²) of a grassland ecosystem, and flows (arrows; g/m²/yr) that occur between the components. Based on data in Woodmansee et al. 1981.

was low or zero and that the volatilization of nitrogen-rich animal wastes was most important. Schimel et al. (1986), however, found that losses of ammonium from urine, animal biomass removal, and nitrous oxide loss totaled only 0.07 g N/m²/yr in a Colorado shortgrass prairie, or about 25 percent of the nitrogen. They calculated a potential loss of ammonium from senescing vegetation (0.26 g N/m²/yr), which, they noted, is an order of magnitude larger than all other losses combined.

Nutrients probably accumulate in grassland ecosystems in most years because additions are typically greater than losses (Clark et al. 1980; Woodmansee 1978). Grassland soils can therefore become quite fertile and are excellent for crop production when water and temperature conditions are suitable for the crop being planted. Of course, the harvesting of crops is itself a nutrient drain; fertilization is required after a few years of cultivation.

Nitrogen-fixation by bacteria in the nodules of legumes and some other plants is an important nitrogen source for many ecosystems. The legumes locoweed, lupine, milkvetch, scurfpea, and vetch are known to have nitrogen-fixing bacteria in shortgrass and mixed-grass prairie (Johnson and Rumbaugh 1981). There is evidence that a few nonleguminous plants, such as Louisiana sagewort and fringed sagewort, may also be capable of nitrogen fixation (Farnsworth and Hammond 1968; Porter 1969). Lichens have been identified as nitrogen-fixers in some desert shrublands (West 1990), but preliminary experiments suggest that this is not true for the most conspicuous lichen in Wyoming grasslands, *Xanthoparmelia chlorochroa* (fig. 5.6; Steven Williams, pers. comm.).

Symbiotic nitrogen fixation is commonly thought to be the single most important mechanism providing nitrogen for plant growth (in the form of ammonium and nitrate), but this is not the case in mixed- and shortgrass prairie, where legume density is low and suitable environmental conditions for fixation exist for only a short time. Rain, snow, and dust are larger sources of nitrogen in western grasslands. Woodmansee (1978) summarized data that suggest that nitrogen fixation

F I G. 5.6. *Xanthoparmelia chlorochroa*, a common lichen in Wyoming grasslands and shrublands.

accounts for only about 7 percent of the annual nitrogen inputs on a shortgrass prairie south of Cheyenne.[9] Much of the nitrogen required for plant growth probably results from the microbial mineralization of soil organic matter. Still, there is considerable evidence that nitrogen can be a limiting factor in the primary productivity of mixed-grass and shortgrass prairie, especially during wetter-than-average years (Lauenroth et al. 1978; Black and Wight 1979; Dodd and Lauenroth 1979; Nyren 1979; Risser and Parton 1982; Wight and Godfrey 1985). This observation has led some scientists to suggest that fertilizing native rangelands might be profitable. Indeed, more forage of higher protein content could result. Major changes in species composition because of nitrogen fertilization are unlikely unless the amounts applied are massive (>150 lb N/acre/yr). Several researchers, however, have observed that cool-season grasses tend to respond more readily to nitrogen fertilization than do warm-season species (Nyren 1979; Wight and Black 1979), probably because they grow when nitrogen, rather than water, could be the primary limiting factor.

Despite the apparently desirable effects of fertilization, several concerns have been identified: (1) fertilizing only part of a rangeland might result in selective overgrazing (Wight 1976; Senft et al. 1985); (2) forage nitrate content could increase to toxic levels (Houston et al. 1973); (3) certain weeds, such as cheatgrass, may be favored (Nyren 1979); and (4) despite increases in forage quality and productivity (or yield), fertilization is not yet economical (Houston and Hyder 1975; Wight 1976; Rauzi and Fairbourn 1983).

Although annual nitrogen inputs may be low, this is of little consequence if losses are also low and if a considerable amount of nitrogen has accumulated in the belowground biomass and soil organic matter. Clark et al. (1980) concluded that most plant needs for nitrogen above ground are met by the translocation of stored nitrogen in the root system, with additional nitrogen being made available by the mineralization of organic matter and microbial tissue. As Clark (1977) noted, "Once a given N atom makes its initial entry into the blue grama plant, there is a greatly increased probability that the atom will again enter new herbage growth in each of several following years." Much (50–80 percent) of the nitrogen in the senescing leaves of western wheatgrass and blue grama is translocated to perennial plant parts, thereby conserving the nutrient (Polley and Detling 1988).

As with productivity, the accumulation and cycling of nutrients is not uniform across the landscape. Many years of natural erosion from ridgetops and accumulation in valley bottoms, combined with pronounced differences in the microclimate and moisture availability, have led to the development of soil catenas (Burke et al. 1987). All aspects of nutrient cycling, as well as the flow of energy and water, vary from one place in the landscape to another. The effect of this patchiness on plant and animal abundance is a major theme in modern ecological research.

THE EFFECTS OF GRAZING BY LARGE MAMMALS

Herbivores of any kind form a small portion of the ecosystem biomass, but they have played an important role in determining the nature of grassland ecosystems (Weaver and Clements 1938; Larson 1940; Ellison 1960). Studying the effects of herbivory provides not only insights relevant to land management, but also interesting examples of how coexisting plants and animals are adapted to each other.

On the surface it appears that any grazing is disadvantageous to plants: aboveground grazing

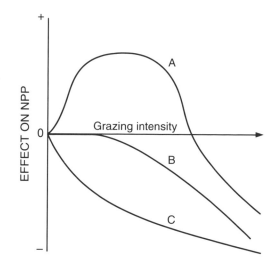

F I G. 5.7   Three hypothesized relationships between large ungulate grazing intensity and NPP. Line A, the overcompensation hypothesis, suggests that grazing stimulates NPP up to some threshold level, after which the effect is negative. Line B is the partial compensation hypothesis, which states that plants are able to maintain NPP despite some grazing on some plants. Line C suggests that any amount of grazing by large herbivores causes a decline in NPP. Adapted from Heitschmidt et al. 1990.

reduces the leaf area available for photosynthesis. Moreover, energy that might be expended for the production of new plants is diverted to replacing the tissue eaten by the herbivore. Some grassland plants, however, seem well adapted to tolerate considerable grazing pressure, and some ecologists maintain that grasslands subjected to light or moderate grazing have higher cover and a higher annual NPP than do comparable grasslands where herbivory is excluded (fig. 5.7). This response, known as overcompensation, has been the topic of considerable research (Lauenroth et al. 1993; Painter and Belsky 1993). With a few exceptions, however, such studies have focused on mammalian herbivores rather than the belowground grazers that account for most energy flow. Scientists can construct fences to exclude large mammals, but excluding nematodes, mites, and other belowground herbivores is a formidable challenge.

How might mammalian herbivory lead to increased grassland productivity? The research suggests various possible mechanisms (McNaughton 1979; Coughenour 1985; McNaughton and Chapin 1985):

1. A more open canopy is created, reducing the potential for leaf shading. If senescent leaves and stems are removed by the herbivore, more light is available for photosynthesis. This mechanism may be especially important on mesic sites, where grassland vegetation is comparatively tall.

2. If water is limiting at the time of grazing, less leaf area for transpiration might increase water availability for the remaining plant tissues, possibly allowing for more photosynthesis per square meter than had occurred before.

3. Herbivores improve nutrient availability for plants by concentrating certain elements in feces and urine, which are decomposed more readily than plant detritus.

4. With the removal of apical dominance, cell division and elongation may accelerate. The new leaves may compensate for the lost leaf area and may be more active physiologically than the older leaves that were eaten.[10]

With such mechanisms in mind, the *compensatory growth hypothesis* was proposed (Owen and Wiegert 1976, 1981, 1982). The hypothesis states that NPP is increased to some optimal level by light-to-moderate grazing (fig. 5.7), after which continued grazing pressure leads to a decline in NPP because of an excessive drain on the energy stored in the root system and less efficient water use (caused by soil compaction that slows the rate of infiltration) (Rauzi 1963; Naeth et al. 1990). Painter and Detling (1981) and Detling and Painter (1983) conducted an experiment on western wheatgrass that showed how, after two days, the photosynthetic rate per unit leaf area was 5–10 percent higher than were the controls for the undamaged leaves of clipped plants. The increased rates of photosynthesis did not completely compensate for the photosynthesis that would have occurred had the total leaf area remained on the plant, but partial compensation is suggested by the results. Several investigators have concluded that few data support the hypothesis that grazing benefits plants (Ellison 1960; Belsky 1986; 1987;

Heitschmidt 1990; Painter and Belsky 1993), but the idea has appeal and will continue to guide future experiments. McNaughton (1985) noted that "compensatory growth did not completely replace the vegetation consumed by herbivores" and that "it is improper to conclude that grazing is strictly advantageous to the plants."

Plant species composition also may be changed by grazing. Some species are called *increasers* because they increase in relative abundance with grazing pressure, while others are classified as *decreasers* because they become less abundant. At some point, grazing pressure, whether from native or domestic herbivores, becomes so heavy that weedy species (*invaders*) increase in abundance. Range managers have historically used the relative abundance of increasers, decreasers, and invaders as an indicator of range condition. Growth-form composition may also be an indicator of grazing history, with shorter growth forms of the same species found more commonly on heavily grazed rangeland (Peterson 1962; McNaughton 1985; Jaramillo and Detling 1988).

Notably, not all grasslands respond in the same way to grazing pressure. Milchunas et al. (1988) compared grasslands over a wide range of climatic conditions and concluded that such grasslands as those occurring in eastern Wyoming—in a semiarid environment and with a long history of grazing by such native ungulates as bison—are tolerant of grazing pressure. In contrast, the taller grasslands of more humid climates, and grasslands in areas with no history of long-term grazing, change rather dramatically with the introduction of livestock. Of course, any grassland can be affected adversely if too many animals are fenced into small areas.[11] Coughenour (1991a, 1991b) described in detail how the nomadic movements of large herbivores, triggered by the greater availability of preferred food nearby, constitute an important mechanism that enables some grasslands to tolerate grazing. Various kinds of livestock grazing systems are now used with this in mind (as, for example, short duration grazing and rest-rotation grazing; Holechek 1983); but there is still considerable debate over their effectiveness (probably because generalizations are difficult). Each ranch operation requires its own management plan based on goals developed with an understanding of the environmental constraints of available land.

It has become natural for ecologists to expect unusual phenomena in African grasslands, where there are twenty-five species of large mammalian herbivores instead of three or four, as in North America. One example is the way in which some herbivores improve food availability for others. Gazelles on the Serengeti appear to depend on wildebeest for their survival because the wildebeest eat the coarser forage, thereby stimulating the growth of the new foliage required by gazelles (Bell 1971; McNaughton 1976). McNaughton found that a month or so after the migratory wildebeest had moved through the area, consuming up to 85 percent of the green biomass, regrowth had created a "grazing lawn" on which the gazelles fed. Rather than competition, one herbivore facilitates the survival of the other.

Such relationships are not unique to Africa. Research on mixed-grass prairie in North America has shown that bison graze on the terrain of prairie dog colonies ("towns") more often than would be expected by chance (fig. 5.8; Coppock et al. 1983a, 1983b; Coppock and Detling 1986; Krueger 1986; Day and Detling 1990). The forage on the colonies is more nutritious than off, apparently because more animal waste products are deposited there. Nitrogen-uptake by the grazed plants is greater, and there are more younger leaves in the available forage (which tend to be higher in nitrogen content). In addition, the regrowth of young shoots is stimulated by grazing. Over time, the bison bring more nutrients to the town, thereby further enhancing soil fertility and forage quality. Krueger (1986) found that pronghorn also used the dog towns preferentially, but only in areas with an abundance of fringed sagewort. Mielke (1977) had earlier suggested a simi-

FIG. 5.8 Mixed-grass prairie in Wind Cave National Park with bison grazing on a prairie dog town (elevation 1,246 m; 4,100 ft). The tree is a ponderosa pine. Photo by James K. Detling, copyright American Institute of Biological Sciences

lar relation between bison and pocket gophers, with the bison improving conditions for the forbs preferred by the gophers. Pocket gophers, he suggested, improve soil fertility by their burrowing and feeding, thereby increasing forage production for the bison. Several other studies also found that the growth forms of western wheatgrass on prairie dog towns were shorter and more prostrate, with higher blade-to-sheath ratios, than those away from the towns, again suggesting that some genotypes of a single species tolerate grazing better than others (Detling and Painter 1983; Detling et al. 1986; Jaramillo and Detling 1988; Whicker and Detling 1988).

The prairie dog–bison interaction is preserved today in Wind Cave National Park (Whicker and Detling 1988), but it must surely have been a widespread phenomenon in presettlement times. McNaughton (1985) suggested that the magnificent carnivores in African savannas are more numerous because of the nutrient and energy flows facilitated by the large grazers. Similarly, North American bison may have been more abundant, along with their predators (grizzly bears and wolves), because of prairie dogs and other burrowing animals. Other animals may have benefited as well. Kamm et al. (1978) found more herbivorous insects on rangeland grazed by cattle than on rangeland that had not been grazed for a long period. Other research suggests that belowground invertebrates are denser in the soil of prairie dog towns than they are away from the towns (Ingham and Detling 1984).

Large herbivores have attracted the attention of most grassland scientists, yet other animals also play significant roles in the complex prairie ecosystem. Ingham et al. (1985) found that plants growing in soil with bacteria and bacteria-feeding nematodes grew faster than where nematocides had been used. Apparently, microbial grazers (such as nematodes) improve nutrient availability for vascular plants just as their much larger aboveground counterparts.

Disturbances and Succession in Grasslands

A common theme for ecological research is *secondary succession:* the gradual changes that occur as an ecosystem "recovers" from a disturbance. All ecosystems are subjected to various kinds of disturbance, and although often viewed as unfortunate, such events are natural phenomena that allow for the coexistence of a larger number of species than otherwise would occur. Also, nutrient cycling is more rapid, and levels of primary productivity are often higher following disturbances. As is commonly observed, the suppression of disturbances can lead to undesirable consequences.

Potential disturbances in western grasslands, aside from plowing for crop production, are drought, fire, periodic heavy grazing by domestic and native ungulates, insect outbreaks, and the burrowing of small mammals and harvester ants. Each disturbance has the potential to modify species composition somewhat, as well as to change productivity and nutrient cycling rates. Some disturbances, such as fire and drought, are largely physical phenomena, whereas herbivory and burrowing are biotic. Grassland recovery from predisturbance conditions often occurs within a few months to a year if the damage has been only above ground and if the energy stored in roots is adequate for new growth. In such cases, there is hardly any disturbance at all.[12]

The effect of natural grassland disturbances is to add to an already high level of spatial variability, such as found by McNaughton (1985) in the Serengeti: "The savanna-grassland environment is characterized by continual, stochastic fluctuations of rainfall, grazing, nutrient availability, and fire." These fluctuations generate pulses of primary productivity in different parts of the landscape, causing a nomadic way of life for large herbivores and humans alike. Moreover, because of this nomadism, plants in any particular place are not grazed continuously (Coughenour 1991). Today in North America it is difficult to observe animal movements and patchiness over large areas because of fencing, the suppression of fires, plowing, and urban development.

FIRE

Natural grasslands are characterized by frequent fires because the xerophytic leaves and stems accumulate rapidly, producing large amounts of flammable fuels. Vegetation structure influences fire spread, with bunchgrass communities requiring more fuel than grasslands with a more continuous biomass.[13] Grazing and drought can greatly diminish the rate of fuel accumulation, but such influences tend to be patchy or short-lived. Before the land was settled by Europeans, grassland fires would have burned for weeks or months until extinguished by cold or wet weather, or until a fire break was reached (a river, a ridge, or an area of inadequate fuel).

The source of ignition was usually lightning strikes, which occur in grasslands as readily as in forests. Higgins (1984) calculated an annual average of twenty-five to ninety-two lightning-caused fires per 10,000 km² in different parts of the northern Great Plains; Komarek (1964) found that lightning was the major cause of grassland fires north of Douglas (fig. 5.9). Most fires occur during July and August, when fuels tend to be drier and thunderstorms more frequent. Thun-

derstorms often produce enough rain to extinguish such fires, but storms with no or little rain are common.[14]

Native Americans also started fires, sometimes to facilitate their hunting (Moore 1972). Gruell (1985) reviewed 145 historical accounts of fire by forty-four observers in the Rocky Mountain region, concluding that fires set by native Americans were most common in the lowlands and that they could have been annual events in some areas (though probably the same grassland did not burn two years in a row). Wright and Bailey (1980) summarized the fire literature for much of the Great Plains, concluding that mean fire intervals were from 2–25 years, with longer intervals occurring in areas of rough topography. Grassland fires in western Nebraska probably occurred at intervals of 15–30 years (Wendtland and Dodd 1992). Bragg (1982) estimated that fires occurred every 4–6 years in central Nebraska from 1850 to 1900, after which they became less common because of fire suppression. Bragg made this estimate using fire scar data on such trees as ponderosa pine, which grow on the fringes of grasslands. Wyoming grasslands probably burned less frequently owing to slower fuel accumulation in a more arid climate.

The effects of fire on grasslands can be considered from several perspectives (Daubenmire 1968; Vogl 1974; Wright and Bailey 1980, 1982). First, some species are more easily killed by fire than others. Even though most perennating buds in grasslands are at the soil surface or below and the heat of a grassland fire tends to be above ground, fuel sometimes accumulates to the point where buds and root crowns in the soil burn along with the shoots (Antos et al. 1983). Furthermore, fire suppression may lead to fuel accumulation and hotter-than-normal fires, thus causing more mortality than might otherwise be expected. Plant mortality apparently depends on the season in which the fire occurs as well as on its intensity and other site-specific conditions (Wright and Bailey 1980; Rennick 1981).

Generalizations about the effects of fire on specific plants are not easy to derive (Wright and Bailey 1980), but warm-season species tend to be more tolerant of fire than cool-season species (Schacht and Stubbendieck 1985; Wright and Bai-

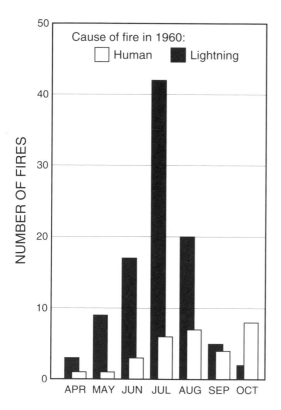

F I G. 5.9   The number of recorded human- and lightning-caused fires in the grasslands and shrublands of the Powder River Basin north of Douglas, Wyoming, in 1960. Adapted from Komarek 1964.

ley 1980; Whisenant and Uresk 1990), probably because they evolved at the time of year when fires are more frequent. Some species increase following fires, not because the fire is directly beneficial to the plant, but because additional water and nutrients are available following the death of other species or because a more favorable environment is created by the removal of detritus from the soil surface (Knapp and Seastedt 1986).

Fires also affect nutrient cycling and primary productivity. In some cases, the productivity following fire is higher, with greener, more palatable nutrient-rich stems and leaves resulting. Burned grasslands initially appear black and totally devastated, but it is common to see herds of bison, antelope, and cattle grazing on recently burned grasslands after plant regrowth begins (Coppock and Detling 1986; D. Uresk, pers. comm.). Indeed, Indians probably burned grasslands intentionally in order to attract game to smaller areas, where they could be more easily hunted. The animals were probably lured by more nutritious forage, resulting from the sudden availability of nutrients after accumulated detritus was burned. Combustion, like microbial decomposition, converts organic matter into the inorganic nutrients needed for plant growth. As a result, a more favorable carbon-to-nitrogen ratio develops, stimulating both microbial and vascular plant metabolism (Hobbs and Schimel 1984; McNaughton 1985).

Another influence of burning, especially in taller grasslands, is to remove the insulating detritus, thereby allowing the soil to warm up sooner and plants to grow earlier in the spring, when water is available (Knapp and Seastedt 1986). Though some nitrogen is volatilized during fires, it represents only a small portion of the nitrogen available in most grassland soils (Woodmansee et al. 1978). Furthermore, the amount lost is probably replaced within a few years by precipitation (see fig. 5.5).

Because of these influences, primary productivity is sometimes higher after a fire in tallgrass prairie (Knapp and Seastedt 1986), leading some managers to encourage the burning of grasslands as a regular practice, at least during wet years. Mixed-grass prairie, however, is water-stressed more often during the year than tallgrass prairies are, and water stress can be aggravated by fire if

moisture that would have been used for plant growth during the short growing season evaporates from the warmer, blackened soil surface. Redman (1978) found more water stress and lower primary productivity a year after a fire on mixed-grass prairie in Saskatchewan, and Engle and Bultsma (1984) suggested that increases in NPP should not be expected following burning in drought years in western South Dakota. Wenger (1943) observed that "burning shortgrass pastures is always accompanied by a temporary reduction in yield and vigor of grass." The word *temporary* is used because most grassland ecosystems are well adapted to fires, and prefire levels of productivity can be expected within a few years.

The effects of grazing and fire are interactive. Grazing can be so intense that there is little detritus on the soil surface, which leads to lower flammability. If, however, too much detritus accumulates owing to fire suppression or very little grazing, then much of the water in short summer showers can be intercepted by the aboveground biomass and evaporated, with little infiltration into the soil, where it can contribute to plant growth. The negative hydrologic impacts of detrital interception may be balanced by the increased trapping of snow and lower evaporation from the soil surface, but there is inadequate information to draw conclusions on such relations.

DROUGHT

Though less sudden and spectacular than fire, drought also can be viewed as a disturbance in grasslands. The continental climate of the western Great Plains typically has great fluctuations in annual precipitation, and it is common to have several consecutive years with well-below average precipitation (Mock 1991). Predictably, NPP and rangeland carrying capacity are much lower during droughts (Newbauer et al. 1980), but just as important, there can be considerable shifts in plant cover and, to a lesser extent, species composition.

Some research suggests that shifts in species composition commonly attributed to heavy grazing could be the result of changes in climate (Reed and Peterson 1961; Hyder et al. 1975; Branson and Miller 1981; Branson 1985). Some plant species (such as Sandberg bluegrass, pricklypear cactus, blue grama, and threadleaf sedge) increase during drought years in Wyoming grasslands (at least initially), while others decrease (junegrass, little bluestem, needle-and-thread grass, red threeawn, silver sagebrush, and western wheatgrass; Ellison and Woolfolk 1937; Whitman et al. 1943; Weaver and Albertson 1956; Reed and Peterson 1961; Albertson and Tomanek 1965; Tomanek and Hulett 1968; R. F. Miller et al. 1993). Yet the effects of drought can be complicated. Several investigators have reported that extended drought alone could cause the formation of shortgrass prairie dominated by blue grama and buffalograss, where mixed prairie had existed previously (Albertson and Weaver 1944; Newbauer et al. 1980). Others observed a high mortality of blue grama and buffalograss due to drought in southeast Montana (Ellison and Woolfolk 1937; Reed and Peterson 1961). One of the most successful species during a drought is Sandberg bluegrass, which often increases at the expense of blue grama and buffalograss (Reed and Peterson 1961). With extended drought even this species begins to decline. Species that survive or increase during a drought do so partially because of reduced competition from less drought-tolerant species.

The changes associated with drought may be in terms of total biomass and plant cover. Plant cover, for example, decreased from 69 percent to 2 percent during a drought in eastern Colorado (Branson and Miller 1981), and from 28 to 2 percent in southeastern Montana (Reed and Peterson 1961). Such drastic changes might be indirectly the result of grasshopper outbreaks, which tend to coincide with drought years or periods of heavy grazing.

Unfortunately, the drought years of the 1930s followed a fifty-year period of grassland plowing, spurred on in some areas by large markets for wheat and the notion that mixed-grass and short-grass prairie could be used for crop production in a manner similar to the tallgrass prairies to the east. The consequences were staggering dust storms.

Recovery from extended drought in a mixed-grass prairie may require several years or more, with some species reinvading from the shelter of clumps of pricklypear cactus or other unpalatable plants that become more abundant during droughts. As Weaver and Albertson (1956) noted, the spiny cacti provided refugia against grazing as well as a more shaded environment. Cacti are not usually appreciated by livestock managers, but spiny plants appear to play a role in protecting other grassland plants against drought and heavy grazing (Turner and Costello 1942; Houston 1963). Forage availability increases dramatically after killing cactus through burning, not because of greater forage productivity but simply because forage is more accessible with the removal of cactus spines (J. Dodd, pers. comm.).

With time, native climax species become reestablished following drought. Seedling establishment is more probable with increased precipitation, but sprouting from surviving rhizomes also occurs. Weaver (1943, 1954) suggested that some native perennial plants survive droughts in a dormant or near-dormant state, even for periods of five to seven years. Newbauer et al. (1980) observed that above-average rainfall for thirteen years changed a shortgrass prairie dominated by blue grama to a mixed-grass prairie with needle-and-thread grass, prairie junegrass, and western wheatgrass as well as blue grama. Livestock carrying capacity also increased.

GRASSHOPPERS, PRAIRIE DOGS, AND OTHER HERBIVORES

Drought disturbances are frequently accentuated by outbreaks of grasshoppers (Whitman et al. 1943; Reed and Peterson 1961; Cook and Sims 1975).[15] The causes of outbreaks during drought are not well understood, but Rodell (1977) noted that bacterial and fungal grasshopper diseases are more common under mesic conditions. This led him to conclude that grasshopper populations are limited more by climatic factors than by food. Such a conclusion is of great practical significance. Watts et al. (1982), however, cautioned that even if there is a climatic correlation with grasshopper abundance, "There can be no assertion as to whether it is a direct action of weather on the grasshopper's physiology, an indirect action on the food plants, a differential effect on their predators, parasites, and diseases, or a measure of each." Some of the most interesting and complex ecological questions pertain to the population dynamics of insects.

Favorable conditions for grasshopper population outbreaks have also been attributed to heavy grazing, whether by bison or livestock (Ellison 1960; Anderson 1961; Holmes et al. 1979; Quinn and Walgenbach 1990). The amount of detritus is reduced, thereby exposing more soil to warming by the sun. The soil can also become drier because of less infiltration—the result of less detritus and a more compact surface soil. Again, a warmer, drier soil results. Watts et al. (1982) suggested that the patchy vegetation that may be created by grazing favors grasshoppers, since denuded areas are good for egg laying and vegetated areas provide food for nymphs.

The amount of energy that flows to grasshoppers can be small or large. At the Central Plains Experimental Range in northern Colorado, Rodell (1977) estimated that grasshoppers consumed only about 0.2–0.4 percent of NPP during years with normal populations. Comparable estimates for outbreak years have not been calculated, but Hewitt et al. (1976, 1977) found that during some years a 63 percent forage loss in Montana might result, with about twenty grasshoppers per square meter. Also, Allred (1941) examined the effects of grasshoppers in the Powder River Basin during the 1930s drought, noting: "In 1936 . . . grasshoppers swarmed in such hordes that they destroyed all edible vegetation. . . . The insects occupied this area . . . in numbers from 50 to 100 per square foot." Much earlier, in 1864, Gen. Alfred Sully noted that "the only thing spoken about here [Montana] is the grasshopper. They are awful. They actually have eaten holes in my wagon covers and in the tarpaulins that cover my stores" (quoted in Pfadt and Hardy 1987). Curiously, some areas are not affected even during the worst outbreak years (Pfadt and Hardy 1987), which illustrates another way that disturbances can cause patchiness in landscapes.

Interestingly, the apparent vegetation damage is not a good measure of the amount eaten by grasshoppers. It has been estimated that they cut from two to twenty-five times the amount of forage that is eaten, thereby creating considerable detritus (Rodell 1977; Hewitt and Onsager 1983).

Other herbivores, most notably bison and domestic livestock, may also cause substantial changes in vegetation. Apparently, bison can graze the rangeland heavily and might have thereby reduced the amount of forage available to the livestock of early settlers (Larson 1940; Roe 1951; Lauenroth et al. 1993). Early herds of livestock, widespread in the western states by 1890, could have done the same. Explorers passing through Fort Laramie observed what they thought was overgrazed rangeland (Davis 1959). One noted that the area was "barren country abounding with prickly pears." Another traveler reported in the *New York Times* that "125,000 head of cattle have during the past season been driven upon the range lying between the North Platte and Powder rivers, west of Fort Laramie, in Wyoming Territory. As a consequence much of the land is as devoid of grass as the streets of New York." Presumably, the cause was livestock abundance. Davis visited the same area about a hundred years later, observing that the grassland had recovered. McGinnies et al. (1991) observed considerable resiliency to drought in the grasslands of eastern Colorado. The rate of recovery may be slow or rapid, depending on plant species composition, weather conditions, and prior intensity and duration of grazing (Laycock 1991; Cid et al. 1991). A question that cannot be answered with confidence is whether the rangeland "recovered" because of better livestock management or because climatic conditions had improved.

Some plant ecologists attribute pricklypear cactus abundance to excessive livestock grazing, but Lewis and Clark observed in 1805, long before the first domestic livestock, that "the prickly pears are so abundant that we could scarcely find room to lye [sic]" (quoted in Branson 1985). Various investigators have observed how cactus increases with drought and decreases following wetter than normal years, possibly because of greater susceptibility to insects. There is little or no good evidence that cactus biomass increases because of grazing pressure alone.

Travelers in Wyoming often wonder whether the grasslands and shrublands of today are identical to those observed in presettlement times. The photographic record suggests that few changes have occurred (Johnson 1987). But are all the species still there? Were any plant species driven to extinction by grazing during the cattle boom of the 1880s? Such questions are difficult to answer, as changes probably would have occurred even if cattle and sheep had never replaced the bison. Mack and Thompson (1982) concluded that Great Plains plants are much more tolerant of domestic livestock grazing than plants in the Great Basin. They attributed this difference to a much larger population of bison on the Great Plains in presettlement times, with the result that plants there have evolved in the presence of substantial grazing pressure.[16] Long-term studies south of Cheyenne suggest that climate affects plant cover and species composition far more rapidly than livestock grazing does, and though heavy grazing may cause declines in the abundance of some palatable species, moderate grazing (50 percent removal of plant biomass) did not cause a decline in range condition (Klipple and Costello 1960; Grant 1971; Milchunas et al. 1989).

The mechanism by which heavy grazing affects grassland vegetation is generally well known and was described, perhaps most succinctly, by Weaver and Clements (1938):

> The more palatable species are eaten down, thus rendering the uneaten ones more conspicuous. This quickly throws the advantage in competition to the side of the latter. Because of more water and light, their growth is greatly increased. They are enabled to store more food in their propagative organs as well as to produce more seed. The grazed species are correspondingly handicapped in all these respects by the increase of the less palatable species, and the grasses are further weakened by trampling as stock wanders about in search of food. Soon bare spots appear that are colonized by weeds or weedlike species. The weeds reproduce vigorously and sooner or later come to occupy most of the space between the fragments of the original vegetation. Before this condition is reached, usually the stock are forced to eat the less palatable species, and these begin to yield to the competition of annuals. If grazing is sufficiently severe, these, too, may disappear unless they are woody, wholly unpalatable, or protected by spines.

Ellison (1960), who provided one of the early, more comprehensive reviews of the effects of grazing on rangelands, recognized that North American rangelands have been subjected to grazing for millennia and commented: "Obviously grazing—as opposed to overgrazing—did not have this depleting effect." He did, however, make an important observation with regard to different kinds of rangeland disturbance, writing: "Fires may be rare because the fuel is eaten by grazing animals, denudation by plowing may be sporadic because of marginal returns and recurrent drought; but overgrazing, although causing less complete denudation in any one season, is important because it is widespread year after year." In some areas, rangelands probably are still recovering from the disturbances created by livestock in the late 1800s and early 1900s, prior to the advent of modern management practices.

Burrowing mammals are also important herbivores in the plains and basins, and sometimes they cause prominent changes. Foster and Stubbendieck (1980) found that pocket gophers reduced forage production by 18–49 percent on some sites in western Nebraska, and that gopher mounds could cover 25 percent of the soil surface (though 5–15 percent was more common). Laycock (1958) studied the revegetation of pocket gopher mounds.

Another burrowing mammal capable of altering the grassland ecosystem in significant ways is the prairie dog (fig. 5.8). Two species of prairie dogs are found in Wyoming, the blacktailed prairie dog on the Great Plains to the east and the whitetailed prairie dog in the basins (Tileston and Lechleitner 1966). One of the most obvious effects of prairie dogs is the extensive burrowing that characterizes their towns. Large volumes of soil are moved, improving infiltration, hastening the incorporation of organic matter, facilitating nutrient cycling, and increasing spatial heterogeneity (Archer et al. 1987; Whicker and Detling 1988).[17] Sharps and Uresk (1990) concluded that prairie dogs (and probably other burrowing animals) maintain biological diversity by creating habitat for other organisms through their small-scale disturbances.

In addition to extensive burrowing, the blacktailed prairie dogs on the Great Plains keep the surrounding vegetation clipped close to the ground, presumably to improve their ability to detect stalking predators. This clipping gives the impression of an overgrazed rangeland. Some scientists believe, however, that large prairie dog populations are a symptom of heavy livestock grazing (Koford 1958). Prairie dog populations might increase with heavy grazing, they reason, if predator success were diminished when larger areas around the town have little vegetative cover. Moreover, the increased abundance of some forbs in heavily grazed rangeland provides more food for the prairie dogs. Hansen and Gold (1977) cautioned that, while heavy livestock grazing might lead to increases in prairie dog populations where the vegetation is fairly tall, this is less likely to happen in shortgrass prairie, where visibility is already good.

Like the beaver in riparian zones (see chapter 4), prairie dogs influence Wyoming grasslands to an extent that far exceeds the biomass they constitute in the ecosystem. Nutrient cycling is facilitated (Whicker and Detling 1988; Holland and Detling 1990), which may play a role in attracting other herbivores, such as bison, to the town (as discussed previously). Plant species diversity is increased by the small-scale disturbances caused by the diggings of prairie dogs (Hansen and Gold 1977; Coppock et al. 1983a; Whicker and Detling 1988; Bonham and Lerwick 1976), and animal species diversity may increase because of the habitat provided for the badger, rattlesnake, burrowing owl, black-footed ferret, and cottontail (and bison, pronghorn, and elk in presettlement times) (Krueger 1986; Whicker and Detling 1988; Coppock and Detling 1986; Agnew et al. 1986; Sharps and Uresk 1990). Moreover, the burrowing of the prairie dogs is greatly magnified when a badger enlarges the den through its vigorous digging in pursuit of prey.

Prairie dog towns are analogous to beaver colonies in other ways. Just as the beaver colonies move up and down a stream, eventually creating meadows or shrub thickets over much of the stream course, prairie dog towns move about the grassland. Though they appear permanent, the towns expand and contract in size. Probably, much of the grassland would be affected during the course of a thousand years or more—a short time in the history of any ecosystem.[18]

Ground squirrels are another group of small mammals that can have an effect on vegetation through herbivory and burrowings. They do not form towns as prairie dogs do, but their populations oscillate between nearly zero and several hundred per hectare (Nancy Stanton, pers. comm.). Causes of ground squirrel population fluctuations are not known.

Another common feature of Wyoming plains and basins are the mounds created by the harvester ant *(Pogonomyrmex occidentalis)* (Rogers 1987; Kirkham and Fisser 1972; Sneva 1979). These ants are often viewed as destructive because they denude areas around their dome-shaped mounds (fig. 5.10). Hull and Killough (1951) estimated that ants had denuded 33,500 ha in the Bighorn Basin, and Scott (1951) estimated that more than 6 million mounds could be found in the Wind River Basin. The common reaction to such statistics is to lament the loss of forage for livestock. Yet like other burrowing animals, ants mix the soil, and the vegetation surrounding denuded areas is often more vigorous, perhaps because of improved water and nutrient availability caused by ant burrowing and defecation. Birkby (1983) estimated that increased growth around ant hills more than compensated for ant denudation in southeastern Montana.

Burrowing mammals and ants are just another kind of disturbance that add patchiness to the grassland mosaic—above ground and below. These disturbances lead to continual change in the grassland community as species adjust to different environmental conditions. The native flora and fauna are well adapted to these disturbances; their diversity is high in large part because of them. As in other ecosystems, however, the effect of people has often been to modify the frequency of grassland disturbances or to increase their intensity, resulting in a more homogeneous landscape with lower species diversity. In many areas the natural disturbances have been replaced with new ones, like plowing. Plowing on a large scale is a disturbance with which the native grassland species did not evolve. Little wonder then that introduced weedy species, originating in landscapes with a much longer history of agriculture, have done so well in the comparatively new North American environment known as cropland.

F I G. 5.10   Like other burrowing animals, western
harvester ants alter vegetation structure on the grasslands and shrublands of the western Great Plains.
Plants on the edge of the disturbed area around ant
mounds typically grow taller because of reduced competition for water and nutrients.

CHAPTER 6

Sagebrush Steppe

The intermountain basins to the west of the Great Plains are characterized by a mosaic of shrublands. Wyoming big sagebrush is the most widespread shrub, but shallow soils and windswept ridges often have black sagebrush or communities of cushion plants. Gardner saltbush and greasewood are typical where soils are fine textured and alkaline, such as in depressions or on some floodplains. Soils along ephemeral streams have basin big sagebrush or silver sagebrush. The transitions between these various components of the vegetation mosaic are sometimes sharp, owing to abrupt changes in soil conditions or moisture availability. Many of the associated plant species are also found in grasslands and include western wheatgrass, junegrass, needle-and-thread grass, Sandberg bluegrass, pricklypear cactus, scarlet globemallow, and rabbitbrush. Common mammals include the pronghorn antelope, blacktailed jackrabbit, badger, ground squirrel, and vole (see table 5.1 and R. F. Miller et al. 1993).

The distribution limits of any species are an interesting ecological puzzle, and big sagebrush is no exception. What accounts for the rarity or absence of sagebrush on the Great Plains at lower elevations? Why is it absent from parts of the Laramie and Shirley basins? Definitive answers are not yet available, but climatic and edaphic factors are relevant (Houston 1961; West 1983, 1988; R. F. Miller et al. 1993). Notably, winter precipitation is a larger proportion of annual precipitation in the western half of the state, where sagebrush is most common (see chapter 3). Water accumulates in the snowpack, and when melting does occur, more water infiltrates to greater depths in the soil. The presence of shrubs increases the leaf area beyond what might be expected in a grassland, and therefore shrubs occur where there is a more dependable source of water. In drier environments or toward the eastern fringes of its range, big sagebrush is found only on the lee sides of ridges or in ravines that are mesic because of snowdrifting (fig. 6.1). Elsewhere, big sagebrush forms an extensive, uniform cover (fig. 6.2).

Plant distribution is determined more by seedling requirements and tolerances than by the characteristics of adult plants. In Wyoming, sagebrush seedlings are commonly observed in some years but only rarely in others; cooler- and wetter-than-average early summers seem to be required.[1] Lower temperatures minimize the rate of water loss from drought-sensitive seedlings, providing more time for roots to grow down to a dependable water supply. Though the roots may not grow more than a few centimeters in the first year, the presence of deep soil water contributes to a more mesic environment in the surface soil because of capillary movement from below, or possibly because of condensation in the surface soil of water vapor that emanates from the deeper soil water (Hennessy et al. 1983). Once established, the adult plants tolerate the occasional years when little winter precipitation occurs.

Variation in the Sagebrush Mosaic

With twenty-three species of woody shrubs and forbs in Wyoming (Beetle and Johnson 1982; Dorn 1992), the genus *Artemisia* is found from the lowlands to the alpine zone above treeline (figs. 6.3, 6.4). The following species merit closer examination when considering sagebrush steppe: big sagebrush, silver sagebrush, black sagebrush, and low sagebrush.[2]

Most prominent of the sagebrush species is big sagebrush, a species characteristic of deep, well-drained soils (Goodwin 1956; Thatcher 1959; West 1988). The species can be observed from eastern Wyoming all the way to the Cascade Mountains of Oregon and Washington, and from Canada south to Arizona and New Mexico (West 1988). It occurs in habitats ranging from the warm Utah deserts at 900 m to mountain shrublands at 3,000 m or higher. Soils with big sagebrush are typically classified as Aridisols (see table 3.1) and have pH values ranging from 6.6 to 8.5 (Thatcher 1959).

As is usually the case with species occupying such diverse environments, big sagebrush exhibits considerable genetic variation. Three varieties are now commonly recognized (Beetle and Johnson 1982): basin big sagebrush, Wyoming big sagebrush, and mountain big sagebrush. Basin big sagebrush is found at lower elevations (fig. 6.3) and is usually restricted to comparatively moist ravines or valleys (Barker and McKell 1986). It grows taller than any other species of *Artemisia* (up to 2 m or more). In Wyoming it is especially common in the Bighorn Basin and in the southwestern quarter of the state.

Wyoming big sagebrush is the most common shrub of the intermountain basins. It is normally less than 0.5 m tall and occupies the drier uplands, with the taller basin big sagebrush sometimes occurring in adjacent ravines. Shumar and Anderson (1986) observed that basin big sagebrush was more common on sandy soils, which are more mesic because of the inverse texture effect, and Wyoming big sagebrush was more common on fine-textured soils (with lower infiltration rates and consequently lower water availability). Stands with both subspecies occurred where soil conditions were intermediate.

Mountain big sagebrush is found in the foot-hills at higher, cooler elevations (fig. 6.3). It is commonly 1 m tall and often occurs in foothill shrublands adjacent to mountain forests (up to 3,000 m). Winward (1970) observed that water stress did not usually develop until September on sites dominated by mountain big sagebrush, whereas it occurred in mid-July on Wyoming big sagebrush sites and late July or early August where basin big sagebrush was common.

Silver sagebrush and black sagebrush are two other prominent species of sagebrush in the lowlands. Plains silver sagebrush,[3] a subspecies of silver sagebrush, usually occurs on more mesic sites east of the Continental Divide (Thatcher 1959; Hazlett and Hoffman 1975; Walton 1984; White and Currie 1984). Sometimes it occurs in ravines or on floodplains in areas where Wyoming big sagebrush dominates the upland vegetation (figs. 6.5, 6.6). Silver sagebrush also occurs on sandy soils. Unlike most woody species of *Artemisia*, silver sagebrush is capable of sprouting from the root crown following fire or other disturbances (White and Currie 1983).

Black sagebrush is usually less than 0.5 m tall and occurs on drier, coarser textured, shallower soils than either silver sagebrush or big sagebrush (see fig. 6.3; Thatcher 1959). It is found on calcareous soils throughout the intermountain basin region (Zamora and Tueller 1973; Baker and Kennedy 1985). Soils, however, may become even too coarse or shallow for black sagebrush. For example, some areas just east of Laramie have small patches of grassland on shallow soils adjacent to black sagebrush steppes, where the soils are slightly deeper.

Low sagebrush is usually less than 25 cm tall and is found only in the western part of the state—for example, in the lowlands of Jackson Hole and Grand Teton National Park, where it is interspersed with mountain big sagebrush (Sabinske and Knight 1978). Typically the soils have a dense layer of fine-textured clay at a shallow depth that impedes drainage in the spring. Patches of low sagebrush several meters across are often found in shrublands otherwise dominated by mountain big sagebrush (see chapter 15). Alkali sagebrush is also usually less than 25 cm tall and forms a similar pattern with basin or Wyoming big sagebrush (as in North Park, Colorado). As its

F I G. 6.1 Shrublands dominated by Wyoming big sagebrush form a mosaic with saltbush desert shrubland in this area west of Rawlins (elevation 2,052 m; 6,750 ft). Big sagebrush is restricted to ravines, where drifting snow accumulates, thereby providing more water for plant growth. The adjacent shrubland is typically dominated by birdfoot sagewort, Gardner saltbush, western wheatgrass, and winterfat. This area is one of the driest in the state (see fig. 3.4).

name implies, however, it is found on more alkaline soils. Usually there is an abrupt transition between big sagebrush steppe and the drier shrublands that are dominated by low sagebrush and alkali sagebrush (Robertson et al. 1966).

The vegetation mosaic of the intermountain basins is highly variable. Burke et al. *(Topographic control,* 1989) concluded that the main control over vegetation distribution south of Rawlins is the effect of wind on snow redistribution. They found black sagebrush on windswept ridges or slopes with shallow soils; Wyoming big sagebrush on adjacent slopes with more snow accumulation and deeper soils (but minimal salt accumulation); mountain big sagebrush in moist ravines with deep snow; and desert shrubland in basins where salt accumulation is high. Where snow accumulation is exceptionally deep, only grasses and forbs can survive (nivation hollows; fig. 6.7). Big sage-

brush is generally thought to be intolerant of soils that are frequently wet or have high salt concentrations, unlike greasewood or Gardner saltbush. In desert shrublands with saline soils, sagebrush is sometimes found only in the ephemeral washes or in ravines with less saline soils (probably from more frequent leaching). In the Bighorn Basin, grasslands typically occur on shallower soils, and big sagebrush steppe on deeper soils (fig. 6.8).

A second common pattern occurs where ravines with Wyoming big sagebrush are surrounded by grasslands or by sagebrush steppe. The shrubs are more dense and taller in the ravines, creating dark stringers of sagebrush (or taller sagebrush) across the landscape (see fig. 6.1). The sagebrush in the ravines and on the uplands may or may not be of the same variety. A third pattern occurs where big sagebrush predominates on the upland

F I G. 6.2   Wyoming big sagebrush steppe south of
Pinedale (elevation 2,189 m; 7,200 ft). Aspen groves
occur on the distant hills, where more water is provided by deeper snow accumulation or groundwater
seepage to the surface.

but is absent in ravines, perhaps because of slow
drainage, which prevents sufficient soil aeration
for sagebrush roots (see fig. 6.5; Lunt et al. 1973;
Ganskopp 1986).

Clearly, sagebrush distribution in the landscape depends on the responses of the species or
variety to soil moisture, salinity, depth, and texture, as well as to climatic factors. The patterns are
accentuated over short distances because of wind,
topography, and abrupt changes in soil conditions
(Smith 1966; Welsh 1957; Myers 1969; Baker and
Kennedy 1985; Burke et al., *Topographic control,*
1989).

SAGEBRUSH ISLANDS

Wyoming big sagebrush sometimes occurs in circular, oval, or lens-shaped patches that are 3–15
m across and are commonly referred to as sagebrush islands (fig. 6.9). These islands may in fact
be patches of sagebrush that are taller than the
surrounding sagebrush shrubs, or they may be
patches of sagebrush in a matrix of grassland or
desert shrubland.

Several explanations are plausible for sagebrush islands. First, small silt dunes or sand dunes
commonly form on the lee side of taller shrubs.
These dunes may have been initiated by a period
of wind erosion, possibly due to heavy grazing by
bison or livestock during a dry period. The accumulated silt and sand create conditions more
favorable for shrub growth by improving snow
accumulation and infiltration. Consequently, additional shrubs become established nearby. A
larger number of taller shrubs increases the deposition of windblown particles and snow, with the
result being the development of small dunes that
may become 2 m deep or more. Such dunes are
known as coppice dunes and typically have spiny

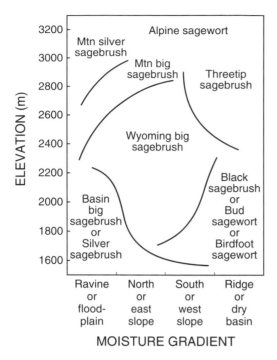

F I G. 6.3   The approximate distribution of different species of *Artemisia* (sagebrush and sagewort) in relation to elevation and topographic position.

caused snowdrifting on the lee side of the shrub, creating conditions favorable for the establishment of additional shrubs—perhaps from seed produced by the original plant. In addition to snow, the taller shrubs could also lead to the accumulation of blowing organic matter as well as some soil. This accumulation, combined with nutrients brought up from the soil by the shrubs and deposited on the surface with litterfall, could create "islands of fertility" (Garcia-Moya and McKell 1970; West and Klemmedson 1978; Allen and MacMahon 1985). Such islands are found where water and other resources are inadequate to support a uniform sagebrush steppe. As West (1988) suggested about sagebrush-dominated vegetation in the Great Basin of Utah, any activity that alters sagebrush islands reduces the productivity of the ecosystem as a whole. Other sagebrush islands appear to be associated with mima mounds (see chapter 8), and Thatcher (1959) observed sagebrush islands occurring in "pockets of deeper soil" surrounded by vegetation dominated by threetip sagebrush. As is usually the case, a vegetation pattern can have several explanations.

hopsage, rubber rabbitbrush, green rabbitbrush, and basin wild rye in addition to Wyoming big sagebrush. Lundberg (1977) observed coppice dunes near the town of Green River dominated by spiny hopsage and big sagebrush that were surrounded by a "desert pavement" dominated by bud sagewort. Spiny hopsage seems to be an indicator of sandy and probably more mesic soils in the semiarid intermountain basins.

Elsewhere, the sagebrush islands are not associated with windblown material and are less easily explained. They are commonly characterized by numerous small mammal burrows, which can disrupt shallow hardpans of calcium carbonate (caliche) or in other ways create soil conditions that favor infiltration and taller shrub growth. Perhaps mammals are attracted by the shade provided by the shrubs, but how did the shrubs become established there in the first place? One explanation is that the patch began with a single shrub that became established by chance (perhaps adjacent to small mammal burrows). Its presence

Sagebrush Adaptations

Understanding vegetation ecology requires information on the requirements and tolerances of the dominant species. Many scientists have studied big sagebrush, probably because it is a conspicuous feature of the landscape over large areas in the West and because many ranchers would like to reduce its abundance. Big sagebrush is poor livestock food, and it competes with preferred grasses for water and nutrients. In contrast, wildlife biologists recognize sagebrush as important browse for deer, elk, and pronghorn antelope during the winter. It also provides an important habitat for sage grouse and other birds. Some tourists view the sagebrush steppe as a barren landscape; others appreciate its unique aroma—especially after a summer rain—and the bluish-green cast it gives to the open landscape.

Big sagebrush is adapted for surviving in an environment where water becomes limiting by mid- to late summer.[4] The sagebrush root system is deep enough to make use of deep soil moisture,

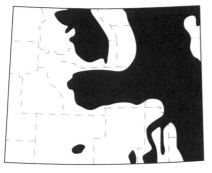

Plains silver sagebrush

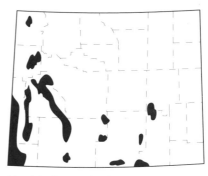

Mountain silver sagebrush

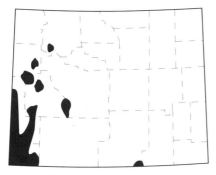

Alkali sagebrush

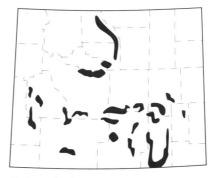

Black sagebrush

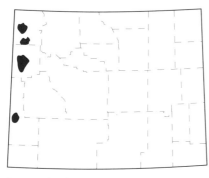

Low sagebrush

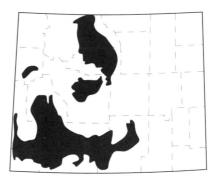

Basin big sagebrush

Mountain big sagebrush

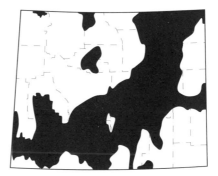

Wyoming big sagebrush

F I G. 6.4   Distribution maps for eight species of
sagebrush *(Artemisia)* in Wyoming. Adapted from
Beetle and Johnson 1982.

F I G. 6.5   Mixed-grass prairie intermingles with Wyoming big sagebrush steppe in the Powder River Basin. In this area, the sagebrush cannot grow in the ravines, probably because they are too wet for too long during the year. Some of the shrubs closest to the ravine are plains silver sagebrush, which is more tolerant of wet soils. Elevation 1,459 m (4,800 ft).

F I G. 6.6   Big sagebrush and silver sagebrush are easily distinguished by examining leaf shape. The longer leaves on big sagebrush are ephemeral, dropping off early in the summer, whereas the shorter leaves remain on the shrub for a full year. Both species are evergreen. Drawing by Judy Knight.

extending down 2 meters or more, but the shrub also has shallow roots that enable it to use summer rainwater (Tabler 1964; Campbell and Harris 1977; Sturges and Trlica 1978). The root system extends laterally to a distance of 1.5 m (Sturges and Trlica 1978). Sturges (1977b) observed that the plants use surface moisture early in the growing season and progressively deeper water as the summer continues. Significant soil water recharge usually occurs only once a year, during the snowmelt period.

Several mechanisms allow the efficient use of water by big sagebrush. First, the stomata close rapidly as water stress develops during the day, thereby limiting transpiration until the next morning. The plant reequilibrates at night, when water uptake exceeds losses (Caldwell 1979). Stomatal closure limits photosynthesis as well, but prolonging the growing season by conserving water seems to be adaptive. The plant is evergreen, retaining

F I G. 6.7 Wyoming big sagebrush steppe south of Rawlins. Nivation hollows form where drifted snow persists on leeward slopes until June, thereby creating a wet environment that big sagebrush cannot tolerate. Elevation 2,250 m (7,400 ft).

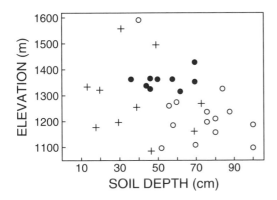

F I G. 6.8 The distribution of grasslands (+), black sagebrush steppe (•), and big sagebrush steppe (o) in relation to soil depth and elevation. Data from the Bighorn Basin.

most of its leaves for one year (Beetle and Johnson 1982), and acclimation to changing temperatures occurs readily, even down to 0°C (Caldwell 1979). Evergreen plants fix carbon whenever water and temperature conditions are within tolerance ranges, with no lost time in the production of new leaves, thereby lengthening the growing season. Some research suggests that big sagebrush is capable of photosynthesis in the early spring and even during warm winter days (Depuit and Caldwell 1973, 1975; Caldwell 1979).

Water is most readily available in the spring, when plants have the largest leaf area for photosynthesis. Two leaf types are produced: ephemeral leaves in early spring, which fall as water stress develops during the summer, and evergreen (or overwintering) leaves, which remain on the plant until the following spring (see fig. 6.6; Beetle and Johnson 1982; Miller and Schultz 1987). Essentially the plant is evergreen, but the loss of ephemeral leaves by early summer allows for some reduction in the total leaf area, thereby providing another mechanism for conserving water and extending the length of the growing season. Shadscale also produces both ephemeral and persistent leaves (West 1983).

The carbohydrates produced by photosynthesis are translocated throughout the plant wherever they are needed for energy. A large portion of these energy-rich compounds, however, is stored in the twigs. Thus, not only are leaves available on the shrub all winter, but the twigs are a good source of energy for herbivores. Furthermore, the shrub is tall enough to be above the snow during most winters. Little wonder then that historically the shrubs have been subjected to considerable browsing pressure and that they have evolved a mechanism to minimize this herbivory. The characteristic scent of sagebrush—produced by volatile oils known as terpenes—is probably a defense against herbivores. Without this adaptation, the shrub may not have been able to survive the intense browsing that might have resulted from large herbivore populations controlled most directly by the winter food supply.[5]

Unlike most other associated plants, big sagebrush lacks the capacity to sprout from roots or the root crown. Consequently, longevity and seed production are especially important for the per-

F I G. 6.9   Wyoming big sagebrush sometimes grows in patches ("islands"), such as in this area southeast of Rock Springs (elevation 2,189 m; 7,200 ft). Often the patches are associated with coppice dunes.

sistence of the species. Sagebrush produces annual rings in the wood similar to temperate zone trees, and the age of some shrubs has been estimated at more than a hundred years (Passey and Hugie 1963; Ferguson 1964). Shrubs that are forty to fifty years old are common (Tisdale and Hironaka 1981).

Seed germination and seedling establishment are especially important processes for plants lacking the ability to reproduce vegetatively. In most years, big sagebrush produces thousands of tiny seeds (Goodwin 1956; Weldon et al. 1959; Robertson et al. 1966; Choudhuri 1968; McDonough and Harniss 1974; Daubenmire 1975; West et al. 1979; Cawker 1980; Romo 1984; Eckert et al. 1986). As with other plants found in the intermountain basins, germination occurs in late winter or early spring, when moisture is available. No cold period is required to break seed dormancy (stratification), although some investigators have found that this treatment causes higher germination rates in mountain big sagebrush. Germination can occur in the dark, but light does have a stimulatory effect. Even very low salt concentrations in the soil are sufficient to inhibit big sagebrush germination (Choudhuri 1968), an observation that may explain why the plant is less common on saline soils.

Although 80 percent or more of the seeds may germinate, few if any seedlings become established in most years—apparently because of seedling sensitivity to water stress, soil salinity, and competition from already-established plants (Robertson et al. 1966). Seedlings become established only during favorable years or following a disturbance that reduces the competitive ability of neighboring plants (Lommasson 1948; Robertson et al. 1966). The age distribution of a sagebrush population is therefore discontinuous, with several age classes represented (Daubenmire 1975; West et al. 1979; Cawker 1980). Cawker (1980) suggested that conditions for high growth rates early in the summer are essential so that the seedlings can develop a root system adequate to cope with late-summer droughts, but that low temperatures during the growing season and winter could also be a cause of mortality.

In some respects, big sagebrush has the characteristics of a weed. It produces many seeds that often lead to plant establishment on disturbed sites, and its initial growth rates can be rapid. Once established, however, its competitive abilities allow it to persist until fire or other disturbances kill the plant.

The Sagebrush Ecosystem

Compared to grasslands, the distinguishing features of sagebrush ecosystems are the presence of a conspicuous shrub and a larger proportion of the annual precipitation occurring in the winter. Otherwise grasslands and sagebrush steppe are similar: plant growth (NPP) is limited by water availability and the length of the growing season, and most of the biomass and herbivory is below ground. Potential evapotranspiration is usually greater than the annual precipitation, and consequently nutrient leaching is rare; fire, drought, and burrowing animals are common disturbances; plant and animal species have evolved to minimize competition by using different resources and by using them at different times of the year; and coexisting herbivores undoubtedly interact in complex ways to affect nutrient cycling and productivity. The discussion on sagebrush ecosystems in this chapter focuses on the differences caused by adding a large, evergreen, aromatic shrub that is unable to sprout. Considering these differences is interesting from an ecological perspective and is relevant to land management because landowners often use fire or herbicides to reduce big sagebrush cover in order to increase grass for their livestock.

HYDROLOGY AND PRODUCTIVITY

Because wind and snow are characteristic features of the plains and basins, snow often accumulates on the lee side of shrubs (fig. 6.10). Drifted snow

F I G. 6.10   Snow frequently accumulates on the leeward side of shrubs, which probably increases the amount of water that infiltrates into the soil.

can be a significant supplement to soil water recharge, especially on comparatively level sites, where accumulation is not affected by topography (Sturges 1975). By removing big sagebrush, total water availability and NPP could be reduced. Ranchers know that there is more livestock forage following reductions in sagebrush. Still, removing sagebrush may reduce ecosystem NPP. Harniss and Murray (1973) noted that the NPP for sagebrush steppe is higher than in grasslands because the shrubs tap resources unavailable to most grasses and forbs. Sagebrush steppe has a range in ANPP of 800–2,500 kg/ha/yr (West 1983), which is generally higher than for mixed-grass prairie (Singh et al., *Vegetation*, 1983).

Water availability to plants may also be influenced by shrubs through the "blackbody effect" (Robertson 1947; Sonder 1959). This phenomenon would be especially important where the vegetation is more or less uniformly covered with snow except for the tops of shrubs—a common occurrence. Solar energy is absorbed by the shrub tops, causing snowmelt around the shrub and creating a depression or *well* into which more snow can drift during the next storm. In this way the shrubs facilitate the percolation of meltwater into the soil or deeper into the snowpack while creating new snow accumulation potential. The depressions around the shrubs may be created and filled several times during a winter, thereby augmenting moisture input above that which would be expected if the shrubs were not there.

F I G. 6.11   The growth of grasses and forbs increases greatly after big sagebrush steppe is burned, as in this area east of Saratoga (elevation 2,280 m; 7,500 ft). The forests in the distant Medicine Bow Mountains are dominated by lodgepole pine and aspen.

If sagebrush is reduced by fire or other means, grass and forb cover increases (fig. 6.11). Yet because of fewer deep roots in the newly created grassland, total water consumption is lower. The NPP is also lower, suggesting again how NPP correlates with total evapotranspiration in dry climates (see fig. 5.4). Sturges (1977a) reviewed the hydrology of sagebrush steppe, noting that the greatest effects of sagebrush control occur when soils are deep. Otherwise the grasses are able to tap the same volume of soil as the shrubs.

ENERGY FLOW

The food web of the sagebrush steppe must be similar to that of grasslands. Less of the annual precipitation, however, comes during the summer, which limits the success of warm-season species such as blue grama, and much of the productivity is concentrated in the less palatable big

sagebrush. Consequently, the carrying capacity for large ungulates might well be less than on the grasslands of the Great Plains. As Mack and Thompson (1982) suggested, summer precipitation and warm-season C_4 species are comparatively infrequent in the intermountain basins, which may explain the fewer numbers of bison found there (see chapter 5).[6] Sagebrush is known to be an important food for large native ungulates, but primarily during the winter, when other forage is scarce or buried by snow. Bison rely on grasses and forbs more than on shrubs (as do cattle).

The role of fire in sagebrush steppe is still uncertain, but historically, flammable conditions must have developed. The inevitable fire would have removed the nonsprouting big sagebrush, freeing up nutrient and water resources that favored the development of a wheatgrass-needlegrass–Indian ricegrass prairie with a higher carrying capacity for large ungulates. Although animals are constrained by their distance to water and other factors, concentrations of some mammals may have shifted from one area to another according to the amount of sagebrush steppe burned in the preceding few years. Perhaps the optimal habitat for some animal species was a mosaic of old-growth sagebrush intermingled with recently burned sagebrush. Presettlement sagebrush steppe might have been more varied where periodic fires were possible, creating patches of grassland and young sagebrush as well as the old-growth sagebrush that seems so prevalent today. Sagebrush cover might be more uniform today, the result of fire suppression and more extensive livestock grazing.[7]

Primary productivity is enhanced by the presence of big sagebrush (Blaisdell 1949; Harniss and Murray 1973), in part because of its deeper root system, but also because it has leaves throughout the year. Photosynthesis is possible whenever environmental conditions are favorable. Pearson (1965a) suggested that sagebrush evergreenness is an adaptation for lengthening the growing season, as the plants are capable of photosynthesis even at freezing temperatures. He also observed that 65–85 percent of the biomass was below ground. Sagebrush growth was greatest in the fall and early spring, while the growth of needle-and-thread grass was most rapid during the late spring and early summer. Depuit and Caldwell (1973) found that water stress and leaf age could limit the rate of sagebrush photosynthesis, and Joyce (1981) calculated that fall-through-spring precipitation was the best predictor for sagebrush steppe NPP. Fetcher and Trlica (1980) discovered that big sagebrush productivity was highly correlated with spring temperature as well as with spring precipitation.

NUTRIENT CYCLING

Nutrient distribution in sagebrush steppe ecosystems is similar to that of grasslands, with the largest concentrations being in the soil (Mack 1971; Murray 1975; Burke 1987; Carpenter and West 1987). Water is more limiting to NPP than nutrients, although nitrogen may be limiting during wet years (Charley 1977; Doescher et al. 1990). The major source of nitrogen is probably precipitation, as in grasslands, but symbiotic nitrogen fixation occurs locally in microphytic crusts on the soil surface (West and Skujins 1977, 1978; West 1990), in the nodules of lupine and other legumes, and possibly adjacent to the roots of Indian ricegrass (Wullstein 1980). Losses of nitrogen may occur through erosion, denitrification, or the loss of nitrogen gases (NO and N_2O) by diffusion (Matson et al. 1991). Several investigators have provided data on the rates of litterfall and decomposition in shrublands dominated by big sagebrush (Mack 1971; Branson et al. 1976; West 1985; Parmenter et al. 1987), and there is some evidence of nutrient translocation from sagebrush leaves to twigs prior to leaf fall (Mack 1971; Charley and West 1975; Charley 1977). Still, soil nitrogen is higher under shrubs. There is also great variation in soil organic matter and nitrogen availability from ravines with mountain big sagebrush to ridgetops with black sagebrush (Burke et al., *Organic matter*, 1989).

The major effect on nutrient cycling of adding big sagebrush, or any shrub, to the ecosystem is to create a larger amount of biomass above ground, specifically woody biomass. Nutrient accumulation is thereby greater above ground than where grasses and forbs predominate. When the shrub dies, or as twigs and branches die, more of the nutrients are added to the soil surface in the form of wood rather than herbaceous biomass. Also, woody fuels accumulate that can create conditions for a longer, hotter fire.

Another influence of shrubs is to create islands of fertility, as discussed previously. The shrub growth form accumulates windblown debris and nutrients (Charley and West 1975; Charley 1977; Burke 1987), and the shrub's root system extracts soil nutrients from a meter or more beyond the canopy while dropping them as litter in a smaller area. It would be interesting to determine whether such nutrient concentrations persist after a fire and thereby cause the patchiness in the grassland that develops afterward. The death of the sagebrush root system must be important as well, not only in reducing competition for water and nutrients with neighboring grasses and forbs, but also in providing a substantial pool of nutrients through the decomposition of sagebrush roots. The woody root system could also lead to an altered soil structure.

More research remains to be done, but the presence of woody shrubs has significant effects on ecosystem processes. It would be useful to compare energy flow and nutrient cycling in a sagebrush steppe that has been burned to one where the shrubs have simply been cut and removed. Only in this way can the effects of the fire be distinguished from those of removing the shrub.

Disturbances and Succession in Sagebrush Steppe

Various disturbances lead to sagebrush death and ecosystem change. Fire is one that occurs when fuel accumulates. Once started, whether by humans or lightning, fire may burn large areas of sagebrush steppe. James Chisholm explored the Wind River Basin and South Pass City areas in 1868 (Homsher 1960), during the Wyoming gold rush, and observed the rapid spread of at least two human-caused fires across what must have been sagebrush-dominated uplands. He wrote in his journal: "The grass took fire and all our efforts could not extinguish it. This time the situation was really alarming, for there was a prospect that the entire Wind River Valley might go up in a flame. . . . The flame went over the nearest hill with amazing velocity and away, Heaven knows how far. . . . The mountains were black and bare over which we travelled for the rest of that day, and we saw the fire pursuing its way far ahead in several directions, but fortunately away from the valleys." The words *mountains* and *away from the valleys* in this quotation suggest that much of the burned area was in the foothills or higher, where fuel accumulation might have been greater. Evidence was reviewed by R. F. Miller et al. (1993) that suggests that the mean fire-return interval of sagebrush steppe in presettlement times ranged from twenty to a hundred years, depending on the NPP of the site.

Drought, too, can kill sagebrush. Ellison and Woolfolk (1937) noted that the shoots of silver sagebrush were more susceptible than those of big sagebrush to drought in eastern Montana, but that silver sagebrush recovered more rapidly because of its ability to sprout from the root crown. Winter mortality of sagebrush can occur as well, probably from water stress created by frozen soils, low soil water potential, and below-average snowfall. Another cause may be the premature breaking of dormancy during periods of above-average air temperature. Extensive areas of mountain big sagebrush were apparently killed by spring frost damage during the 1976–77 winter in southwestern Wyoming and neighboring states (Hanson et al. 1982; Nelson and Tiernan 1983), but shrub dieback during the mid-1980s in the Great Basin has been attributed to unusually wet conditions that persisted for five years (West 1988; Wallace and Nelson 1990). Big sagebrush, shadscale, fourwing saltbush, winterfat, rabbitbrush, and antelope bitterbrush were all affected.

Allred (1941) observed that the coincidence of drought and an outbreak of grasshoppers caused the death of approximately 50 percent of the big sagebrush in a portion of the Powder River Basin. He wrote: "In 1936 . . . the grasshoppers swarmed in such hordes that they devoured all of the edible vegetation, ate the leaves and bark from the twigs of the sagebrush, and completely girdled the more tender stems." Other insects known to kill big sagebrush include gall midges (Furniss and Krebill 1972), the aroga moth (Henry 1961; Gates 1964), certain beetles (Pringle 1960), and Mormon crickets (Cowan 1929; Young 1978). Sagebrush is commonly infested with gall-forming

flies, but apparently with little detrimental effect (Robert Pfadt, pers. comm.). Snowmolds can be another cause of lost vigor and possibly death (Nelson and Sturges 1986; Allen et al. 1987), and voles may girdle the stems of big sagebrush, causing a significant disturbance when their populations are high (Frischknecht and Baker 1972; Parmenter et al. 1987; R. F. Miller et al. 1993). Anderson and Shumar (1986) found that, in southeastern Idaho during years of high population density, jackrabbits can cause a reduction in winter plant cover. However, no species declined in density, and several grew more vigorously after jackrabbit herbivory, suggesting compensatory growth (see chapter 5).

Big sagebrush is a competitive shrub capable of developing high levels of cover. The plant is easily killed, however, and subsequent grass production may be more than double that of pretreatment levels (Robertson 1947; Alley and Bohmont 1958; Fuller 1958; Sturges 1975, 1986; Eckert et al. 1986). Depending on the intensity of livestock grazing, the increased forage production may last from ten to twenty years (Johnson 1962). Big sagebrush usually regains dominance as new seedlings become established and the shrubs grow to maturity, but for a decade or more the sagebrush steppe appears and functions more like a grassland if most of the shrubs are killed.

Shrub death also means that less water is transpired because the root systems of the surviving grasses and forbs generally do not penetrate as deeply as the sagebrush roots. Total evapotranspiration is reduced by about 15 percent in several subsequent growing seasons following herbicide treatment (Sheets 1958; Tabler 1968; Sturges 1986). In some areas, especially in the mountain sagebrush zone, this lower evapotranspiration may allow for more streamflow from sagebrush-dominated watersheds.

DISTURBANCE BY FIRE

Because big sagebrush does not sprout, changes following fire in sagebrush steppe are more prominent than those following fire in grasslands. Other species may also be reduced by burning (such as Idaho fescue, needle-and-thread grass, pricklypear cactus, and threadleaf sedge), while some are favored (such as bottlebrush squirreltail,

Sandberg bluegrass, and western wheatgrass; Wright et al. 1979).

Susceptibility to fire-caused mortality seems to depend on the amount of fuel that has accumulated above the root crown, the depth of perennating buds below the soil surface, and the stage of growth at the time of burning. Mihlbachler (1986) observed that western wheatgrass vigor and production increased with spring burning but declined with fall burning, and that burning at any time reduced the productivity of needle-and-thread grass. Repeated burning in consecutive years can cause a shift from a community dominated by a mixture of cool- and warm-season species to one dominated primarily by warm-season species, which are generally more fire-tolerant. This shift in dominance can be significant if early-season production is viewed as a critical resource, whether for livestock or big game. Burning sagebrush may be useful, but it should be done only in the context of management plans and after careful consideration of the undesirable changes that could result (Smith et al. 1985).

Rennick (1981) conducted a study of prescribed spring burning of foothill sagebrush steppe in southeastern Montana. Big sagebrush was almost completely killed, but prairie rose, silver sagebrush, skunkbush sumac, western snowberry, and winterfat persisted through sprouting. As expected, perennial grass production on the burned sites was higher than comparable unburned areas even though a few species were reduced in vigor in the first growing season after the fire (in particular, bluebunch wheatgrass and prairie junegrass). Western wheatgrass increased rapidly after the burn. Two weedy biennials also increased—common dandelion and yellow salsify—but Rennick concluded that the abundance of annuals (for example, cheatgrass) can be reduced temporarily if burning occurs when the plants are actively growing in the spring. If the rangeland is in good condition, perennial grasses become dominant after the fire (West and Hassan 1985).

The frequency of fire is another important consideration. Wright et al. (1979) noted that repeated burning may reduce the abundance of perennials and increase the frequency of annuals. Several investigators have also observed that sprouting shrubs, such as horsebrush and rabbitbrush, may be favored over big sagebrush if fires occur more than once every twenty to twenty-five years (Young and Evans 1978; Wright et al. 1979; Tisdale and Hironaka 1981; Young 1983; Bunting et al. 1987).

SUCCESSION FOLLOWING DISTURBANCES

With time, sagebrush often regains dominance on disturbed areas. As observed by Harniss and Murray (1973) on a site in eastern Idaho, mountain big sagebrush production gradually increased over a thirty-year period while that of other shrubs, grasses, and forbs gradually decreased (fig. 6.12). After any natural or artificial disturbance, the rate of big sagebrush reinvasion increases in proportion to the amount of precipitation (Lommasson 1948); the amount of litter on the soil (Beetle 1960); the extent of competition from herbaceous plants (Robertson et al. 1966); the intensity of grazing (Johnson 1969); the number of sagebrush seeds in the soil (Frischknecht and Plummer 1955; Mueggler 1956); and the number of live shrubs that remain (Johnson 1969). With continued protection from grazing and fire, as might occur in exclosures, sagebrush apparently continues to increase in cover (Anderson and Holte 1981; Uhlich 1982). Cooper (1953) reported a decline in sagebrush and an increase in perennial grasses (junegrass, needle-and-thread grass, and Sandberg bluegrass) with protection from grazing, but sagebrush generally persists until the next fire or drought.

Big sagebrush also readily invades abandoned farmland. Lang (1973) monitored permanent plots on old fields in the Powder River Basin from 1943 to 1965, noting that grasses, big sagebrush, and pricklypear cactus all increased in cover during this twenty-two-year period. The cactus and sagebrush increased on both grazed areas and in livestock exclosures. Interestingly, the establishment of big sagebrush on abandoned mine land is often not easily accomplished (E. DePuit, pers. comm.; Allen et al. 1987).

CHEATGRASS IN THE SAGEBRUSH STEPPES OF WYOMING

The burning of big sagebrush, when done in appropriate places and at appropriate times, is viewed as a useful management tool for increasing herbaceous production. It simulates a natural

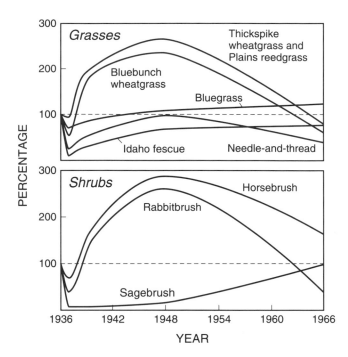

F I G. 6.12   Changes in the canopy cover of several grasses and shrubs after the burning of big sagebrush steppe in eastern Idaho. The burn occurred in 1936. An increase (or decrease) in plant cover is indicated when the line for a species is above (or below) the broken line. The two shrubs that increase, rabbitbrush and horsebrush, are capable of sprouting; big sagebrush declines because it does not sprout. Adapted from Harniss and Murray 1973.

disturbance to which many native species are adapted. A complication with this practice has developed, however, in parts of Utah and Nevada, and it may be a problem in Wyoming as well—at least at the warmer, lower elevations, where cheatgrass is common.

Cheatgrass is an annual grass that was introduced in western North America in about 1889 from the steppes of Europe (Mack 1984). It has become common in some areas, partly because it is a winter annual that begins growth in the fall or early spring and thereby gains a competitive edge over many native species. Although livestock grazing may hasten the spread of cheatgrass, some have observed that this pioneer species can invade climax communities with no grazing pressure or disturbances in the Great Basin (Tisdale et al. 1965; R. F. Miller et al. 1993). Often cheatgrass appears to be a stronger colonizer than the native pioneer species, and once established it is a formidable competitor with native species (though native species may become reestablished with proper management in Utah; West and Hassan 1985).

With cheatgrass as a component of the flora, land managers are confronted with the prospect of increasing the population of an introduced weed while trying to reduce the amount of big sagebrush and increase native perennial grasses and forbs.[8] Furthermore, cheatgrass leads to the rapid accumulation of a highly flammable fuel, shortening the fire-free interval (fig. 6.13). Fires occur more frequently, thereby diminishing the chances of sagebrush reestablishment, causing a decline in some perennial grass species, and favoring cheatgrass expansion still further (Young and Evans 1973; Young et al., *Great Basin*, 1976; Young et al., *Downy brome*, 1976; Billings 1990; McArthur et al. 1990; Whisenant 1990). The economic impact of cheatgrass is moderated by the fact that it is good livestock forage before it flowers. But converting semiarid steppes to vegetation dominated by an introduced annual is a significant change that usually has undesirable consequences (D'Antonio and Vitousek 1992).

The cheatgrass problem may not be so severe over much of Wyoming's basins because of cooler temperatures and more summer precipitation. Cheatgrass is a common weed, however, and along with others, it is probably becoming more widespread as land is disturbed by road building, plowing, and mineral exploration (Beauchamp et al. 1975; Allen and Knight 1984). Weed-free soil

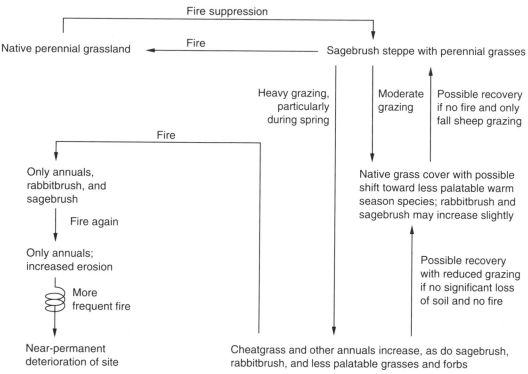

F I G. 6.13   The apparent effect of cheatgrass introduction, heavy livestock grazing, and increased fire
frequency in the big sagebrush steppes of Utah and
Nevada. This scenario may not apply to most high-
elevation steppes in Wyoming, where cheatgrass is less
widespread, but the introduction of exotic species can
cause undesirable ecosystem changes. Adapted from
West 1988.

samples can still be found in remote locations at
some distance from disturbances (Beauchamp et
al. 1975; Allen 1979), but it seems probable that
such areas will become more rare. Several investigators have proposed methods for estimating the
probability of cheatgrass expansion following
Wyoming big sagebrush control (Young et al.,
Downy brome, 1976; Young and Evans 1978; Hassan and West 1986).

The cheatgrass problem is not restricted to land
managed for livestock. Recently, a fire burned a
large stand of big sagebrush in Little Bighorn Bat-
tlefield National Monument in southern Montana. Much of the sagebrush was killed and bluebunch wheatgrass, a co-dominant before the
burn, became less abundant. Cheatgrass became
more common on the burned area, as did yellow
salsify. Managing the vegetation of a national
monument so that it reflects presettlement conditions is a goal that may be impossible once certain

introduced species become established. Further
research is necessary to resolve perplexing ecological problems such as this one.

THE EFFECTS OF GRAZING ON
SAGEBRUSH STEPPE

As in grasslands, the effects of grazing sagebrush
steppe need not be viewed as a disturbance unless
too many animals are retained by fences in too
small an area for too long a time. Bison, pronghorn, elk, deer, and other herbivores are integral
parts of the steppe ecosystem (Mack and Thompson
1982). Ungulate grazing does not generally kill the
dominant big sagebrush, but changes in plant species composition do occur (Lommasson 1948;
Frischknecht and Plummer 1955; Weaver and Albertson 1956; Ellison 1960; Blaisdell et al. 1982; R.
F. Miller et al. 1993). Miller and co-workers concluded that the effects of grazing management on

the plant communities of intermountain shrub-lands are largely unknown, but that heavy live-stock grazing in some areas can cause an increase in the cover of unpalatable woody species, a decrease of perennial forbs and grasses, and an increase in introduced annuals. They also con-cluded, however, that changes in plant composi-tion would have occurred even without livestock grazing, because of plowing, the introduction of cheatgrass and other annuals, altered fire frequen-cies, and even climate change.

One of the most widely mentioned responses to grazing is an increase in big sagebrush cover (Ellison 1960). Historical records, however, sug-gest that sagebrush has been a dominant feature of the Wyoming landscape for hundreds of years and that its abundance should not be considered an artifact of livestock grazing pressure (Vale 1975; Dorn 1986; Johnson 1987). John Fremont wrote in 1845 that

> one of the prominent characteristics in the face of the country is the extraordinary abundance of the "artemisias." They grow everywhere— on the hills, and over the river bottoms, in tough, twisted, wiry clumps; and, wherever the beaten track was left, they rendered the pro-gress of the carts rough and slow. As the coun-try increased in elevation on our advance to the west, they increased in size; and the whole air is strongly impregnated and saturated with the odor of camphor and spirits of turpentine which belongs to this plant. This climate has been found very favorable to the restoration of health, particularly in cases of consumption; and possibly the respiration of air so highly impregnated by aromatic plants may have some influence.

Similarly, Platt and Slater (1852) noted, "Along the Sweet Water, most of the way, are nar-row bottoms of good grass. Adjacent to these bot-toms are large, arid, wild-sage plains, extending to the mountains." The same comments are valid today (fig. 6.14). Sagebrush may have increased on some sites, but it has clearly been a dominant feature of the Wyoming landscape for thousands of years (Beiswenger 1991).[9]

Sagebrush might have been favored by grazing in some areas because of reduced competition

from more palatable species and because of more mineral soil exposure where there was less mulch and standing dead biomass (Weldon et al. 1959). Both conditions facilitate the survival of sagebrush seedlings, which might have happened most commonly on the fringes of the Great Plains (for example, in portions of the Powder River Basin), where the competitive ability of the shrub is re-duced because of less winter precipitation and more summer rain—conditions apparently more favorable for herbaceous grasses and forbs.

Though grazing surely does influence species composition to some extent, drought is another important factor. Several studies have concluded that conditions appearing to have resulted from poor livestock management were actually a result of extended dry periods (Haag 1949; Reed and Peterson 1961). Indeed, short-term shifts in cli-matic conditions can cause great changes in NPP (and biomass) in the semiarid west.[10] As with most ecological phenomena, explaining the ef-fects of grazing lies in understanding the interac-tion of several factors rather than the effects of one. Participants in the debate over whether livestock grazing has greatly increased sagebrush cover must recognize that what actually happened probably varied considerably from place to place. Heated debates over ecological phenomena fre-quently develop when sweeping generalizations are made.

Grazing pressure could also have caused a de-cline in microphytic crusts on the surface of inter-mountain basin soils (Anderson et al. 1982; West 1990). The research on this effect has been done entirely in the Great Basin to the west, but it may be relevant to Wyoming as well. Microphytic crusts are growths of lichens, algae, mosses, fungi, or bacteria that form on the soil surface, thereby minimizing erosion and sometimes creating fa-vorable conditions for seedling establishment (Kleiner and Harper 1972, 1977; Anderson et al. 1982; West 1990; R. F. Miller et al. 1993). The crusts may also be important for nitrogen fixa-tion. Microphytic crusts were possibly less com-mon in Wyoming because of more trampling by bison and antelope, or because temperatures were too low at the higher elevations when adequate soil moisture was available.

The theory of Mack and Thompson (1982) is

F I G. 6.14   This photograph along the Sweetwater River east of Jeffrey City (elevation 1,885 m; 6,200 ft) was taken by William Henry Jackson in 1870. The vegetation, dominated by Wyoming big sagebrush, is essentially the same today (Johnson 1987). The granitic Sweetwater Rocks on the left have limber pine and Rocky Mountain juniper. Photo provided by the U.S. Geological Survey (USGS 292).

again relevant. They maintained that the different climatic conditions of the Great Basin, and possibly much of western Wyoming, were unfavorable to large herds of bison because of summer drought and the scarcity of warm-season C_4 species.[11] Thus, food was less available during the critical early-summer calving period, and population densities were lower than on the Great Plains. Microphytic crusts might have formed under these conditions (or at least during the mid-1800s, when bison populations were low; Urness 1989). When cattle and sheep were in large numbers in the late 1800s, the microphytic crusts would have been destroyed by trampling, causing the soil to become more erodible and more suitable for the invasion of introduced weedy annuals such as cheatgrass. Native species such as big sagebrush may have increased as well (Urness 1989; West 1990, 1991).

Rangeland management today is more refined than it was a half century ago, but the adverse impacts of excessive livestock grazing in those early years probably exist to this day in some areas. Perhaps the sagebrush-dominated rangelands most similar to those of presettlement times are those some distance from frequently used sources of water and where fires have not always been suppressed. Speculation on the past and current effects of bison, antelope, and elk grazing, compared to the effects of domestic livestock grazing, will continue. If polarizing pronouncements can be avoided, such discussions will lead to the sustainable management of rangeland ecosystems.

CHAPTER 7

Desert Shrublands and Playas

In addition to sagebrush steppe, the intermountain basins have a variety of desert shrublands dominated by greasewood, shadscale, fourwing saltbush, Gardner saltbush, winterfat, spiny hopsage, and kochia—all characteristic of the Great Basin deserts to the west (West 1988; Osmund et al. 1990). These shrublands typically occur where the average annual precipitation is less than 25 cm and where the soils have high concentrations of salts (see fig. 1.5).[1] In some cases the salts are so concentrated that the soil surface is white. Salinity combines with moisture availability in determining the abundance of different species (fig. 7.1).

Except for soil salinity, desert shrublands have many features in common with the sagebrush steppe discussed previously: dominance by the shrub growth-form, moisture and nutrient limitations to plant growth, and sensitivity to various kinds of herbivores. Because of salinity and low precipitation, the plant cover of desert shrublands is generally less than in the slightly more mesic grasslands and sagebrush steppe. Soils tend to be deep, but other characteristics vary considerably, creating a high level of patchiness in the vegetation that appears to be topographically and edaphically controlled (fig. 7.2).

Surviving in the Desert Shrubland Environment

The salts that accumulate in desert soils pose a significant problem for plant survival, primarily because water uptake is prevented if the salt (solute) concentration of the soil water is higher than that of the plant cells. Halophytes have evolved adaptations that enable the maintenance of cell solute concentrations at higher levels than those found in the soil solution, thereby enabling water uptake. In most cases an excess of salts accumulates, and this excess contributes to the succulent leaves found in some halophytes, such as Rocky Mountain glasswort and greasewood.[2]

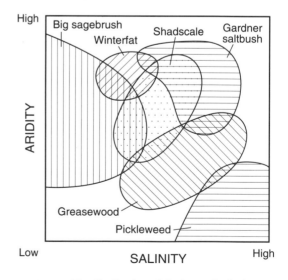

FIG. 7.1 The distribution of six desert shrubs in relation to gradients of salinity and aridity. Pickleweed, the most salt-tolerant species, is found in Utah and Nevada, not Wyoming. Adapted from West 1988.

F I G. 7.2   Sharp transitions among different kinds of desert shrublands are common, such as in this area north of Rock Springs. Mixed desert shrubland occurs on the left and on the distant escarpment (shale, marl, and sandstone in the Green River formation); saltbush desert shrubland, dominated by Gardner saltbush, occurs on the right. Elevation 1,976 m (6,500 ft).

Some of the salts accumulated in plant tissues are discarded each year with leaf fall. Halophytes function essentially as pumps, collecting the salts from throughout the rooting zone and depositing them in a smaller area around the plant through leaf fall. Other halophytes have evolved special adaptations for excreting salts as their concentrations build to high levels (Goodin and Mozafar 1972; Detling and Klikoff 1973). This excretory mechanism involves the development of salt glands on leaf surfaces, such as occurs on saltgrass, saltcedar, and saltbush (fig. 7.3). The microscopic glands enlarge until they burst, after which rainwater washes the salts from leaf surfaces onto the soil. Some ecologists have hypothesized that salt deposition under the canopy is an allelopathic mechanism, but it is probably simply a secondary result of plant adaptation to saline environments.

As in grasslands and sagebrush steppe, the competition for resources is important in determining which desert plants and animals survive. This competition is often minimized because of seasonal differences in activity, or because they tap different parts of the ecosystem for their resources

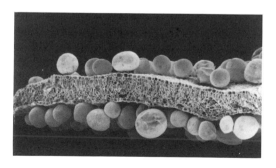

F I G. 7.3   A scanning electron micrograph of salt-secreting glands on the leaves of saltbush. The glands are modified leaf hairs (trichomes) that enable some halophytes to tolerate highly saline soils. Magnification about 33×. Micrograph by Howard Stutz.

(see chapter 5). One study specific to desert shrublands was done by White (1976), who wondered how coexisting winterfat and shadscale differed in their adaptations. He found that winterfat, a C_3 species, was capable of rapid photosynthesis in the spring and then became relatively inactive during the rest of the year, when moisture was limiting. In contrast, shadscale, a C_4 species, had moderate rates of photosynthesis throughout the growing season. Shadscale had photosynthetic rates that were about half that of winterfat in the spring but two times faster in the summer. Minimum soil water potentials for winterfat and shadscale photosynthesis were -5 and -7 MPa, respectively, indicating a great tolerance for water stress. Both were able to fix carbon at low temperatures—a potentially important adaptation for the cool, short growing season that characterizes the desert shrublands of Wyoming. Differences between the two species explain how they are able to grow at different times of the year, thereby reducing competition for resources. Other desert shrubs have similar adaptations (Blauer et al. 1976).

In addition to pricklypear cactus, some palatable desert shrubs are spiny (such as greasewood, bud sagewort, shadscale, and spiny hopsage)—an adaptation to minimize grazing. Also, like big sagebrush, shadscale produces two kinds of leaves: ephemeral leaves and overwintering leaves (West 1983).

The Desert Shrubland Mosaic

The vegetation mosaic of Wyoming deserts is strongly influenced by water availability and topography. Greasewood desert shrubland and saltgrass meadow are characteristic of playas and other comparatively wet depressions. On the upland, the mosaic is composed of mixed desert shrubland, saltbush desert shrubland, and desert grasslands. Shrublands dominated by big sagebrush are intermingled with desert shrublands and grasslands, commonly occurring where soils are less saline, where drainage is not impeded by fine-textured soils, and on the lee sides of slopes, where snowdrifting is greater. Extensive sagebrush steppes, as described in chapter 6, are found at

FIG. 7.4   Greasewood shrubland in the Great Divide Basin. Saltgrass and Gardner saltbush are also common in this area. The soil surface is often white because of salt accumulation. Elevation 1,991 m (6,550 ft).

slightly higher elevations or where annual precipitation is somewhat greater.

Greasewood shrublands occur on the fringes of playas, desert lakes, ponds, rivers, and creeks (fig. 7.4). Though characteristic of the desert shrubland region, greasewood does not tolerate long periods of drought. This deciduous shrub commonly grows to a height of 1 m or more, and total plant cover is greater than for other desert communities (table 7.1). Like other members of the goosefoot family (Chenopodiaceae), greasewood has salt glands adapted for excreting excess salts, and consequently soil salinity is typically higher under the shrubs (Sharma and Tongway 1973). Notably, greasewood seedlings can survive under the parent shrubs, where the salinity is highest (Romo and Eddleman 1985). The seeds, however, usually germinate early in the spring, when water is available near the surface and when the soil solution is less saline because of dilution

T A B L E 7 . 1   Characteristic plants of desert shrublands in Wyoming

	Saltbush Desert Shrubland	Mixed Desert Shrubland	Greasewood Desert Shrubland	Saltgrass Meadow	Basin Grasslands
SHRUBS					
Alkali sagebrush	X	—	X	—	—
Basin big sagebrush	ravines	ravines	—	—	—
Birdfoot sagewort	X	—	—	—	—
Bud sagewort	X	X	—	—	—
Douglas rabbitbrush	—	X	—	—	X
Fourwing saltbush	—	X	—	—	—
Gardner saltbush	X	—	X	—	—
Greasewood	—	—	X	—	—
Horsebrush	—	X	—	—	X
Rubber rabbitbrush	—	X	—	—	—
Shadscale saltbush	—	X	X	—	—
Spiny hopsage	—	X	—	—	—
Winterfat	—	X	—	—	—
Wyoming big sagebrush	—	X	—	—	X
GRASSES					
Alkali sacaton	—	—	X	X	—
Alkaligrass	—	—	X	X	—
Blue grama	—	—	—	—	X
Bluebunch wheatgrass	—	—	—	—	X
Bottlebrush squirreltail	—	—	X	X	—
Foxtail barley	—	—	—	X	—
Indian ricegrass	—	X	—	—	X
Needle-and-thread grass	—	—	—	—	X
Salina wildrye	—	—	—	—	—
Saltgrass	—	—	—	X	—
Sandberg bluegrass	—	X	—	—	X
Western wheatgrass	X	X	X	X	X
FORBS					
Fringed sagewort	—	X	—	—	X
Goosefoot	—	X	—	X	—
Greenmolly summercypress	—	X	—	—	X
Halogeton	X	X	X	X	—
Hood phlox	—	X	—	—	X
Hooker sandwort	—	X	—	—	X
Monolepis	—	—	—	—	—
Pepperweed	—	X	—	X	X
Pricklypear cactus	X	X	X	—	X
Salicornia	—	—	X	X	—
Scarlet globemallow	—	X	—	—	X
Sea blight	—	—	X	X	—
Spiny aster	—	X	—	—	X
Wild onion	—	X	—	—	X
Yellow beeplant	—	X	—	—	X

Note: — = absent or less common

by spring snowmelt and rains (Romo 1984). Greasewood is known to produce root sprouts. Some of the roots grow to depths of 3 m or more (Nichols 1964a).

Hamner (1964) classified the greasewood-dominated vegetation of the Bighorn Basin into four community types: greasewood-shadscale, greasewood-big sagebrush, greasewood-grass, and a greasewood monoculture. These communities were differentiated in part by edaphic features, with the greasewood-shadscale stands occurring on heavy clay-loam soils. Blue grama was more common on somewhat sandier soils, alkali sacaton and saltgrass on siltier soils, and western wheatgrass on soils with high amounts of clay. The soil pH of greasewood-dominated stands was commonly 8 or higher and soil conductivity (salinity) ranged from 2.9 to 8.8 dS/m. Nichols (1964) found that the infiltration rate in greasewood-dominated vegetation was much lower than in stands dominated by big sagebrush.

An anomaly occurs in riparian zones, where greasewood shrublands are often found on terraces and basin big sagebrush occurs in ravine bottoms, closer to the water table (fig. 7.5). In the Bighorn Basin, the soils of ravine bottoms are more permeable and therefore more favorable for big sagebrush (Nichols 1964). Such soil differences might indicate different conditions during sediment deposition, but there is also the possibility that more frequent flooding of the ravine causes more leaching of accumulated salts. For some reason, basin big sagebrush has a competitive advantage in the ravine, closer to the water table, where greasewood might be expected. Greasewood apparently receives adequate water on the terrace, as it does higher on the upland, where groundwater seeps to the surface (such as in some badlands; Brown 1971).

Some saline depressions with shallow groundwater have little greasewood and are characterized by inland saltgrass, alkaligrass, alkali sacaton, and in wetter areas, alkali cordgrass. Several halophytic forbs in the goosefoot family are also found in these meadows. One, Rocky Mountain glasswort, is a small, succulent plant that turns bright red in the fall and seems especially well adapted to the sodic "slickspot" soils—a name derived from the fact that the soil is often slippery when wet because of the soil-particle dispersing effect of sodium. Glasswort commonly forms a concentric band closest to ponds and depressions, with bands of saltgrass meadow, greasewood shrubland, and mixed desert shrubland occurring as the distance from the pond increases (fig. 7.6, table 7.1). This zonation is apparent because of abrupt transitions in dominant plant species, probably caused by their tolerances to soil salinity, soil cracking during dry periods, and depth to available groundwater (Flowers 1934; Flowers and Evans 1966; Steger 1970; Detling and Klikoff 1973; Skougard and Brotherson 1979).

Inland saltgrass has been the subject of several studies because of its value as a forage species

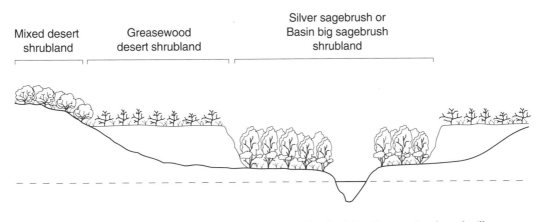

Mixed desert shrubland   Greasewood desert shrubland   Silver sagebrush or Basin big sagebrush shrubland

F I G. 7.5 Along many ephemeral streams, greasewood is common on terraces, and basin big sagebrush, plains silver sagebrush, and willows are common in ravine bottoms.

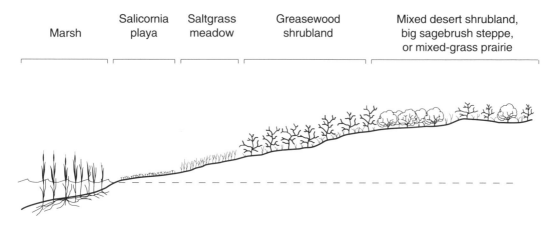

| Marsh | Salicornia playa | Saltgrass meadow | Greasewood shrubland | Mixed desert shrubland, big sagebrush steppe, or mixed-grass prairie |

F I G. 7.6   The vegetation around desert playas or wetlands changes with distance from open water, probably because soil salinity decreases and depth to water table increases. The result can be concentric bands of different vegetation types.

(McGinnies et al. 1976; Bowman et al. 1985). Saltgrass grows normally at low levels of salinity, but it tolerates soils with conductivities ranging from 5 to 25 dS/m and a pH of 7 to 10 (Blaisdell and Holmgren 1984). Saltgrass is much more tolerant of saline conditions than blue grama, though Tiku (1976a) found that it does not concentrate salts in its tissues. Cluff et al. (1983) found that the seed of inland saltgrass germinated best when there was a 20°C diurnal temperature fluctuation and when the seedbed was fairly warm during the day (about 40°C) and wet, with relatively low salinity. They concluded that saltgrass seed germination is an episodic event, occurring only when moisture events coincide with optimal seedbed temperatures and when moisture is adequate to leach sufficient salts to raise soil moisture potentials above -1.5 MPa. The same might be said for many desert plants.

In a study of saltgrass meadows in Colorado (east of the Front Range), McGinnies et al. (1976) identified four soil conditions with somewhat different cover and associated species. Slickspot or alkaline soils were dominated by a sparse cover of saltgrass, whereas mounds were dominated by alkali sacaton and had much higher cover. Level areas had a mixture of saltgrass, blue grama, and alkali sacaton; swales had mostly saltgrass, western wheatgrass, and threadleaf sedge. Sharp boundaries were observed between these variations of saltgrass meadow. Interestingly, the surface horizons of the soils throughout were neither saline nor alkaline. The subsurface horizon, however, was hard when dry and impermeable when wet. Below that, the soil was saline though moist throughout the year. The salt problem apparently occurs in the deeper soil rather than on the surface. McGinnies and co-workers concluded that blue grama was highly sensitive to salinity, but its coexistence with saltgrass on level areas seems to be an anomaly that requires further study.

Mixed desert shrubland occurs where the moisture supply is less dependable.[3] Wyoming big sagebrush is a prominent species, but unlike in sagebrush steppe, it coexists with bud sagewort, shadscale, winterfat, spiny hopsage, greasewood, and Gardner saltbush (table 7.1, fig. 7.7). Shadscale is more common on somewhat sandier, nonsaline, drier soils; spiny hopsage is more common on sandy upland sites, where snow accumulation is higher than usual; and winterfat is more common on dry, nonsaline, silty, and slightly gypsiferous upland soils (Branson et al. 1976; Baker and Kennedy 1985). Typically, big sagebrush occurs in upland ravines, where the soil may be less saline because of more frequent water flow and where water infiltration into the soil in the spring may be higher (figs. 7.7, 7.8).

Although big sagebrush is thought to be intolerant of saline or alkaline soils, Hamner (1964) found it with greasewood on soils having a pH of 8.4 and a conductivity of 3.4 dS/m. Gates et al.

F I G. 7.7　Mixed desert shrubland on the left, with
more dense Wyoming big sagebrush in ravines; near
Green River. Elevation 1,976 m (6,500 ft).

(1956) concluded that big sagebrush could not be
used as an indicator of low salinity but noted that
it tended to grow on soils with a relatively low salt
content. Nichols (1964) found that big sagebrush
soils had higher infiltration rates than other desert
shrublands (fig. 7.8), but his study was done on a
lower stream terrace, where basin big sagebrush
was the dominant subspecies instead of Wyoming
big sagebrush. In contrast, Gates et al. (1956) found
big sagebrush (probably Wyoming big sagebrush)
on "the heaviest textured soils of any species studied" in western Utah.

　The great diversity of soils on which big sagebrush is found might be explained by genetic variation, as noted by Blaisdell and Holmgren (1984),
or by the fact that plant species composition and
growth are functions of climatic conditions as well
as soil characteristics. Topography, aspect, and
parent material may be better predictors of plant
distribution than soil pH or conductivity (Billings

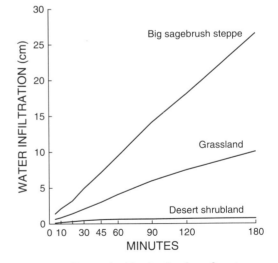

F I G. 7.8　Desert shrublands often have finer textured soils than grasslands or big sagebrush steppe,
which accounts for the much slower water infiltration
observed in desert shrublands (see also fig. 3.9). In
arid regions, less water is available for plant growth in
soils with slow infiltration rates. Adapted from
Nichols 1964.

1949; Gates et al. 1956; Brown 1971; Baker and Kennedy 1985).

Mixed desert shrublands are sometimes modified by the formation of desert pavement and coppice dunes. Silt and sand can easily be eroded by wind in an arid environment, leaving a surface of pebbles adjacent to small dunes, where the finer particles accumulate around shrubs (see chapter 6 and fig. 6.9). Bud sagewort, galleta, greenmolly summercypress, Gardner saltbush, and Indian ricegrass are common on the drier desert pavement soils; big sagebrush, greasewood, Great Basin wildrye, Indian ricegrass, spiny hopsage, and thickspike wheatgrass are common on the dunes. Whether by creating microtopographic relief that favors the accumulation of snow or by having improved infiltration, the coppice dunes apparently provide a more mesic (and perhaps less saline) environment.

Coppice dunes undoubtedly existed in presettlement times, but their formation may have been accelerated by heavy livestock grazing (Wood et al. 1978; Hodgkinson 1983). Trampling can have a significant effect on the kinds of microsites provided for seedling establishment. Eckert et al. (1986) observed a tendency for desirable range species to require less trampled microsites, while heavy trampling favored sagebrush and some weedy forbs.

Further variation in the desert shrubland mosaic and also in the foothill shrubland mosaic is caused by seleniferous soils. Selenium is a chemical element required for proper animal nutrition but that can be extremely toxic to livestock in high concentrations. Selenium indicators, such as woody aster and several species of milkvetch, princesplume, and goldenweed, sometimes occur as distinct communities and have been useful in mapping selenium-rich bedrock. The element is most highly concentrated in certain shales, siltstones, and claystones that are carbonaceous or have high concentrations of iron (Beath et al. 1941; Kolm 1975).

Usually, selenium-rich soils are restricted to small areas, but selenium-bearing bedrock is common in the lowlands and foothills of the West. There has been a continuing concern that certain land management activities, including surface mining and irrigation, could increase the amount of land where selenium accumulates to toxic levels in water, livestock forage, and even meat and plant products intended for human consumption (Harris 1992). Any practice that exposes selenium-bearing rock also increases the density of plants that accumulate selenium. Eventually this selenium is deposited on the soil surface, where it is subject to redistribution by erosion. If this happens, toxic concentrations could develop in places where they would not otherwise be expected.

There are two kinds of plants that accumulate selenium in their tissues. One, the selenium indicators, require selenium for their growth. These include two-grooved milkvetch, tineleaved milkvetch, woody aster, and princesplume. The others, in contrast, are facultative accumulators, which do not require selenium but nevertheless often develop high concentrations. Such plants include woody aster, Gardner saltbush, gumweed, locoweed, mentzelia, snakeweed, spiny hopsage, winterfat, and certain grasses (Trelease and Beath 1949). Other plants do not accumulate selenium even though they are frequently found on seleniferous soils.

Both selenium indicators and facultative accumulators can concentrate large amounts of selenium in leaves and stems without adverse effects on the plant itself. For example, two-grooved milkvetch is known to accumulate 4,000 ppm or more from soils containing only 1.1 ppm (Beath 1937). Grasses and cereal grains with only 10–30 ppm can be lethal or crippling to cattle and horses (Trelease and Beath 1949). Both the indicators and accumulators increase surface-soil selenium concentrations as their shoots decompose in the fall, thereby possibly improving conditions for their own growth. A shortage of water, however, often limits the amount of biomass that can be produced.

Saltbush desert shrubland is perhaps the most arid vegetation type of the intermountain basins.[4] It is characterized by sparse plant cover and the prevalence of Gardner saltbush, a short, evergreen shrub also known as saltsage or Nuttall saltbush (fig. 7.9). Characteristic of this plant community are birdfoot sagewort and bud sagewort (table 7.1). Saltbush desert shrubland occurs commonly on fine-textured soils developed from shale or alluvium. In some areas, the low shrubs are rare,

F I G. 7.9   Saltbush desert shrubland east of Rock Springs (elevation 1,976 m; 6,500 ft). The treeless escarpment is formed from late Cretaceous sandstones and shales.

and the vegetation appears more like a desert grassland, with blue grama common on somewhat sandier soils and western wheatgrass on fine-textured clay soils (Hamner 1964; Hodgkinson 1987). On windswept, shallow soils with low salinity, the dominant species may be bluebunch wheatgrass, needle-and-thread grass, Hooker sandwort, Hood phlox, rock tansy, salina wildrye, and such forbs as Nuttall goldenweed, Hooker sandwort, and several species of cushion plants (fig. 7.10; table 7.1). Some foothill grasslands are similar (see chapter 9).

Several studies have found that saltbush desert shrubland exists on various soil textures, including sandy loam, sandy clay loam, loam, and clay loam (Gates et al. 1956; Vosler 1962; Nichols 1964; Gibbens 1972; Knight et al. 1987). Saltbush desert soils have low infiltration rates compared to sagebrush- and blue grama–dominated communities (fig. 7.8)—probably because of their relatively high clay content or the dispersing effect of sodium on surface-soil particles. A soil pH of 8 or above is common. Yet the soils are not always saline (the conductivity is commonly 1.5 dS/m or less; Vosler 1962; Gibbens 1972; Knight et al. 1987). Gibbens found that soil conductivity was noticeably higher under shrubs than between them. Bud sagewort was dominant in two of Gibbens's study areas, both of which had alluvial soils. The soil pH and conductivity were about the same as for stands dominated by birdfoot sagewort and Gardner saltbush.

Gardner saltbush is a valuable forage species on winter range and is sometimes planted for reclamation purposes, especially when the topsoil has a high pH and salinity. Its roots spread laterally to a distance of 2 m, with taproots penetrating to 1 m or more (Russey 1967; Gibbens 1972). Root sprouting or layering is the major form of reproduction in established stands (Russey 1967; Nord et al. 1969; Blaisdell and Holmgren 1984), though seedling establishment is common under

F I G. 7.10   Windswept desert shrubland in the Great Divide Basin, with cushion plants adapted for conserving water and heat. Two common species are squarestem phlox and Hooker sandwort. Similar cushion plants, but of different species, are found in alpine tundra. Elevation 2,067 m (6,800 ft).

certain conditions (Ansley and Abernethy 1984; Young et al. 1984). Gibbens found that some Gardner saltbush plants were nearly fifty years old.

The Desert Shrubland Ecosystem

There is great variation in desert shrublands, ranging from greasewood shrubland in comparatively moist habitats to arid stands dominated by Gardner saltbush and bud sagewort. Consequently, considerable variation in ecosystem characteristics should be expected. With regard to rates of plant growth (NPP), Branson (1976) ranked a series of desert shrubland types in western Colorado by dominant species, in descending

order: big sagebrush, Gardner saltbush, greasewood, rabbitbrush, shadscale, spiny hopsage, and winterfat. The most productive shrublands undoubtedly have the highest water consumption as well, but hydrologic data are rare. Harr and Price (1972) calculated an annual evapotranspiration of 21–25 cm in greasewood shrublands of central Washington, of which 18–31 percent was transpiration. They also noted that shrub height, canopy cover, and total leaf area were inversely related to groundwater depth. Rickard (1967), also working in Washington, found that deciduous greasewood caused less soil water drawdown during the winter than did evergreen big sagebrush. He observed a more luxuriant growth of cheatgrass in the more mesic greasewood stand.

With the exception of greasewood shrublands, desert ecosystems are characterized by potential evaporation that greatly exceeds annual precipitation. Water stress therefore occurs frequently. Infiltration rates vary considerably and appear to be correlated to the dominance of different shrubs (fig. 7.8). Nutrient leaching is rare, but nitrogen can limit NPP (West 1991). Nitrogen fixation results from species of cyanobacteria, bacteria, or lichens found in microphytic crusts, but such inputs may be balanced by nitrogen losses caused by denitrification when the soil is wet and anaerobic (West 1990, 1991; R. F. Miller et al. 1993). Such losses may be of little consequence since water is usually the primary limiting factor and most nitrogen requirements for new plant growth are probably met by translocation from other plant parts and by the mineralization of soil organic matter. Nutrients other than nitrogen also play important roles in semiarid and arid ecosystems, though less is known about them (West 1991). As in grasslands and sagebrush steppe, herbivores and detritivores increase the rate of nutrient cycling (Whitford 1988; West 1991; see chapter 5), and shrubs may create islands of fertility (Charley and West 1975; Schlesinger et al. 1990; West 1991; see chapter 6).

Except for the effects of livestock grazing, comparatively little is known about plant-animal interactions in desert shrublands (West 1983). Aro (1978) studied a mosaic characterized by patches of winterfat interspersed in big sagebrush shrubland. Jackrabbits preferred winterfat for food, but they seemed to require the taller sagebrush for cover. He hypothesized that the winterfat nearest the sagebrush was subjected to heavier grazing pressure, which in turn facilitated the slow invasion of big sagebrush into the winterfat community. In this way, the jackrabbits had an influence on the mosaic, at least during years of high populations. Some investigators believe that jackrabbits in large numbers can graze rangelands more heavily than livestock or big game can (Currie and Goodwin 1966; Plummer et al. 1968).

Little research has been done on the hydrology, energy flow, and nutrient cycling of Wyoming desert shrublands, but some interesting opportunities exist. For example, just as considering the ecological significance of adding big sagebrush to a grassland ecosystem is helpful (as discussed in chapter 6), it also is helpful to consider the significance of higher salinity (or selenium) on nutrient cycling and other ecosystem characteristics. Does nitrogen become more or less available as salinity increases? Are soils high in sodium more susceptible to erosion? How do patterns of herbivory and decomposition change, if at all, because of higher salt concentrations in the halophytic plant biomass? In the case of greasewood shrublands, with finer textured soils that are wet and possibly anaerobic during much of the year, is most of the herbivory still below ground, as in grasslands and other shrublands?

Disturbances and Succession

The major disturbances in desert shrublands are probably drought and extraordinarily heavy herbivory, whether from grasshoppers, Mormon crickets, bison, or livestock. Fires are less frequent in more arid situations, but they surely occur in stands of greasewood and mixed desert shrubland, where adequate fuel can accumulate, especially after a period of relatively light grazing or where cheatgrass has invaded. Harvester ants can be abundant in some saltbush desert shrublands (Wight and Nichols 1966; Kirkham and Fisser 1972), and sometimes two years or more of drought will cause portions of the shrub canopy to die, if not the entire plant (Allred 1941; Blaisdell and Holmgren 1984). Shadscale is ap-

parently especially sensitive, and dramatic annual variations have been reported for shadscale-dominated shrublands in Idaho—apparently caused by annual differences in spring precipitation and the abundance of a scale insect (the mealy bug) that feeds on roots (Sharp et al. 1990).

Drought is not the only explanation offered for desert shrub mortality. Between 1977 and 1986, the shrubs over large areas in the Great Basin died for no obvious reason. Wallace and Nelson (1990) suggested that four factors were involved—all triggered by several years of unusually high precipitation: (1) increased soil salinity, (2) frequent periods when the soil was anaerobic, (3) the increased susceptibility of shrubs to soil-borne diseases, and (4) the loss of shrub tolerance to the droughts that did occur. Thus far, a similar dieback of desert shrubs has not been reported for Wyoming.

Introduced annuals have been highly successful in desert shrublands, just as they have been in shrublands dominated by sagebrush (see fig. 6.13). Three common species are cheatgrass, halogeton, and Russian thistle. Rickard (1964) suggested that the increase in cheatgrass was an important factor in causing a shift from sagebrush dominance to greasewood dominance in Washington shrublands. The cause, he hypothesized, was higher plant cover following cheatgrass invasion, which led to the interception of more annual precipitation; less deep percolation because of the dense cheatgrass root system near the soil surface; increased fire frequency because of the high cover of the highly flammable cheatgrass; a reduction in the amount of nonsprouting sagebrush following fires, while greasewood was able to maintain its position and even expand because of sprouting; and increased surface-soil salinity caused by the expansion of greasewood that was not favorable for sagebrush. Similar successional patterns could occur in Wyoming if cheatgrass were to become abundant.

Succession in desert shrublands subjected to livestock grazing has been studied in several locations. As expected, the most palatable species tend to decrease (for example, fourwing saltbush, spiny hopsage, winterfat, and some grasses), and less palatable species to increase (rabbitbrush, shadscale, and such weedy species as Russian thistle;

Ellison 1960; Harper et al. 1990; Yorks et al. 1992). The season of grazing can be important in determining rangeland response, with early-winter sheep grazing generally having less impact on desirable species than late-winter (early-spring) grazing (Blaisdell and Holmgren 1984). Ranchers commonly put sheep on desert shrublands in the winter and then transport them to mountain rangelands in the summer. With protection from livestock grazing, palatable species (perennial herbaceous species) increase, and the abundance of introduced annuals may decrease or remain about the same (Blaisdell and Holmgren 1984; Yorks et al. 1992). Harper et al. (1990) found that such desert shrubs as shadscale and winterfat in Utah gradually decline in density following the cessation of grazing, whereas perennial grasses and a few other species increase (such as budsage).

Desert shrublands are second only to sagebrush steppe in the amount of land they occupy in the intermountain basins. Although similar to sagebrush-dominated communities, the plants are adapted to saline soils and many of the shrubs sprout following disturbances. Typically, desert shrublands occur at the lowest elevations of the basins—in depressions, on fine-textured soils with slower infiltration rates, and on soils developed from highly saline shales. The distribution of the dominant shrub species apparently depends on the amount and timing of water availability, temperature at the time water is available, soil texture and salinity, and depth to groundwater. Desert shrublands seem to be more sensitive to livestock grazing than the grasslands of the Great Plains, perhaps because their evolutionary history did not include large numbers of bison.

C H A P T E R   8

Sand Dunes, Badlands, Mud Volcanoes, and Mima Mounds

Sand Dunes

Stretching across the intermountain basins of central Wyoming are several large complexes of active and stabilized sand dunes (fig. 8.1, and see fig. 1.5). Dune location depends on wind patterns and barriers to moving sand. Thus, sand dunes occur in the windward foothills of mountain ranges (for example, the Ferris and Laramie basin dunes; Gaylord 1982); in wind corridors, where blowing sand might be funneled (such as the Killpecker Dunes); and in other areas where wind speeds are insufficient to move the sand farther or where improved growing conditions increase the probability of widespread dune stabilization by plants (such as the Platte River dunes in eastern Wyoming and the Nebraska Sand Hills). Sand dunes create a dramatically different environment in the semiarid lowlands, primarily because they are more mesic and because the shifting sand creates special problems for plant establishment and survival. Ponds frequently occur between the dunes.

The origin of such large volumes of sand is one of the first questions that arise when considering dune ecology. The answer lies in the physical weathering of sandstones over long periods and the power of water and wind to sort and move the resulting particles. The sand grains (0.05–2 mm diameter) are gradually separated from coarser gravel and much of the finer silt and clay. Under appropriate geologic conditions, the sand accumulates, first along creeks and rivers. For example, most sand in the Killpecker dunes originated from the appropriately named Big Sandy and Little Sandy creeks, which drain the Wind River Mountains (Ahlbrandt 1972). These creeks were much larger during glacial melting, bringing large volumes of sand to the area northeast of Rock Springs. Westerly winds in the Holocene blew much of this sand along the ground (saltation) to the east, over the Continental Divide, and across the Great Divide Basin to the Ferris and Seminoe mountains. Lighter silt and clay particles would have been dispersed more widely. The presence of different-aged sand grains in now-dormant parts of the Killpecker dunes suggests that sand movement occurred in this area at various times over the past twenty thousand years (Gaylord 1982, 1983). Indeed, it continues to this day.

The kinds of plants present on dunes depend on the extent of sand stabilization as well as on temperature, moisture, and the amount of organic matter in the sand. Only a few plant species can survive on the active dunes in Wyoming, such as blowout grass, Indian ricegrass, needle-and-

F I G. 8.1   The Killpecker sand dunes northeast of Rock Springs (fig. 1.5), with Boars Tusk (an old volcanic neck) in the background. Dune colonizers in the foreground include Indian ricegrass, slimflower scurfpea, and alkali wildrye. These species, plus basin big sagebrush, rabbitbrush, rusty lupine, skeletonplant, and spiny hopsage, are common on the stabilized dune. Alkali cordgrass is common in the wet swales found at the base of many dunes. Elevation 2,128 m (7,000 ft).

thread grass, prairie sandreed, rusty lupine, salina wildrye, sand lovegrass, sand muhly, sandhill muhly, and scurfpea.[1] These species increase in cover with dune stabilization. Missing from Wyoming is American beachgrass, commonly found on inland dunes to the east and along the Atlantic and Pacific coasts, and yellow wildrye, a common colonizer in the dunes of southeastern Idaho (Chadwick and Dalke 1965). The drier, coarser soils of dune tops typically have different species than the finer textured soil between dunes (Barnes and Harrison 1982), where wetlands dominated by alkali cordgrass may occur if the water table is close to the surface.

Succession in dune ecosystems is clearly a process of soil stabilization. The continued growth of pioneer species gradually leads to more organic matter in the sand and more soil binding by the root systems. The mere presence of the plants causes slight but significant reductions in wind velocity, thereby diminishing the extent of sand movement. This stabilizing process continues as plant cover increases, and eventually the entire dune may be covered with vegetation. During stabilization, many other grasses, forbs, and shrubs become common, including antelope bitterbrush, big sagebrush, rabbitbrush, silver sagebrush, spiny hopsage, and others. The pioneer species become less common, but they persist in the community.

Stabilized dunes are subject, however, to the same kinds of periodic disturbances important elsewhere—drought, fire,[2] burrowing, heavy grazing, and human traffic. Such disturbances lead to a decline in plant vigor, subjecting small areas of the dune to rapid wind erosion. The resulting blowout, created in the short time of a few days or weeks, is eventually stabilized again through the persistence and expansion of the pioneer species. Significantly, the suppression of fires is thought to have greatly increased the proportion of some dune fields that are now in a stabilized condition, such as in the Nebraska Sand Hills (Weaver and Albertson 1956; Burzlaff 1962; Bleed and Flowerday 1990). This tendency toward stabilization in the absence of fire may be counteracted, however, by heavy grazing or the recreational use of motorized dune buggies.

Compared with surrounding grasslands and shrublands, the dune environment is more favorable for the survival of many plants and animals. The annual production of new plant biomass can be more than double the adjacent sagebrush steppe (Pearson 1965b). Even with limited precipitation, rapid infiltration through the coarse sand provides good water storage (the inverse texture effect), and sometimes snowdrifts on the lee sides of dune crests are buried by drifting sand in the spring. It is not uncommon to find buried snow in midsummer, which provides a source of water for ponds and perennial streams as it melts.[3] The landscape pattern is diversified further by the shrubs and small trees that sometimes occur along such streams, for example, basin big sagebrush, common juniper, currant, oregongrape, Rocky Mountain juniper, water birch, wild rose, and willow.

Good water storage creates a more mesic environment for plants, but it also contributes to the formation of small ponds between the dunes. Many animals are attracted to the dunes because of water availability, including deer, elk, waterfowl, and wild horses. The ponds provide habitat for various aquatic organisms, including salamanders and freshwater shrimp. Because dunes are an oasis in the semiarid basins and plains, livestock sometimes concentrate there, creating management problems similar to those encountered in riparian landscapes. Early humans were also attracted to the dunes, undoubtedly because of water and food availability (Bleed and Flowerday 1990). Artifacts of native Americans, estimated at more than ten thousand years old, have been found in the Killpecker Dunes, commonly in association with bison bones (Ahlbrandt 1972).

Shifting sands provide a poor substrate for seedling establishment, and once established a plant is always threatened with being buried by saltating sand. Such pioneer dune plants as blowout grass, Indian ricegrass, and scurfpea often survive burial because of rhizomes that grow up or down to maintain an appropriate depth. As important, the rhizomes are long, frequently branched, and have numerous shoots (figs. 8.2, 8.3)—all characteristics that diminish the probability that the entire plant will be buried to excessive depths. Adventitious roots also form on some plants, even as seedlings. Maun (1981) studied the seedling establishment of prairie sandreed, observing that the first internode after germination of the small seeds could be up to 6 cm long and have adventitious roots along its entire length. Most initial growth is in the root system, which

F I G. 8.2   The linear growth pattern of alkali wildrye in the Killpecker sand dunes is caused by rapidly extending rhizomes that help the plant survive when covered by sand. Other grasses have the same adaptation (for example, Indian ricegrass and blowout grass).

F I G. 8.3   Scurfpea, like other dune-colonizing plants (see fig. 8.2), also has elongated rhizomes. Note the prominent taproots, which are exposed here because of wind erosion. Killpecker sand dunes; elevation 2,128 m (7,000 ft).

soon becomes infected with mycorrhizal fungi that increases the efficiency of the root system in binding the soil, as well as in the uptake of water and nutrients (Koske and Polson 1984). Maun found, however, that the principal cause of seedling mortality is desiccation, which suggests that new seedlings become established only in relatively wet microsites or during comparatively wet periods. This requirement seems typical for other plants in the intermountain basins.

Dune plants encounter other problems as well, such as the sandblasting of their leaves and stems. Perhaps for this reason they are herbaceous, with temporary shoots that require maintenance for only a few months. Unlike in the coppice dunes discussed in chapter 7, woody plants invade only after the sand dunes have become partially stabilized. Also, the warm winds of dunes accelerate

transpiration, thereby causing plant water stress even though comparatively more water is available. Many dune plants are water-efficient C_4 species, and a large portion have both shallow and deep roots (>2 m) (Potvin and Harrison 1984). Barnes and Harrison (1982) found that deep-rooted dune plants in Nebraska have stomata that are less sensitive to drought than most grassland plants, probably because of continued water availability in the dunes.

Extensive root systems must also be important for obtaining nutrients, as sand is an infertile substrate. Mycorrhizae facilitate nutrient acquisition (Koske and Polson 1984), and some dune plants also have the capacity for nitrogen fixation. In fact, two common dune colonizers in Wyoming are nitrogen-fixing legumes (scurfpea and lupine). Another is Indian ricegrass, which apparently promotes the growth of nitrogen-fixing microorganisms near its roots (Wullstein et al. 1979; Wullstein 1980). With dune stabilization, other nitrogen-fixing species (such as bitterbrush) may invade. Soil organic matter accumulates and nutrient availability improves. But much of the organic matter is fragmented with the next blowout and is blown downwind as it comes to the surface. The newly active dune is again left with a deficiency of nutrients.

Badlands

Some parts of the lowlands are nearly devoid of vegetation because of naturally high rates of erosion. Such areas are known as badlands and are some of the most fascinating landscapes in the Rocky Mountain region (for example, Hell's Half Acre, Honeycomb Buttes, Adobe Town, the Powder River Breaks ridges, Wind River Badlands, and Grizzly Buttes; fig. 8.4). Badland topography, usually angular with sharp ridges, occurs in arid climates on shales or mudstones. The clay-rich soils expand and contract with wetting and drying—conditions that typically prevent plant establishment and facilitate erosion. Flash floods are common because of intense thunderstorms and slow infiltration.

Because badlands are subject to rapid erosion, they are favorite locations for fossil hunters. Each year new specimens are exposed on the surface.

F I G. 8.4   The Wind River Badlands near Dubois, one of several badlands found in Wyoming. Here the badlands have developed from the shales, claystones, and sandstones of the Eocene Wind River formation (2,100 m; 6,888 ft).

The fossils (such as crocodiles, primates, and tapirs) often reflect the subtropical climate of the Cretaceous and early Tertiary periods (McGrew et al. 1974). Some badlands are now protected as national monuments because of their value for paleontological studies as well as their inherent beauty (such as Fossil Butte National Monument, west of Kemmerer and Badlands National Park in South Dakota).

Badland vegetation is sparse on the arid, rapidly eroding slopes and is typically composed of species that are common on adjacent grasslands and shrublands. Depending on the location, common plants include big sagebrush, bottlebrush squirreltail, Gardner saltbush, greasewood, shadscale, silver sagebrush, tufted evening primrose, juniper, and western wheatgrass (Knight et al. 1976). Halophytes are common. The ravines sometimes have more plant cover, including such riparian species as basin big sagebrush, basin

wildrye, greasewood, narrowleaf cottonwood, silver sagebrush, and various species of willow. Plant species composition and cover vary considerably with elevation, substrate, and precipitation. As erosion is a periodic if not continuous process, there is little opportunity for the successional development of a stable vegetative cover on the steeper slopes.

Because of the abrupt, finely dissected topography, the fauna of badlands can be more unusual than the flora. Hawks, eagles, swallows, and swifts find suitable nesting sites on exposed cliffs, and larger animals use the badlands as shelter on cold, windy days.

Perhaps because the vegetation is so sparse and there are few unique plant species, the badlands have not been well studied botanically. Brown (1971) studied the Powder River Breaks in northern Wyoming and southeastern Montana, which were formed from weakly consolidated shales interspersed with lignite seams in the Fort Union formation.[4] He identified seven types of shrublands and woodlands in the area: greasewood, shadscale-sagebrush, big sagebrush–shadscale–western wheatgrass, big sagebrush–western wheatgrass, skunkbush sumac–western wheatgrass, juniper–western wheatgrass, and ponderosa pine–juniper. Mixed-grass prairie can be found on mesa tops capped by scoria (Larson and Whitman 1942). Greasewood is found on the upland, where seepage to the surface occurs, and saltbush desert shrubland is common on steep, fine-textured soils. Plant cover is sometimes contoured in narrow bands that parallel beds of gray clay, buff-colored silt, and lignite—an indication of strong edaphic control.[5]

Pine and juniper are usually restricted to escarpments that are comparatively resistant to erosion. Without this feature, erosion occurs before mature trees have a chance to develop. Landslides, however, can cause significant changes. For example, a photograph taken in 1889 in Fossil Butte National Monument shows an escarpment dominated by Douglas-fir (figs. 8.5, 8.6). Today, the same area has a grove of narrow-leaf cottonwoods, an unusual situation because the cottonwoods are not located near a stream as they usually are. This anomaly was probably initiated when the conifers were cut for railroad ties by the nearby Union Pacific Railroad, which destabilized the slope and allowed for a significant landslide (R. D. Dorn, pers. comm.). Apparently, small basins were created by the slide, thereby trapping water during spring snowmelt and providing a sufficiently moist substrate to support the cottonwoods on this upland site.

mits and then dried. Similar mounds are found in northern Albany County (L. Munn, pers. comm.)

The geologist and explorer Ferdinand Hayden was the first scientist to discover these unusual landforms when he visited the area in 1877. According to his report (Hayden 1879), the mounds at that time had pools of muddy water at the top. Upon observing that a bubble of gas would periodically rise to the surface of the pools, he and his men conducted an experiment: "A rifle-ball shot down vertically into one of the openings produced a sudden eruption of the whole mass. Water and mud were thrown to a height of about 10 feet, covering the luckless experimenter from head to foot. From a safer distance the trial was several times repeated and almost always followed by the same result. . . . Crude as this test may be, it shows the presence of gas at some depth, held there under mechanical pressure."

Today, the mud volcanoes appear to be dormant. There has been surprisingly little weathering of the mounds, suggesting that the springs became inactive just recently. A possible cause is the pumping of groundwater in the Great Divide Basin for livestock and industry, which could be relieving some of the water pressure that must have been important for their creation. It is possible, however, that the mounds are rejuvenated by periodic, rather than continuous, "eruptions."

Hayden also observed deep pools of muddy water in the area, covered by a thin layer of crusty silt that gave the appearance of solid ground: "Innumerable bones of animals, who here sought to quench their thirst, prove the treacherous character of the soil." A nearby pond was christened "Death Lake" by the explorers.

Mud Volcanoes

Perhaps the most remarkable landscape of the intermountain basins is the small area of mud volcanoes or mud springs found in the lowest part of the Great Divide Basin (the Chain-of-Lakes area north of Wamsutter). These conical mounds are 1–5 m high and are surrounded by mudflats or playas with only a few plants (fig. 8.7). The cones seem to provide better conditions for plant growth. At one time, a mud slurry oozed from their sum-

Mima Mounds

A peculiar feature of some lowland grasslands and shrublands in eastern Wyoming is circular or oval mounds that are usually less than 0.5 m high and 4–8 m in diameter (fig. 8.8). The cause of these mounds, which are especially common south of Laramie, west of Cheyenne, and north of Medicine Bow, has been the subject of controversy for many years. Equally puzzling mounds occur in Colorado, Saskatchewan, Minnesota, coastal

F I G. 8.5 A photograph taken in 1889 of Fossil Butte in Fossil Butte National Monument, west of Kemmerer. The trees at the base of the escarpment, composed of Green River claystones, are Douglas-fir (2,250 m; 7,400 ft). The Green River formation is a lake deposit that contains many fossil fish (see fig. 2.3). Photo courtesy of the Wyoming State Archives and Robert D. Dorn.

F I G. 8.6 The same part of Fossil Butte shown in fig. 8.5. By 1984 the Douglas-fir woodland had been converted to a narrowleaf cottonwood woodland. See text for explanation. Photo courtesy of Robert D. Dorn.

126

F I G. 8.7   Mud volcanoes in the Chain-of-Lakes area of the Great Divide Basin. Plants on the mounds include greasewood, rabbitbrush, and saltbush. A mud flat (playa) with little or no vegetation surrounds the mounds. A small lake is visible in the background. Elevation 1,981 m (6,515 ft).

Texas and Louisiana, northern Mexico, California, and western Washington, as well as in Kenya (Cox 1984; Cox and Gakahu 1987), South Africa (Cox et al. 1987), and Argentina (Cox and Roig 1986). Sometimes they occur at high elevations (above 3,600 m; Cox et al. 1987). Known as prairie mounds, pimple mounds, and biscuit land, they are most commonly referred to as mima (pronounced mīma) mounds because those of the Mima Prairie in Washington were the first described (Dalquest and Scheffer 1942; Washburn 1988).

Mima mound topography raises two questions: what is the effect of mima mounds on plant and animal distribution, and what is the explanation of their origin? The effect of the mounds on plants is subtle in some areas, with no obvious differences between mound and intermound vegetation. Elsewhere, the vegetation on the mound is taller or is composed of different species. For example, the mounds appear to be one cause of sagebrush islands, where clumps of taller big sagebrush occur in a matrix of grassland or shorter shrubland. In Colorado, McGinnies (1960) found a higher NPP on the mounds, and Hansen (1962) observed that Idaho fescue was more common between mounds. Some investigators have also noted that burrowing mammals (pocket gophers and ground squirrels) appear to be more common on the mounds. Through their burrowing, they create soils that have a lower bulk density, higher infiltration capacity, and improved nutrient availability (Arkley and Brown 1954; McGinnies 1960; Hansen 1962; Mielke 1977; Brotherson 1982; Cox 1984). Vegetation differences can be expected when mound and intermound soils are different, but whether the differences are due to the soil or to associated fauna is still unknown.

The more perplexing question about mima mounds pertains to their origin. In reviewing the

F I G. 8.8   Ground view of mima mounds in the Laramie Basin (elevation 2,250 m; 7,400 ft). Jelm Mountain is in the background. The vegetation is high-
elevation mixed-grass prairie, dominated by blue
grama, fringed sagewort, june grass, and Sandberg
bluegrass.

literature, Cox (1984) and Washburn (1988) identified four prevailing hypotheses: wind deposition
(coppice dunes), erosion, frost, and burrowing
rodents. In Wyoming it is easy to distinguish
mima mounds from wind-caused coppice dunes,
based on mound form and soil characteristics.
Certainly, windborne sediment deposition could
occur on mima mounds with taller vegetation,
but other explanations seem necessary. The erosion hypothesis suggests that the tops of mounds
were once the location of a stabilizing object, such
as a tree or tall shrub, with the soil in between
more susceptible to erosion (Cain 1974). Mound
distribution patterns are often similar to tree distribution patterns in woodlands and savannas,
but there is little or no supporting evidence for
this explanation in Wyoming.

The frost hypothesis has received more attention than the erosion and deposition hypotheses.
From the air, the mound pattern gives the impression of sorted polygons that are common in the
arctic tundra (fig. 8.9), an observation that suggests that the mounds were formed about twenty
thousand years ago, when there were glaciers in
the mountains and the climate was colder. Indeed,
Mears (1981, 1987) has described Wyoming intermountain basins as being a windswept permafrost
environment with relatively little snow cover. In
such environments, soil cracks 1 m deep or more
would be formed by the contraction caused by
either freezing or desiccation. These cracks were
often filled with blown soil particles, and today
their approximate form and depth are commonly
seen when the soil profile is exposed (see fig. 2.8).
Depending on the cause of the cracks, the distinct
patterns in the soil profile are referred to as fossil
ice wedges or sand wedge relics, the latter having
been formed in desiccation cracks.

Spackman and Munn (1984) dug trenches
through mima mounds in the Laramie Basin and
determined that the fossil wedges described by
Mears were a key factor in mound formation.

F I G. 8.9   Aerial view of mima mounds in the Laramie Basin (elevation 2,204 m; 7,250 ft). Similar mounds are found in other intermountain basins.

They concluded that the mounds can be attributed to "cryostatic pressure created from water entrapped between a layer of permafrost or bedrock and a downward-freezing frost layer from the surface. Pressure was ultimately released through planes of weakness created by large fossil permafrost sand wedges." The evidence for their hypothesis included the observation that the mounds were composed of gravel and sand mixtures that were formed, they reasoned, when the gravels were forced to the surface through sand-filled wedges. They observed that the mounds are usually underlain by a gravel substrate and are regularly distributed, as were the soil wedges. Spackman and Munn suggested further that the fossorial rodent hypothesis could not apply to the Laramie Basin mounds because mound-forming animals tend to bury surface-soil horizons. Instead, they observed that the soil strata followed the curvature of the mounds, as though it had been pushed up from below. McFaul (1979) concluded that the

mima mounds in Washington were also formed in association with permafrost and ice wedges.

Investigators of mima mound origin in other areas have supported the burrowing rodent hypothesis, most often invoking pocket gophers as the cause. In one of the earliest papers on the origin of mima mounds, Arkley and Brown (1954) proposed the origin of the mounds in California as follows:

The method by which the gopher accumulates soil into a mound is explained by his tendency to place his nest in a well-drained spot where the soil is deepest; thus generation after generation of gophers may keep building nests near the crest of any high spot in the land surface, or over a window in a hardpan. When a gopher is tunneling, he moves the soil beneath his body, and forces it backward to a surface opening already established. . . . Thus, over a long period of time, the gopher, by digging outward from his nest, tends to move the soil toward it. . . . The mound rises very gradually over a considerable period of time and perhaps many generations of gopher occupation. The fine gravels are moved with the soil, but cobbles and boulders are undermined until they are exposed on the surface in the intermound, or sink to the base of the mound forming a thin layer of cobbles.

In Colorado, Branson et al. (1965) considered both the frost and the burrowing hypotheses for the mima mounds on Rocky Flats, west of Denver. They acknowledged that a periglacial environment would have existed during glaciation, but unlike Spackman and Munn (1984), they observed that the mounds were randomly distributed, which they did not think would occur if the cause was frost or geologic. Pocket gophers seemed like a more probable explanation, especially since the water table on Rocky Flats in March is only 10 cm deep between the mounds and 30 cm deep in the mounds. Another study in Colorado suggested that the soils between mounds were too shallow to support pocket gophers (Hansen 1962). Apparently, the problem was not a shallow water table, as suggested by Branson and co-workers, but that the gophers could still improve their habitat through mound building (creating, perhaps, a warmer, better-insulated winter nest; Cox et al. 1987).

The most recent analyses of mima mound topography are by Washburn (1988) and Cox and his associates (Cox 1984; Cox and Gakahu 1986; Cox and Allen 1987a, 1987b; Cox et al. 1987). Like Arkley and Brown (1954), Cox reasoned that the gophers slowly move soil into mounds through their digging, leaving pebbles and stones too large for the gophers to move in the intermound area. Moreover, Cox and Allen (1987a) were able to prove that this was happening by placing soil plugs with small metal fragments near the pocket gopher mounds they studied in California. Using a metal detector, they observed that the metal fragments were gradually moved in the direction of the mound. Especially interesting is the observation by Cox and Allen that the "intensity of moundward translocation is greatest where the soil becomes shallowest and most poorly drained." Based on such evidence and the observation that none of the other hypotheses could account for the lack of larger rocks on the mounds,

Cox and his associates concluded: "Fossorial rodents are the builders of Mima mounds, the largest and most widespread landscape features produced by any mammal other than man." In addition to pocket gophers, mound building has been attributed to ground squirrels, badgers, toads, and the mole rats of Kenya.

The rationale of Cox, suggesting that all mounds worldwide were created by burrowing, still needs to be reconciled with the evidence obtained by Spackman and Munn (1984) for the frost hypothesis. Berg (1990) added a new dimension to the debate by concluding that most mima mounds are the result of seismic vibrations in the earth's crust in areas with unconsolidated fine sediments on a relatively rigid, flat substratum. Cox (1990a), however, was not convinced by Berg's evidence and persisted in arguing that small mammals are the cause of mima mound formation. Perhaps there is room for more than one explanation for the origin of these widespread, puzzling landscape features that influence plant growth and animal distribution.

IV

Foothills and Mountains

CHAPTER 9

Escarpments and the Foothill Transition

The lowlands of Wyoming are interrupted by a dozen or more mountain ranges (see fig. 1.4) that are flanked by foothills formed largely from eroding sedimentary rocks and glacial moraines. In some ways the foothills are like the riparian zone. Both are linear features that add diversity to the plains and basins, usually providing more cover and a more mesic environment for plants and animals. They are important components of big game winter range, providing food and cover at a critical time of the year. The foothills have a more moderate climate than the cooler mountains above or the drier lowlands below. Away from the mountains are numerous escarpments that can be compared to islands (fig. 9.1). They provide distinct environments on the plains and increase the diversity of plants and animals that can survive. As important, these topographic features contribute immeasurably to the aesthetic appeal of the region.

Foothills and escarpments are more moderate winter habitats than the lowlands because of their abrupt topography. Wind-driven snow accumulates on lee slopes, which provides more moisture for plant growth the following summer than on the surrounding plains (Burke et al., *Topographic control*, 1989). The lee slopes also serve as windbreaks for elk, deer, antelope, and other animals, thereby moderating stress from wind chill. Nearby, windward and south-facing slopes are usually

snow-free, providing easier access to forage. Shelter and forage are thus nearby—unlike the mountain forests, where there is abundant shelter but little forage in the winter (owing to continuous snow cover), or the surrounding lowlands, where both shelter and forage may be hard to find.

The vegetation mosaic of escarpment and foothill landscapes consists of shrublands, grasslands, woodlands, and windswept ridges (fig. 9.2). Shrublands dominated by mountain-mahogany are abundant on many rock outcrops with little or no soil development. Skunkbush sumac, antelope bitterbrush, mountain big sagebrush, serviceberry, and snowberry occur on deeper soils (table 9.1). Black sagebrush or threetip sagebrush commonly occurs on windswept ridges or plateaus with shallow soils (ibid.), and nivation hollows with tufted hairgrass and other species occur where deep snowdrifts develop. Foothill grasslands are characterized by bluebunch wheatgrass. The woodlands have varying densities of juniper, limber pine, ponderosa pine, aspen, Douglas-fir, and—in one or two localities in extreme south-central Wyoming—pinyon pine or Gambel oak. Windswept ridges, characterized by bluebunch wheatgrass and a variety of cushion plants, have low plant cover because of shallow soils and little snow accumulation.

In general, the foothill-escarpment mosaic is patchy, a reflection of abrupt changes in geologic

F I G. 9.1 The coarse, rocky soil of Cretaceous sandstone escarpments in the Powder River Basin supports
little bluestem, ponderosa pine, Rocky Mountain juniper, skunkbush sumac, and other foothill plants.

Wyoming big sagebrush, blue grama, and western
wheatgrass dominate the shrubland on the deeper,
fine-textured soils. Elevation 1,520 m (5,000 ft).

substrate, snow accumulation, microclimate, soil
depth, and inverse texture effect. Vegetation banding is sometimes observed, such as in the Centennial Valley, west of Laramie, where Myers (1969)
found vegetation differences between resistant,
ridge-forming sandstone and limestone, and the
intervening, parallel, nonresistant formations with
shales, siltstones, and claystones. Linear aspen
groves can occur at the contact between the impervious granite of the mountains and the over-
lying sedimentary strata (fig. 9.3; Hanna 1934;
Myers 1969), where runoff from the mountain
encounters deeper soils. Trees also occur on ridges
because the groundcover there is usually insufficient to fuel a fire. Thus, trees have a greater
chance of becoming established than in the adjacent grassland or sagebrush steppe, where the
competition for water from grasses and forbs is

more intense and where the time between fires is
shorter (States 1968; Wells 1970b).

Plant Adaptations and Vegetation Dynamics

MOUNTAIN-MAHOGANY SHRUBLANDS

One of the more conspicuous foothill shrublands
is dominated by mountain-mahogany (*Cercocarpus* spp.). This shrub forms dense thickets on
rocky or shallow soils from the western Great
Plains (1,500 m) up to elevations of 2,400 m in the
mountains (fig. 9.4, table 9.1). Two species of *Cercocarpus* occur in Wyoming: the deciduous true
mountain-mahogany in the Black Hills and across
the southern half of the state, and the evergreen
curlleaf mountain-mahogany in the foothills of

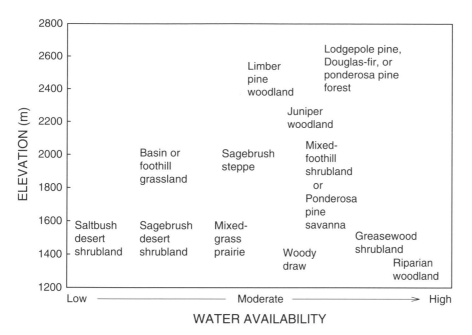

F I G. 9.2   The distribution of foothill grasslands, shrublands, and woodlands in relation to gradients of water availability and elevation in the Bighorn Basin. Vegetation types that occur above and below the foothills are shown for comparison.

the Bighorn Mountains and to the west and south (fig. 9.5; Miller 1964). There is little overlap in the distribution of the two species,[1] but both occur on a variety of bedrock types, including sandstones, limestones, and shales (Johnson 1950; Medin 1960; Brooks 1962; Miller 1964; Brotherson et al. 1984; Knight et al. 1987). Mountain-mahogany dominance on poorly developed soils is undoubtedly facilitated by the nitrogen-fixing ability of both species (Vlamis 1964; Youngberg and Hu 1972; Lepper and Fleschner 1977).

The two species of mountain-mahogany are browsed frequently by deer in the winter (Medin 1960; Duncan 1975; Austin and Urness 1980), and they respond by growing lateral branches or buds. The more heavily the shrubs are browsed, the more spiny the branches become, making further browsing more difficult. In this way the plants develop their own defense mechanism.

Some ecologists have observed the invasion of adjacent communities by curlleaf mountain-mahogany following a period of fire suppression, suggesting that frequent fires have restricted mountain-mahogany to rocky sites (Young and

Bailey 1975; Gruell et al. 1985; Arno and Wilson 1986). Duncan (1975) observed curlleaf mountain-mahogany stems that ranged from 5 to 85 years old in Montana, but she cited other studies that reported ages of more than 150 years. Johnson (1950) observed sprouting after fall burns, and Miller (1964) noted some potential for layering.[2] Large seed crops, however, are common (though often on a two- to ten-year cycle), and new stands of curlleaf mountain-mahogany often become reestablished by seedlings (Ken Stinson, pers. comm.).

JUNIPER WOODLANDS

A characteristic feature of some foothills and escarpments is the stands of picturesque woodlands dominated by Utah juniper or Rocky Mountain juniper (figs. 9.6, 9.7). Associated species include true and curlleaf mountain-mahogany, black sagebrush, big sagebrush, limber pine, and ponderosa pine (table 9.1). In eastern Wyoming and the northern Great Plains, Rocky Mountain juniper occurs in ravines or where summer precipitation

TABLE 9.1   Characteristic plants of foothill vegetation types in Wyoming

	Mountain-mahogany Shrubland	Juniper Woodland	Mixed Foothill Shrubland	Aspen Woodland	Conifer Woodland	Oak Woodland	Woody Draws
TREES							
Aspen	—	—	—	X	—	—	—
Boxelder	—	—	—	—	—	—	X
Bur oak	—	—	—	—	—	X	—
Douglas-fir	—	—	—	—	X	—	—
Gambel oak	—	—	—	—	—	X	—
Green ash	—	—	—	—	—	—	X
Limber pine	X	X	—	—	X	—	—
Ponderosa pine	—	—	—	—	X	—	—
SHRUBS							
American plum	—	—	—	—	—	—	X
Antelope bitterbrush	—	—	X	—	X	—	—
Black sagebrush	X	X	X	—	—	—	—
Mountain big sagebrush	—	—	X	X	X	—	—
Silver sagebrush	—	—	X	—	—	—	X
Threetip sagebrush	—	—	—	—	—	—	—
Wyoming big sagebrush	X	X	X	—	—	—	—
Chokecherry	—	—	X	X	—	—	X
Common juniper	—	—	—	X	X	X	—
Rocky Mountain juniper	—	X	X	—	X	X	X
Utah juniper	—	X	—	—	X	—	—
Curlleaf mountain-mahogany	X	X	—	—	—	—	—
True mountain-maghogany	X	X	—	—	—	—	—
Snowbush	—	—	X	—	—	—	—
Douglas hawthorn	—	—	—	—	—	—	X
Douglas rabbitbrush	X	X	X	—	—	—	—
Rubber rabbitbrush	—	X	X	X	X	X	X
Mountain snowberry	—	—	X	X	X	X	—
Western snowberry	—	—	—	—	X	X	X
Saskatoon serviceberry	—	—	X	X	—	—	X
Utah serviceberry	—	—	X	X	—	—	—
Skunkbush sumac	—	—	X	—	—	—	X
Wax currant	—	—	X	—	X	—	—
Woods rose	—	X	X	X	X	X	X
GRASSES							
Basin wildrye	—	—	X	X	X	—	X
Blue grama	X	X	X	—	—	—	X
Bluebunch wheatgrass	X	X	X	—	X	—	X
Danthonia	—	—	—	X	—	X	X
Idaho fescue	—	—	—	X	—	—	—
King spikefescue	—	X	X	X	X	—	—
Little bluestem	—	—	X	—	—	X	X
Junegrass	X	X	X	—	X	X	—
Sideoats grama	—	—	X	—	X	X	X
FORBS							
Arrowleaf balsamroot	—	—	X	—	—	—	—

TABLE 9.1 (*Continued*)

	Mountain-mahogany Shrubland	Juniper Woodland	Mixed Foothill Shrubland	Aspen Woodland	Conifer Woodland	Oak Woodland	Woody Draws
Fringed sagewort	X	X	X	X	X	X	X
Hairy goldenaster	X	X	X	—	—	—	—
Hooker sandwort	X	X	—	—	—	—	—
Lambert locoweed	X	X	X	—	—	—	—
Lupine	—	—	X	X	—	—	—
Mouse-ear chickweed	X	X	X	—	—	—	—
Oregongrape	—	—	X	X	X	X	X
Pricklypear cactus	X	X	X	—	—	—	X
Pussytoes	X	X	X	X	X	—	—
Sulfurflower buckwheat	X	X	X	—	X	—	—
Yarrow	X	X	X	X	X	X	X

Note: — = absent or less common

FIG. 9.3 Aspen groves, in the foothills of the Medicine Bow Mountains, occur in moist ravines and along the contour, where the impermeable granite of Centennial Ridge meets the permeable soils that have developed on sedimentary rocks (elevation 2,584 m; 8,500 ft). Sagebrush steppe, dominated by threetip sagebrush, occurs to the east (right); Douglas-fir and limber pine occur just above the aspen. Forests at higher elevations are dominated by Engelmann spruce, lodgepole pine, and subalpine fir.

F I G. 9.4   True mountain-mahogany shrubland, west of Cheyenne, occurs on shallow, rocky soils adjacent to mixed-grass prairie on the deeper soil. Elevation 2,280 m (7,500 ft).

F I G. 9.6   Juniper woodland usually occurs on escarpments with coarse soils, such as this area southeast of Lander. Rocky Mountain juniper is more common in the eastern half of Wyoming, Utah juniper in the more arid parts of the western half. Both species are found near Lander; elevation 1,794 m (5,900 ft). See also fig. 9.7.

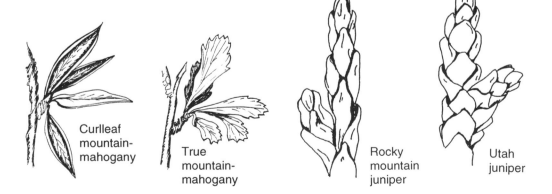

F I G. 9.5   True mountain-mahogany is easily distinguished from curlleaf mountain-mahogany.

F I G. 9.7   The small, scalelike leaves of Rocky Mountain juniper are longer than the leaves of Utah juniper. Drawings by Judy Knight.

is higher. Yucca, skunkbush sumac, and ponderosa pine are common associates. Utah juniper occurs on escarpments in the more arid basins of western Wyoming (for example, in the Wind River Canyon and Bighorn Basin).

Like mountain-mahogany shrublands, juniper woodlands are usually found on shallow or rocky soils. Yet the cause of juniper distribution is a puzzle yet to be solved, because juniper is absent from many habitats that seem appropriate. Billings (1954) suggested that juniper on the upper or northerly edge of its range, as in Wyoming, occurs only in warmer thermal belts, where frosts are less frequent at night. Like mountain-mahogany, the two species of juniper occur on a variety of geologic formations (Wight and Fisser 1968; Knight et al. 1987). Juniper, however, is more capable than mountain-mahogany of invading adjacent rangelands on deeper soils. Differences in juniper and mountain-mahogany distribution may be the result of differences in the microclimate, relative ability to invade adjacent habitats, or soil-bedrock features that have not yet been studied in sufficient detail.

The juniper woodlands of the Bighorn Basin have been the subject of several studies under the direction of H. G. Fisser (Robinson 1966; Wight and Fisser 1968; Wamboldt 1973; Hanson 1974; Spaeth 1981; Waugh 1986). Juniper woodland distribution in this area ranges from about 1,100 to 1,900 m (overlapping but generally lower than mountain-mahogany shrublands). Wight and Fisser (1968) observed that the lower limits in specific areas seemed to be where shallow rocky soils meet deeper alluvial or colluvial soils. The upper limit is probably determined by a climatic factor.[3] Mountain-mahogany was rather uncommon in the southern Bighorn Basin, with Wyoming big sagebrush, black sagebrush, and broom snakeweed accounting for most shrub cover, but curl-leaf mountain-mahogany was commonly found with juniper northward toward the Montana border (especially on steeper and more barren sites).

Juniper invasion into adjacent grasslands or shrublands is one of the most often discussed ecological topics in the intermountain west (Waugh 1986; Everett 1987; R. F. Miller et al. 1993). Using tree ring analysis to study stand history, Waugh (1986) found that the juniper in his study area in

the Bighorn Basin had been restricted to a limestone outcrop until about ten years after the introduction of large cattle herds—a time that also marks the initiation of fire suppression.[4] Juniper has since expanded onto adjacent sites. Waugh noted that more than 90 percent of the juniper seedlings were becoming established under sagebrush, which could have been functioning as nurse plants, and he hypothesized that an increase in sagebrush cover following livestock grazing had created a more favorable environment for juniper invasion. A wide range of juniper ages has been reported in the Bighorn Basin, from very young to more than five hundred years old (Wight and Fisser 1968; Waugh 1986).

The role of nurse plants for juniper establishment in semiarid environments is widely accepted, but other factors may be important as well (as reviewed by Waugh 1986). In particular, livestock grazing could have reduced the fuels available for fire. Fires at thirty- to sixty-year intervals prevented juniper invasion in the Great Basin of Utah (Burkhardt and Tisdale 1976; Tausch and Tueller 1977; Young and Evans 1981), and others have noted that burning can reduce the abundance of sagebrush and juniper (Wright et al. 1979). Woodlands with old junipers are often restricted to rocky ridges, where fires are less frequent. Fire, however, may not be the only factor involved. Waugh (1986) suggested that heavy livestock grazing could reduce the rate of evapotranspiration from the soil, thereby creating a more mesic environment, which favors sagebrush and juniper invasion. Reduced evapotranspiration in combination with an unusually wet period may have triggered the invasion episode observed by Waugh, which ended after about fifty years (in 1940). Others have suggested that drought, in combination with grazing and fire suppression, favors juniper invasion (R. F. Miller et al. 1993).

Herbaceous plant production is greatly reduced with the establishment of juniper in a grassland or shrubland, probably owing to competition for water, light, and nutrients, as well as to the production of toxic chemicals by the juniper (Jameson 1966, 1970). Fires may become less frequent if the shrubs are widely spaced, which perpetuates conditions favorable to the nonsprouting juniper. The woodland, however, can become highly flammable as juniper density increases (Wu 1991). Fire is used to reduce juniper cover, but Spaeth (1981) recommended against burning if the potential existed for the rapid expansion of introduced annuals, such as cheatgrass (see chapter 7). Another caution comes from wildlife managers who have noted that the elimination of juniper could have an adverse effect on deer and other wildlife by reducing the amount of food and thermal cover important for winter survival. Animals stay warmer at night under the juniper canopy because the trees retain heat near the ground. Shade during the summer may also be important for those animals that stay in the foothills.

A voluminous literature on the ecology of pinyon-juniper woodlands exists that is relevant to Wyoming (Gifford and Busby 1975; Everett 1987; R. F. Miller et al. 1993)—though pinyon pine is nearly absent from the state (occurring only near Flaming Gorge Reservoir in the foothills of the Uinta Mountains). Small populations occur in Idaho west of Kemmerer and in Colorado at Owl Canyon (north of Fort Collins, highway 287) (Wright 1952). Across the Colorado Plateau and Great Basin, pinyon pine and Utah juniper are common associates, with pinyon pine found more often at higher elevations and Utah juniper at lower elevations (Welsh 1957; Foster 1968). Though pinyon pine is essentially absent from Wyoming, it is common to see juniper with limber pine and ponderosa pine in the foothills.

PONDEROSA PINE–LIMBER PINE–DOUGLAS-FIR WOODLANDS

Ponderosa pine, limber pine, and Douglas-fir commonly occur in the foothills and on escarpments (fig. 9.8). Growing in association with grasslands and various kinds of shrubland, the tree-dominated woodlands often appear to be increasing in density and expanding their range—probably the result of fire suppression (Progulske 1974; Johnson 1987; Veblen and Lorenz 1991). The encroachment of ponderosa pine into foothill grasslands can result from fire suppression and is a problem for ranchers because the pine leaves are toxic to cattle. Also, the trees reduce the rate of herbaceous forage production and may increase evapotranspiration, thereby reducing streamflow (Orr 1975).

Ponderosa pine is restricted in Wyoming to comparatively warm areas with higher summer precipitation (see fig. 1.5), such as occurs in the Black Hills, at lower elevations in the Bighorn Mountains, on the east slope of the Laramie Mountain Range, in a few localities around the Medicine Bow and Seminoe mountains, and on lowland escarpments scattered throughout the western Great Plains (Tolstead 1947; Potter and Green 1964; Weaver 1965; Brown 1971; Hansen and Hoffman 1987; Hoffman and Alexander 1987). Species with a distribution pattern similar to ponderosa pine include skunkbush sumac, Rocky Mountain juniper, sideoats grama, and little bluestem (table 9.1). Ponderosa pine is essentially absent from western Wyoming, perhaps because of a growing season that is too short and dry, but it does occur in Dinosaur National Monument in Colorado and the foothills of the Uinta Mountains in Utah (Welsh 1957).

Although ponderosa pine is often common on escarpments in eastern Wyoming, where deep water percolation is presumably possible, some geologic substrata provide better conditions for tree growth than others. States (1968) found that the growth of ponderosa pine was best on Mowry shale, where the trees had wider and less variable annual ring widths. Low annual variability in ring width suggests that a consistently good environment for growth was provided by this marine shale, despite the climatic variations that occurred. In contrast, States found that trees on the Pine Ridge and Cloverly sandstones had narrower and more variable annual rings, even though they occurred near the Mowry shale and surely had the same climatic conditions.[5] The association of some foothill species with fracture lines in the bedrock can be striking (see fig. 3.3)

Limber pine sometimes occurs with ponderosa pine, but it is more typical where the climate is drier. Unlike ponderosa pine, it extends upward to the alpine treeline (McNaughton 1984). Lepper (1974) studied limber pine over much of its geographic range and concluded that they were restricted to rocky soils and ridges because the seedlings were not competitive with other species. Limber pine is commonly found on the lee side of boulders, where the seedlings of the now-mature trees would have been protected

F I G. 9.8 Escarpments and foothills on the eastern plains typically have ponderosa pine woodland, such as along the Hat Creek Breaks north of Lusk (Oligocene and Miocene sandstones and claystones; elevation 1,520 m [5,000 ft]). Mixed-grass prairie is also common, sometimes with silver sagebrush and Wyoming big sagebrush.

from excessive wind damage. Once established, however, limber pine occurs on some of the windiest sites in the region (fig. 9.9). Indeed, the unusually "limber" branches that give this tree its name must be an adaptation to minimize breakage during windy periods. Associated species include common chokecherry, ground juniper, king spikefescue, mountain big sagebrush, oregongrape, and western snowberry.

Foothill woodlands in the western half of Wyoming are dominated by Douglas-fir rather than ponderosa pine (fig. 9.10). The best examples occur west of the Continental Divide, such as in the Sunlight Basin, Jackson Hole, and the Greater Yellowstone Area. Precipitation patterns west of the Continental Divide, with a larger portion occurring in the winter, apparently favor Douglas-fir. Soils with Douglas-fir may also be more fertile (Despain 1973, 1990). Limber pine sometimes oc-

curs with Douglas-fir in western Wyoming (Bissell 1973), such as in Fossil Butte National Monument.

Douglas-fir population dynamics appear to be much the same as those of ponderosa pine, with encroachment into grasslands commonly occurring with fire suppression (Gruell 1983). In a Montana study, Arno and Gruell (1986) found that before 1890 fires occurred every few decades, confining the trees to rocky sites or the lee sides of slopes. Fisher et al. (1987) obtained comparable results for ponderosa pine woodlands in Devil's Tower National Monument, where they estimated that from 1770 to 1900 the average number of years between fires was only fourteen. Since 1900, fires have occurred much less frequently (about once every forty-two years), owing to fire suppression and the cessation of ignitions by native Americans.

F I G. 9.9   Limber pine occurs on rocky escarpments from low elevations near Pine Bluffs (1,538 m; 5,045 ft) up to alpine treeline in the Medicine Bow Mountains. This photo, taken in the Laramie Range, shows widely scattered limber pine in a high-elevation (2,432 m; 8,000 ft) shrubland dominated by blue grama, fringed sagewort, junegrass, mountain muhly, Sandberg bluegrass, and threetip sagebrush. The riparian shrubland has baltic rush, beaked sedge, Bebb willow, Nebraska sedge, Pacific willow, and water birch. Ponderosa pine is found on escarpments toward the east at lower elevations.

MIXED FOOTHILL SHRUBLANDS

Sagebrush steppe extends into the foothills in many areas, intermingling with mountain-mahogany shrublands and pine–Douglas-fir–juniper woodlands. Species composition changes, however, as the environment becomes cooler and more mesic. Mountain big sagebrush becomes the dominant variety instead of Wyoming big sagebrush, and other shrubs are common: antelope bitterbrush, common chokecherry, serviceberry, skunkbush sumac, snowberry, snowbrush ceanothus, and wild rose (fig. 9.11, table 9.1). Great Basin wildrye and antelope bitterbrush occur in ravines and on sites where snow accumulates (Walker and Brotherson 1982). Common grasses and forbs include arrowleaf balsamroot, bluebunch wheatgrass, hairy goldenaster, junegrass, and lupine.[6]

Serviceberry is a conspicuous but widely scattered shrub in some mixed foothill shrublands. Two species are common in Wyoming: Saskatoon serviceberry, which occurs throughout the state, and Utah serviceberry, which is found in drier foothill habitats to the west and south.[7] Both species are deciduous, capable of root sprouting, and important food for deer and elk (White 1968; Williams 1976; Tisdale and Hironaka 1981; Young 1983). At lower elevations, where the frost-free period is longer (at least 120 days; Martin 1972), skunkbush sumac is common, especially on the east side of the Bighorn and Laramie mountains.

The mesic environment of the foothills allows for greater plant cover, which enhances the probability of both lightning- and human-caused fires. Periodic fires are required in many areas to prevent a gradual succession from mixed foothill shrubland to dense shrublands or woodlands dominated by junipers or other conifers. Temporary shifts in species composition may occur following a fire because some species lack the ability to reproduce vegetatively (such as mountain big sagebrush and juniper), but most other shrubs are capable of sprouting (Blaisdell and Mueggler 1956; Young 1983). Fires commonly expose mineral soil (Sherman and Chilcote 1972; Clark et al. 1982), thereby facilitating the reestablishment or addition of some species. Antelope bitterbrush, snowbrush ceanothus, and skunkbush sumac, like mountain-mahogany, are capable of nitrogen fixation because of the actinorhizae formed by bacteria in the genus *Frankia* (Wagle and Vlamis 1961; Nelson 1983; Righetti et al. 1983). This adaptation enables the establishment of these species on relatively infertile soils. Snowbrush ceanothus is capable of stem sprouting (Young 1983), but in addition, seed dormancy in this species is broken by the heat of a fire, and seedlings commonly invade burns where the species existed previously (Zavitkovski and Newton 1968).

Sometimes greasewood occurs in the foothills, an interesting anomaly since it is usually found where groundwater is near the surface on saline soils in desert shrubland landscapes. The presence of greasewood on escarpments in southeastern Montana has been attributed to porous lignite seams that cause groundwater to move horizontally to the surface on hillsides (Brown 1971). Seeps or perched water tables, with concomitant salt accumulations, typically develop on such sites.

F I G. 9.10 Douglas-fir, lodgepole pine, and aspen dominate the north slopes of the Gros Ventre Mountains on the east side of Jackson Hole. The treeless south slopes have a mixed foothill shrubland with bluebunch wheatgrass, fringed sagewort, mountain big sagebrush, needle-and-thread grass, saskatoon serviceberry, skyrocket gilia, stonecrop, yarrow, and other plant species. Mountain big sagebrush dominates the shrublands in the foreground (elevation 2,128 m; 7,000 ft).

FIG. 9.11   Saskatoon serviceberry, antelope bitterbrush, rabbitbrush, and mountain big sagebrush dominate this mixed foothill shrubland in the Medicine Bow Mountains (elevation 2,462 m; 8,100 ft). Aspen and alder occur in the ravine, limber pine on the distant ridge.

FOOTHILL GRASSLANDS

Foothill grasslands often develop on windy slopes or plateaus, where snow does not accumulate in large quantities, where soils are too shallow for various shrub species, or where summer rainfall is higher. Bluebunch wheatgrass is the most characteristic plant, occurring most often on relatively warm, dry sites; Idaho fescue is more typical of the higher, more mesic grasslands (>2,121 m; Wright and Wright 1948; Heady 1950; Beetle 1956; Fichtner 1959; Williams 1963; Ludwig 1969; Mueggler 1975; Despain 1990). A dwarf species of *Artemisia,* threetip sagebrush, occurs with Idaho fescue on windswept, shallow soils (see fig. 9.9; Fisser 1962). Little bluestem is sometimes found on the slopes of buttes and other escarpments (Whitman and Hanson 1939; Hansen and Hoffman 1987).

Bluebunch wheatgrass is a highly preferred forage species of livestock and big game. Though it is usually viewed as a bunchgrass, Passey and Hugie (1963) found that it can reproduce by rhizomes on mesic sites. Williams (1961, 1963) surveyed the distribution of bluebunch wheatgrass in the Bighorn Basin and found it to be characteristic of the 25–50 cm precipitation zone in the foothills. At lower, warmer elevations, it is usually found on north slopes, if it occurs at all. Various investigators have observed that bluebunch wheatgrass decreases after hot fires or heavy grazing in spring or summer. Cheatgrass is a common invader following such disturbances, but bluebunch wheatgrass increases again with succession or if grazing is confined to when the plant is dormant (Sauer 1978).

Some foothill grasslands occur on extraordinarily windswept plateaus or slopes with shallow soils. The grasses are scattered and small forbs are more common. Many of the forbs have the "cushion plant" growth form, which keeps the stems and leaves densely aggregated at the soil surface, where the air is warmer. Cushion plants are typical of windswept areas in alpine tundra as well as in the lowlands, though the species composition is completely different (table 9.1; Welsh 1957; Baker and Kennedy 1985; Knight et al. 1987).

BROADLEAF WOODLANDS

Four other components of the foothill-escarpment mosaic are woodlands characterized by either aspen, bur oak, chokecherry, or Gambel oak—all of which are deciduous and occur on mesic sites with deeper soils. The additional moisture apparently comes from a higher soil water–holding capacity, more snow accumulation (from drifting), more summer precipitation, or runoff from the slopes above.

Chokecherry woodlands (also known as hardwood draws) often occur as part of a mosaic comprising grassland and ponderosa pine woodland on the low, eastern plains (fig. 9.12). In addition to chokecherry, common shrubs include American plum, skunkbush sumac, silver sagebrush, and western snowberry (MacCracken et al., *Plant community variability*, 1983; Uresk and Boldt 1986). Trees sometimes found with the shrubs are American elm, boxelder, green ash, and hawthorn (table 9.1). Except for the lack of cottonwood, chokecherry woodlands have the characteristics of some riparian woodlands. There is more plant biomass than on adjacent shrublands, providing good cover for wildlife as well as good forage and shade for livestock.

Aspen-dominated foothill woodlands tend to occur on the upper fringes of the intermountain basins, but they are still separated from the aspen forests found in nearby mountains. Aspen woodlands frequently occur in foothill riparian zones, where moisture is plentiful and cool-air drainage moderates the temperature, or just below large

F I G. 9.12   Ravines in the foothills of the Bear Lodge Mountains and Black Hills near Sundance typically have chokecherry, skunkbush sumac, wild rose, and other shrubs. Mixed-grass prairie occurs on the up-land, with ponderosa pine on some ridgetops with coarse, poorly developed soils. Elevation 1,186 m (3,900 ft).

snowdrifts or seeps on hillslopes. The understory is dominated by a variety of shrubs, grasses, forbs, and sedges typical of the montane aspen forests (table 9.1).

Aspen woodlands in the foothills often form a crescent on the lower sides of snowdrifts. Sometimes they completely surround the drift, forming a doughnut-shaped grove that can be referred to as an *aspen atoll* (fig. 9.13; Burke et al., *Topographic control,* 1989). If trees become established on the windward side of the atoll, they cause a further accumulation of snow in the center. Only herbaceous plants capable of surviving in a short growing season can persist where the snow is deepest, since snow cover may last until mid-July.

Plants found in the central snowglade include various mountain meadow species such as alpine bistort, mountain silver sagebrush, shrubby cinquefoil, and tufted hairgrass. Surrounding the snowglade is a band of shrubby, malformed aspen. Here the average snow accumulation is sufficiently shallow to allow the aspen to persist, but the weight of the snowpack causes breaking and bending that prevent the trees from assuming their normal growth form. Only on the fringes of the "atoll" does the aspen develop into trees.

Oak woodlands are found only in small areas of the northern and eastern slopes of the Black Hills (bur oak) and on the west side of the Sierra Madre (Gambel oak) on the Wyoming-Colorado

FIG. 9.13   Aspen groves sometimes occur on the lee side of the Continental Divide south of Rawlins where deep snowdrifts develop during the winter, which provide adequate water for shrubs and trees in an area that elsewhere is windswept sagebrush steppe (elevation 2,310 m; 7,600 ft). This photo shows a portion of a doughnut-shaped grove, known as an aspen atoll, where the center is dominated by shrubby cin-quefoil and tufted hairgrass. Deep snow accumulation prevents tree establishment in the center. The shrubs in the foreground and on the left are aspen that are regularly broken by the heavy snowpack. Trees develop just beyond the shrubby aspen, where snow accumulation is less; some lodgepole pine are found downwind just beyond the aspen trees.

F I G. 9.14   Bur oak forms a distinctive foothill woodland in the northern Black Hills and Bear Lodge Mountains (elevation 1,277 m; 4,200 ft). Mixed-grass prairie with an occasional skunkbush sumac occurs below the oak; ponderosa pine grows on the ridges. Gambel oak forms a similar woodland on the southwestern foothills of the Sierra Madre (and southward into Colorado).

border (fig. 9.14). Neilson and Wullstein (1983) studied the tolerances and distribution of Gambel oak in relation to climate, concluding that spring frosts and summer drought have limited its spread northward into Wyoming by preventing seedling establishment. Summer precipitation is higher southward into Colorado and Utah (see chapter 3), and late-spring frosts are less likely to occur. No comparable studies have been done on bur oak, but it is interesting that this species occurs only in the northeastern corner of Wyoming, where the elevation is relatively low and annual precipitation is comparatively high (see fig. 3.3). If it were not for late-spring frosts in Wyoming, oak would probably be a co-dominant of foothill woodlands with ponderosa pine, as is common in Colorado (for example, near Colorado Springs).

Oak might also occur with aspen and choke-cherry. Harper et al. (1985), in their review of Gambel oak, observed that it often occurs in northern Utah where temperature inversions regularly create warmer thermal belts in the foothills, with an average of ninety frost-free days a year. In Utah, Gambel oak is usually found on sloping upland sites between 1,675 and 2,286 m, above the juniper-pinyon zone but below ponderosa pine and aspen.

Oak woodlands can become highly flammable, but both bur oak and Gambel oak resprout vigorously and may increase in density after fires (Harper et al. 1985; Tiedemann et al. 1987). With fire suppression, they are invaded by less fire-adapted species, such as Rocky Mountain juniper and ponderosa pine (and in northern Utah, pinyon pine, white fir, and canyon maple; Harper et al. 1985). Both kinds of oak woodland provide winter forage for big game, and their energy-rich acorns are consumed by deer, elk, wild turkey, and squirrel. With age, forage production under the oak canopy declines because of shading and competition for water. Fire can be used to stimulate the growth of more forage, but ranches, summer homes, and resorts located in the foothill environment usually prevent this practice.

Plant-Animal Interactions in the Foothills

ASPEN, ELK, AND FIRE

The management of foothill aspen groves has been one of the most interesting and highly debated natural resource issues in the region (Beetle 1968, 1974a, 1974b; Krebill 1972; Gruell and Loope 1974; DeByle 1979; Olmsted 1979; Boyce 1989; Coughenour and Singer 1991). The controversy began in the mid-1950s, when some of the aspen groves in northwestern Wyoming appeared to be dying (fig. 9.15). Aspen is unique because the numerous "trees" in a grove may result from the root sprouting of only one or a few plants.[8] Seedling establishment is thought to be rare in the foothill environment, but when establishment is successful, the root system sprouts additional aboveground stems that grow into new trees—a classical case of cloning that is more common in herbaceous plants. Like other species of *Populus*,

F I G. 9.15   An aspen grove in northwestern Wyoming that appears to be dying, possibly because of excessive browsing by elk. Mountain big sagebrush, arrowleaf balsamroot, and other foothill plants occur around the dying trees; lodgepole pine and limber pine occur in the background. Elevation 2,158 m (7,100 ft).

each stem is comparatively short-lived, usually dying by the age of a hundred years. The root system, however, is thought to live for centuries, producing new stems around the edges of groves or where older stems are damaged.

The reason for the apparent loss of aspen is attributed to two factors: fire suppression and the excessive browsing of sprouts by large mammals, particularly elk. Browsing was the first factor to be debated.[9] Many of the so-called decadent aspen groves in northwestern Wyoming occurred along known elk migration routes, and many were located adjacent to winter feeding grounds—notably, the National Elk Refuge. Though elk are grazers in the summer, aspen sprouts are an important food for these animals in the winter and spring. Too much browsing for too many consec-

utive years reduces the number of sprouts, or as some have thought, eliminates them altogether. Yet when aspen groves are fenced to exclude large herbivores, small aspen sprouts often grow into trees, even in the decadent stands (fig. 9.16).

Heavy aspen browsing by elk in the Jackson Hole area occurs at least in part because of the blockage of migration routes to traditional winter ranges by urban and agricultural developments. Recognizing this, and with a desire to maintain and even increase elk numbers for tourists and hunters, winter feeding programs were authorized on the National Elk Refuge and elsewhere. As a result, elk winter mortality was greatly reduced and the population increased. The effect on aspen was predictable. Fortunately, aspen groves away from winter feeding grounds are often in good condition.

An interesting perspective on decadent aspen stands was presented by DeByle (1979), who suggested that one of the causes is political pressure to maintain elk populations at the same high level year after year. DeByle reasoned that in presettlement times the number of elk fluctuated considerably—resulting largely from higher mortality in some winters than others—and that aspen sprouts had a better chance of developing into trees during periods when the browsing pressure was low (a natural rest rotation). Elk populations still fluctuate, but probably to a lesser extent because of winter feeding programs.

Wildlife managers are sometimes sensitive about elk receiving so much of the blame for aspen groves that appear to be in bad condition. They argue (correctly in some cases) that livestock frequently congregate in aspen groves because of the abundant forage that develops in the relatively mesic environment, where aspen is typically found. Cattle and sheep can cause considerable damage to aspen through browsing, trampling, and bedding.

Fire is another factor that cannot be ignored in explaining the apparent decline in some aspen groves. Veblen and Lorenz (1986), working in Colorado, suggested that widespread burning in the late 1800s or early 1900s, for whatever reason (lightning- or human-caused), might have created many aspen stands of approximately the same age. Such stands are now more subject to

F I G. 9.16   Aspen groves sometimes recover quickly when protected by a fence from browsing, such as in this area in northwestern Wyoming. Aspen exists outside the fenced exclosure, to the right, but only as small, heavily browsed sprouts.

disease or senescence, and managers are concerned because the problem (if it is one) seems to be widespread. In fact, periodic aspen senescence over a large area might have occurred naturally before European settlement.

The aspen issue is confounded further by fire suppression. With less frequent fires in the foothills and mountains, many aspen stands have become quite old. Such stands might be more susceptible to diseases that kill the entire tree, including the root system. Possibly more important, conifers have become dominant through natural succession, with the effect that the diminishing number of aspen groves are subjected to more browsing than before. Large-scale fires could stimulate the development of new aspen forests (DeByle 1979), which might then remove some of the browsing pressure on any particular grove in the foothills. It is noteworthy that many new aspen seedlings became established in the Greater Yellowstone Area after the 1988 fires (see chapter 14).

Whatever the cause, too many participants in the debate are prone to make sweeping generalizations about the relative importance of fire suppression and damage by domestic and wild ungulates. In fact, the magnitude of the problem is not well known. What proportion of the aspen groves in different mountain ranges is in a degraded condition? Is there any basis for thinking that some aspen groves might have been in "poor condition" long before European settlement? Research at the scale of landscapes is necessary to answer such questions—and to enable wildlife advocates, livestock owners, and federal and state agencies to develop sustainable management plans acceptable to both groups.

WINTER RANGELAND FOR DEER AND ELK

Widely dispersed in the mountains during summer and fall, elk and mule deer spend much of the remaining part of the year (usually November to April) concentrated on winter ranges in or near the foothills. Often the winter ranges are heavily affected by large numbers of animals concentrated in small areas—more now than in presettlement times as the land area available for winter range has diminished because of human activities. Though year-long habitat is fundamental, wildlife biologists generally agree that the availability and condition of the winter range are the most important factors in determining sustainable herd sizes. Many winter ranges are on private land where wildlife sometimes competes with domestic livestock. Consequently, an important goal for wildlife agencies is to cooperate with landowners and in some cases to purchase winter ranges so that they can be preserved and managed in a way favorable to native ungulates.

The survival of antelope, deer, elk, and moose in the winter depends largely on there being adequate food available and on the animals having sufficient fat reserves. In the Rocky Mountains, snow depth causes an increase in the energy required to obtain food through pawing and simple movements, and severely limits the amount of accessible food. The bark and twigs of trees and shrubs, though generally not preferred, are consumed during this time because they are visible above the snow. Browse lines develop on junipers, willows, and other tall shrubs (fig. 9.17).

By the time spring arrives, winter ranges may seem to be heavily grazed and browsed, yet the plants typically recover quickly because the herbivory occurs largely when the plants are dormant. Depending on livestock management practices

F I G. 9.17   Browse lines, such as on this Rocky Mountain juniper adjacent to Yellowstone National Park, are formed when adequate food is not available for elk during the winter. Although possibly indicating excessive browsing, such browse lines might have existed for many centuries near elk and deer winter ranges. Elevation 1,824 m (6,000 ft).

and the extent of overlap between the summer plant preferences of livestock and the winter preferences of big game, winter range plants may have a full summer of rest and become senescent before they are again subjected to large ungulates. The seasonal movements of animals have resulted in a natural rest rotation system (Coughenour 1991a). More than one year of rest may result if the herd selects a different part of the foothills for wintering or if high mortality during a particularly severe winter leads to fewer animals in the following winter.

PINE SEED DISPERSAL BY BIRDS

Another interesting plant-animal interaction in foothill and escarpment landscapes is that between certain seed-eating birds and limber, white-bark, and pinyon pines. Unlike most conifers, these pines have comparatively large nutritious seeds; the cone scales tend to hold the seeds in place, even after the cones have opened. Not surprisingly, several studies have shown that animal seed dispersal is important for these species (Vander Wall and Balda 1977, 1983; Ligon 1978; Tomback 1983; Arno and Hammerly 1984).

One of the best-studied examples of this relationship involves the Clark's nutcracker, a common bird in the foothills and mountains (fig. 9.18). Working in Arizona, Vander Wall and Balda (1977) observed that a single nutcracker was able to carry up to ninety-five pinyon seeds in its sublingual pouch for a distance of 22 km. Such behavior is associated with seed caching in late summer. The seeds are typically buried on south-facing slopes, which are often snow-free during late winter. Many of the seeds are eaten in the spring, a time coinciding with the raising of young and when food for seedeaters is in short supply, but many seeds are not relocated and develop into new trees (and sometimes into clumps of trees).

As might be predicted, there is a great overlap in the distribution of both North American and European nutcrackers with the occurrence of pines having large, wingless seeds (Tomback 1983; Arno and Hammerly 1984; Vander Wall 1991). Seed-caching birds such as the nutcracker (and

F I G. 9.18 Clark's nutcracker buries the seed of limber and whitebark pine in the soil, often on warm south slopes. Many seeds are not recovered and grow into new trees. Wyoming Game and Fish Department photo by LuRay Parker.

certain species of jay) are continually attracted to the pines because of their edible and nutritious seeds, and the plants benefit by having their wingless seeds more widely dispersed and even planted. Seeds that simply fall to the ground are more likely to be eaten by small mammals, and the seedlings are less likely to become established on the forest floor, a less favorable seedbed than that provided by burial in mineral soil. Moreover, the mortality of seedlings that originate from fallen seed may be higher because of competition with the already well-developed parent tree. This coadaptation between birds and some pines is further illustrated by the observation that the nutcracker distinguishes between viable and nonviable seeds. There is little advantage for either the bird or tree if less nutritious, nonviable seeds are cached.

Bird dispersal appears to be one of the factors causing the patchy distribution pattern of limber and whitebark pine. Typically, these trees are found on ridges or sunny slopes that are frequently snow-free in winter (Bamberg 1971; Hutchins and Lanner 1982; Arno and Hammerly 1984), conditions that might reflect an instinctive preference of the birds. Seeds cached on such sites would be more accessible in the winter, when food is scarce. Similar distribution patterns have been observed for stone pine in the European Alps, which also depends on bird dispersal (Tranquillini 1979).

Bird dispersal affects the three pines in other ways also. For example, limber pine is sometimes a multiple-stemmed tree. Tomback (1983) found, however, that the stems are genetically different— a clear indication that the "tree" became established from several cached seeds. Furthermore, the birds probably facilitate the establishment of whitebark and limber pines in burned areas, especially if the birds are attracted to the habitat by perches, such as dead trees or boulders. The establishment of whitebark pine in the Greater Yellowstone Area is especially significant because grizzly bears are known to build up fat reserves for the winter through the consumption of large numbers of whitebark pine seeds in the fall (see chapter 14).

The exposed bedrock of foothill and escarpment landscapes creates sharp environmental boundaries and a patchy vegetation mosaic. Interesting ecological phenomena occur wherever boundaries are found. Conflicts over proper land use will surely continue, primarily because of differing management priorities for private and public lands and because foothill and escarpment landscapes provide important habitats for people as well as an exceptional diversity of plants and animals. Multicolored escarpments with scattered woodlands or shrublands provide grand vistas of the adjacent plains. As important, the benefits of a moderate lowland climate are found close to steep, sheltered slopes, and in the case of foothills, close to the majestic mountain landscapes above. For these reasons, foothills and escarpments are highly valued real estate.

CHAPTER 10

Mountain Forests

Rising above the plains, basins, and foothills are mountains that greatly diversify the landscapes of the region (table 10.1). Most are anticlinal or fault block in origin and have cores of Precambrian granite, gneiss, and quartzite. In many areas the Precambrian rocks are still covered by sedimentary strata formed at the bottoms of ancient seas that spread across Wyoming millions of years ago, before mountain building began. More than fifty peaks now rise above 3,900 m (mostly in the Wind River Mountains). Some of the mountain ranges are massive, while others occur as smaller islands in the intermountain basin landscape (for example, the Granite Mountains; see fig. 1.1). In many areas the mountains have been shaped by glaciers, and the soil has developed on glacial till.

Large or small, the mountain ranges are important economically. Most of the forests are located there, and most of the water used by agriculture, industry, and municipalities originates in the mountains as snowfall. They are also important for outdoor recreation. Ecologically, mountains are interesting because of their diverse array of ecosystem types, the contrasts of mountain ecosystems with those of the surrounding lowlands, and the ways in which mountains affect the surrounding lowlands, for example, through the development of alluvial fans and rainshadows. To no small extent, the landscape patterns observed in the foothills and lowlands are caused by the nearby mountains.

Spatial variation in mountain forests can be attributed partially to climatic conditions that change with elevation and topography. Soils vary in relation to slope steepness and the nature and depth of the underlying bedrock (which is usually near the surface). Other factors causing spatial patterns in high-elevation forests are uneven snow accumulation and melting, with the result that some patches of ground are snow-free much earlier than others. Some species, such as dwarf huckleberry, cannot tolerate snow cover beyond the first week of July (Knight et al. 1977).

The most obvious environmental gradient in mountains is a decrease in temperature with elevation (see fig. 3.6). Though it is tempting to adopt a standard cooling (adiabatic lapse) rate of, for example, 6°C for every 1,000 m of elevation (3°F/1,000 ft; MacMahon and Anderson 1982), such rates vary greatly with the time of year. For example, Callison and Harper (1982) reported cooling rates of 1.3°, 6.4°, 8.9°, and 13.1°C per 305 m (1,000 ft) in January, December, April, and May, respectively, along the same elevational transect in the Uinta Mountains. This variability probably depends on differences in mountain size, cold-air drainage, snow cover duration, the potential for temperature inversions, and other climatic and topographic factors.

Other environmental factors also change with elevation (Daubenmire 1943b). The snow-free period and length of growing season typically decrease with elevation, but precipitation generally increases—at least to midelevations, where

TABLE 10.1   Characteristics of mountains in or adjacent to Wyoming

| Mountain Range | Elevation (m) | | | Geologic Origin | Predominant Surface Rocks | Glaciated[c] |
	Highest Peak	Surrounding Lowlands[a]	Relief[b]			
Absaroka Range	4,009 Francs	1,853	2,156	Eroded remnant of a volcaniclastic deposit that formerly filled the Bighorn and Wind River basins	Volcaniclastic	Yes
Beartooth Range	3,901 Granite	1,548	2,353	Fault block	Granite and metamorphic	Yes
Bighorn and Pryor mountains	4,013 Cloud	1,219	2,794	Anticlinal uplift	Large areas of Paleozoic sedimentary rocks; central area of granite and metamorphic rocks	Yes
Black Hills and Bear Lodge Mountains	2,207 Harney	1,219	988	Domal uplift	Same as Bighorn Mountains	No
Granite Mountains[d]	3,059	1,935	1,124	Downfaulted anticlinal uplift (graben)	Granite and metamorphic	No
Gallatin Range	3,350 Electric	1,615	1,735	Anticlinal	Mostly Paleozoic and Mesozoic sedimentary	Yes
Gros Ventre Range	3,561 Doubletop	2,012	1,549	Anticlinal	Mostly Paleozoic sedimentary	Yes
Hartville Uplift	1,724	1,484	240	Anticlinal	Mostly Paleozoic sedimentary	No
Laramie Range	3,131 Laramie	1,745	1,386	Anticlinal	Granite, anorthosite, and metamorphic	No
Madison Range	3,440 Koch	1,573	1,867	Anticlinal uplift	Mostly Paleozoic and Mesozoic sedimentary	Yes
Medicine Bow Mountains	3,662 Medicine Bow	2,094	1,568	Anticlinal	Granite and metamorphic	Yes
Owl Creek Mountains	3,009	1,582	1,427	Anticlinal	Mostly Paleozoic sedimentary	No
Park Range	3,712 Mount Zirkel	2,255	1,457	Anticlinal	Granite and metatamorphic	Yes
Sierra Madre	3,354 Bridger	2,051	1,303	Anticlinal	Granite and metamorphic	Yes

TABLE 10.1 (*Continued*)

Mountain Range	Highest Peak	Surrounding Lowlands[a]	Relief[b]	Geologic Origin	Predominant Surface Rocks	Glaciated[c]
Teton Range	4,197 Grand Teton	2,002	2,195	Fault block	East face granite, quartz monzonite, and metamorphic; west, Paleozoic sedimentary	Yes
Uinta Range	4,041 Kings	1,868	2,173	Anticlinal	Mostly metasedimentary; some granite	Yes
Wasatch Range	3,008 Doubletop Mountain	1,433	1,575	Fault block	Mostly Paleozoic sedimentary	Yes
Washakie Range	3,815 Washakie Needles	2,005	1,810	Anticlinal (partially buried by volcaniclastic rocks of Absaroka Range)	Mostly Paleozoic sedimentary	Yes
Wind River Range	4,207 Gannett	1,956	2,251	Anticlinal	Granite and metamorphic	Yes
Wyoming, Hoback, Salt River, Snake River and Caribou Ranges[e]	3,463 Wyoming	1,969	1,494	Thrust masses of Paleozoic and Mesozoic sedimentary rocks	Paleozoic and Mesozoic sedimentary	Northern part only

[a] Estimated by averaging evaluations of nearby towns in adjacent lowlands.
[b] Estimated as the difference between the elevations of the highest peak and the adjacent lowland.
[c] Glaciers may have been widespread or restricted to small areas.
[d] Includes Ferris, Seminoe, and Shirley mountains; the highest point is the summit of the Ferris Mountains.
[e] Known collectively as the thrustbelt ranges.

snow accumulation is greatest. Snow accumulation, combined with lower evaporation rates caused by cooler temperatures, provides an environment considerably more mesic than in the surrounding lowlands (see fig. 3.4). Countering the effect of cooler temperature is lower atmospheric pressure, which allows rapid evapotranspiration at higher elevations (Smith and Geller 1979; Smith and Knapp 1990). Midmountain environments, however, are still more mesic than any other upland environment in the region. Primary productivity is comparatively high there as well (see chapter 11).

The effects of topography are just as important as those of elevation (Daubenmire 1943b). South slopes are warmer and drier, and north slopes are cooler and more mesic than might be expected at any particular elevation. Thus, species that are typically found high in the mountains may occur at low elevations in ravines or on north slopes, and species requiring a warmer environment may be found at unusually high elevations on south slopes (fig. 10.1).

Soils can also have a pronounced effect on the landscape mosaic. For example, meadows occur in the Bighorn Mountains on comparatively dry, fine-textured soils derived from sedimentary shale and limestone, even at high elevations, where forests might be expected (see chapter 12). Douglas-fir or spruce-fir forests occur on fine-textured

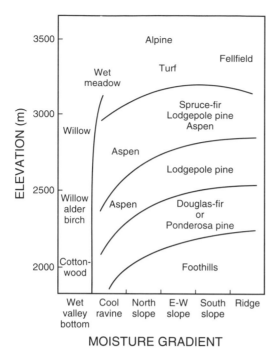

F I G. 10.1   Approximate distribution of different forest types in relation to elevation and moisture availability. Note that each forest type typically occurs at the lowest elevations in cool, moist ravines.

soils (Inceptisols and Alfisols), where moisture stress is less severe in late summer (such as on north slopes), and lodgepole pine is typically the most common tree dominating forests on coarse, less fertile soils derived from glacial till or granite. Still, generalizations are difficult. Weaver and Perry (1978) found lodgepole pine forests on both sandstones and shales in the Bridger Range of south-central Montana.

The vegetation patterns that develop in montane landscapes are determined by disturbances as well as by environmental gradients. The effects of disturbances are important in all landscapes, but in montane forests they can occur over larger areas and their impacts last longer because the usually dominant conifers are unable to sprout. Fires, windstorms, and insect outbreaks—along with human-caused disturbances[1]—often lead to tree death over many hectares (fig. 10.2). A forest develops again only after the successful establishment of new seedlings or with the rapid growth of small trees that had been suppressed by larger

trees prior to the disturbance. This is in contrast to grasslands, where the primary disturbance—fire—does not kill most of the plants and new shoots are produced within a year by sprouting (see chapter 5).

The frequency of disturbance varies with elevation and other environmental factors. For example, forest fires are less common at high elevations because fuels accumulate more slowly in the cooler environment and because the fuels that do accumulate are generally moister and consequently less flammable.

Disturbances in forests are long-lasting and give the impression of devastation or poor management, even though the causes of the disturbance are natural. For example, the insects and pathogens causing tree death in the Rocky Mountain region are usually native species that have coexisted with the plants for thousands and even millions of years.[2] Moreover, at least in forests at lower elevations, fires were more common in previous centuries than they are today (Habeck and Mutch 1973; Loope and Gruell 1973; Arno 1980; Romme 1982; Habeck 1988). One of the challenges for ecologists is to distinguish between landscape patterns caused by disturbances from those caused by environmental changes. Peet (1988), in a review of Rocky Mountain forests, concluded that "the vegetation is perhaps best thought of not as a uniform and stable cover but rather as a mosaic, with the character of each tessera [patch] frequently changing and borders being periodically redefined."

Surviving in the Mountains

One of the fundamental challenges for plant survival in the Rocky Mountains is the short, cool, sometimes dry growing season, which makes it difficult for plants (especially large plants) to fix sufficient energy by photosynthesis for survival and reproduction during the rest of the year. Several adaptations help to maximize the length of the growing season. First, photosynthesis is known to occur in montane plants at temperatures near freezing or below (Freeland 1944; Parker 1953; Perry 1971; Smith and Knapp 1990), and many plants have chlorophyll in their leaves or stems all

F I G. 10.2   Severe winds of short duration can topple trees over large areas, such as occurred in this 6,000-ha area (15,000 acres) in 1987 in the Teton Wilderness. Jackson Hole Guide photo by Bill Conradt.

year (evergreen or wintergreen plants). This ability to acclimate to low temperatures, combined with the continuous presence of chlorophyll in leaves, enables plants to extend their growing season into late fall and early spring. Evergreen coniferous trees are excellent examples, but other plants have similar adaptations. For example, deciduous plants such as aspen and dwarf huckleberry have chlorophyll in their stems, which permits photosynthesis when they have no leaves. The green stems of dwarf huckleberry may account for up to 45 percent of the plant's annual photosynthesis (Kyte 1975).

Some shrubs and herbaceous plants with green leaves or stems under the snow are capable of photosynthesis and growth as soon as the snow is shallow enough to permit light penetration, despite freezing temperatures (Mooney and Billings 1960, 1961; Kimball et al. 1973; Salisbury 1984; Hamerlynck 1992). The capacity for photosynthesis at cold temperatures is an important adaptation, but simply being able to survive cold temperatures is another. Sakai and Weiser (1973) compared the low-temperature tolerances of numerous tree species and found that, unlike the trees from some other climatic regions, those of the Rocky Mountains could tolerate temperatures of −60°C (−76°F).

Another problem for plants in mountain environments is acquiring and conserving nutrients from relatively young, coarse-textured soils. All of the trees have mycorrhizae (fig. 10.3). Also, the mycorrhizal root systems are concentrated in the top 10–15 cm of soil, where decomposition is most active and limiting nutrients, especially nitrogen, are likely to be available. Once incorporated into plant tissues, the nutrients are retained for extended periods simply by the longevity of leaves and twigs. Lodgepole pine leaves, for example, persist for five to eighteen years, depending upon environmental conditions (Knight 1991; Schoettle 1991). When the leaves do fall, whether from an evergreen or deciduous plant, limiting nutrients in the leaves are first partially reabsorbed by the twig. Thus, plants in nutrient deficient environments have evolved mechanisms for retaining limiting nutrients.

Short, cool growing seasons also present problems for seedling establishment since new seedlings are the most susceptible to water, tempera-

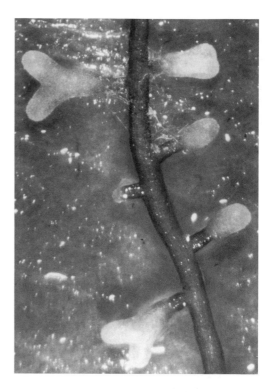

FIG. 10.3   A microphotograph showing the swollen mycorrhizal root tips of lodgepole pine. The fungi facilitate the uptake of water and nutrients by roots. Photo by Steven L. Miller.

ture, and nutrient stresses. Germination often occurs despite cool temperatures in the spring, thereby allowing the maximum time possible for seedlings to develop the stored energy and cold-hardened tissues needed for the following winter. Moreover, the formation of mycorrhizal roots typically occurs within the first few weeks (Harvey et al. 1987). Despite these adaptations, seedlings often succumb to summer drought, late-spring frosts, and other factors (Cui 1990). Because seedling establishment is so precarious, nearly all mountain plants are perennials.

Fire and insect disturbances are common in many vegetation types. Some conifers have a thick bark that reduces mortality following a surface fire (such as ponderosa pine and Douglas-fir), but most are easily killed by either surface or crown fires. Similarly, insect outbreaks can sometimes be curtailed by abundant resin production. Sooner or later, however, the trees become susceptible. Consequently, fire and insects create distinctive openings in the forest canopy that persist, often for several decades, until they are hidden by new trees. Many forest grasses, forbs, and shrubs survive disturbances because of their sprouting ability, but species differ in this regard. Species with deep roots or rhizomes (such as lupines and milk-vetch) are better able to survive forest fires than species with stolons or shallow root systems (bearberry, heartleaf arnica, pyrola, and twinflower; McLean 1969; Hungerford et al. 1991).

Despite their inability to sprout, conifers usually become reestablished after disturbances because of the abundance of seed they produce and the often favorable environment for seedling establishment created by the disturbance. Mineral soil, which is often exposed after the forest floor has burned, is an especially good seedbed for the establishment of pines, spruce, and Douglas-fir (Alexander 1987a). As important, the seedlings are often able to compete with the sprouting plants that survive disturbances. In fact, by trapping snow in the winter and providing some shade on warm summer days, other plants and fallen logs improve the environment for tree seedling establishment (Knapp and Smith 1982)—especially in areas that might otherwise be blown free of snow.

Different species tolerate different levels of environmental stress—a fundamental principle in considering the causes of landscape mosaic anywhere—and stress levels change rapidly with changes in elevation, soils, and topographic position. Consequently, species composition often changes abruptly in the mountains. Variation in species composition is further complicated by single species that exhibit a wide range of genetically determined tolerances (ecotypic variation). Several studies have shown genetic differentiation over short environmental gradients in Rocky Mountain conifers (Rochow 1970; Pearcy and Ward 1972; Linhart et al. 1981; Knowles and Grant 1983; Mitton 1985; Rehfeldt 1985; Schuster et al. 1989), and aspen groves at higher elevations must be genetically different from those in the foothills. Considerable genetic variability, one aspect of biodiversity, can be expected in landscapes with rapid spatial and temporal changes.

Animals have an equally diverse range of adaptations for surviving in mountain landscapes. Some migrate to warmer or less limiting environ-ments at lower elevations (or farther south) in the fall, thereby avoiding cold temperatures, deep snow, and food shortages during the winter. Other species are able to survive in special shelters or beneath the snow, and in some cases by becoming dormant through hibernation. Energy stored during the preceding summer in fatty tissue or food caches is especially critical for animals that are year-round residents.

Variation in the Forest Mosaic

Vegetation zonation on mountain ranges has been of interest to scientists for more than a century. In Wyoming and adjacent states, foothill grasslands, shrublands, and woodlands grade into forests dominated by ponderosa pine and Douglas-fir at lower elevations, lodgepole pine at midelevations, and Engelmann spruce and subalpine fir at higher elevations (Daubenmire 1943b; Hayward 1952; Langenheim 1962; Patten 1963; Ream 1964; Gartner 1967; Jacoby 1971; Despain 1973; Reed 1976; Weaver and Perry 1978; Veblen and Lorenz 1986; Habeck 1987; Peet 1988; Green and Conner 1989). There is considerable overlap in the elevational distribution of these tree species, and in some areas the foothill vegetation grades directly into lodgepole pine forest or Douglas-fir, as, for example, on the east slopes of the Medicine Bow Mountains and around Jackson Hole and the Sunlight Basin. Ponderosa pine is not found in western Wyoming, where Douglas-fir forests usually border the foothill vegetation.

Intermingled with the coniferous forests are mountain meadows (or parks), ribbon forests, snowglades, aspen groves, various kinds of riparian vegetation, and shrublands or woodlands dominated by mountain big sagebrush and limber pine. The upper treeline, where the mountains extend that high, is characterized by short, wind-pruned Engelmann spruce, subalpine fir, and limber pine in most Wyoming mountain ranges, but whitebark pine is typical of the treeline in northwestern Wyoming and the Northern Rockies. In central Colorado, bristlecone pine is common at treeline and also forms subalpine woodlands in some areas (Baker 1992a). Throughout the region, various alpine communities form interesting mosaics (see chapter 13).

The distribution of the major vegetation types is determined by environmental conditions that affect water stress and the length of the growing season. An interesting challenge for plant ecologists is to understand why different species occupy different habitats. For example, why is ponderosa pine restricted to the foothills of eastern Wyoming, lodgepole pine to the midelevations, and subalpine fir and Engelmann spruce to the cooler habitats of higher elevations? The answers to such questions are based on the physiological characteristics of the plants, and more specifically, on the physiological characteristics of seedlings—the most delicate phase of a plant's lifetime. Universally, seedling establishment depends on the availability of seed, seedbed characteristics, and the environment (including herbivory) during germination and establishment. All these factors vary greatly across the landscape at all spatial scales, causing considerable variation in plant community composition.

Rocky Mountain forests are easily classified into several major types (table 10.2). Forest ecologists, however, have found considerable variation within each type, variation that is important when considering management alternatives. The most thorough classification has been made using Daubenmire's habitat-type approach, where forests are classified according to the dominant tree in the climax forest and one or more understory plants that are indicators of special environmental conditions.[3] This classification system has been helpful in evaluating a particular site for wood production, wildlife habitat, and other forest values. Using this approach, Wyoming forests have been divided into a total of 9 series and 110 habitat and community types, including 9 different kinds of ponderosa pine forest, 16 kinds of lodgepole pine forest, and 28 kinds of subalpine fir forest. Such detail is beyond the scope of this book, but the general characteristics of each major forest type are the subject of the following sections. Chapter 11 examines forests from the perspective of energy, water, and nutrients. (See table 10.3.)

PONDEROSA PINE FOREST

Ponderosa pine occurs at low elevations on the east slopes of the Rocky Mountains, where summer precipitation is higher and the growing season is warmer and longer (see fig. 3.4) The importance of summer precipitation is suggested by the absence of ponderosa pine on the southern end of the Bighorn Canyon (in the semiarid Bighorn Basin) and its presence at the same elevation on the northern end of the canyon, where summer rainfall is much higher (49 cm compared to 18 cm). Ponderosa pine is also found on the lower southwest slopes of the Uinta Mountains in Utah, where, as on the Great Plains, a larger portion of the annual precipitation occurs in the summer (Harper et al. 1980). The importance of warmer temperatures was indicated by a laboratory study done by Cochran and Berntsen (1973), who found that ponderosa pine seedlings were more sensitive to cold temperatures than were those of the lodgepole pine, which is found at higher elevations. They also suggested that cold sensitivity explains the absence of ponderosa pine from depressions or frost pockets where cold air accumulates.

The most extensive ponderosa pine forests are found in the Black Hills and on the east slopes of the Bighorn and Laramie mountains (especially near Casper Mountain and Laramie Peak, on the western edge of the Great Plains; see fig. 1.5). As with ponderosa pine forests elsewhere, tree density is a function of fire frequency, with open forests and savannas maintained by frequent surface fires that kill young seedlings and saplings. The older trees are not killed by surface fires, owing to their thick bark, and the understory shrubs and herbaceous plants resprout readily. Canopy fires were probably rare in such forests before European settlement, but fire scar data suggest that surface fires occurred on average every ten to fifteen years (Fisher et al. 1987). Since 1900, fire suppression has lengthened the mean fire-return interval to forty-two years.

Photographs taken in the 1800s in the Black Hills, the Colorado Rockies, and elsewhere show how ponderosa pine and Douglas-fir forests were once more open than they are now (Progulske 1974; Gruell 1983; Veblen and Lorenz 1986, 1991; Johnson 1987). With fire suppression, ponderosa forests have become more dense and are more subject to intense canopy fires than before (figs. 10.4, 10.5, 10.6). Another factor contributing to higher tree density may be livestock grazing, which can improve conditions for tree seedling

TABLE 10.2 Characteristic plants of mountain forests and woodlands in Wyoming

	Limber Pine Woodland	Ponderosa Pine Forest	Douglas-fir Forest	Aspen Forest	Lodgepole Pine Forest	Spruce-fir Forest	Whitebark Pine Woodland
TREES							
Aspen	—	X	X	X	X	X	—
Douglas-fir	—	X	X	—	—	—	—
Engelmann spruce	—	—	—	X	X	X	X
Limber pine	X	X	X	X	X	X	—
Lodgepole pine	—	—	—	X	X	X	—
Ponderosa pine	X	X	X	X	—	—	—
Subalpine fir	—	—	—	X	X	X	X
Whitebark pine	—	—	—	—	X*	X*	X*
SHRUBS							
Antelope bitterbrush	X	X	—	—	—	—	—
Birchleaf spiraea	—	X	X	X	X	X	—
Chokecherry	—	X	X	X	—	—	—
Common juniper	X	X	X	X	X	X	X
Dwarf huckleberry	—	—	—	—	X	X	X
Mountain gooseberry	—	—	—	X	X	X	X
Mountain snowberry	—	—	X	X	X	X	—
Ninebark	—	X	X	—	—	X	—
Pachistima	—	—	X	X	X	X	—
Rocky Mountain juniper	X	X	X	X	—	—	—
Rose	X	X	X	X	X	X	—
Russet buffaloberry	X	X	X	X	X	X	X
Saskatoon serviceberry	X	X	X	X	—	—	—
Skunkbush sumac	X	X	X	—	—	—	—
Wax currant	X	X	X	X	—	—	—
GRASSES							
Bluebunch wheatgrass	X	X	X	—	—	—	—
Bluejoint reedgrass	—	—	—	—	X	X	—
California brome	—	—	—	X	—	—	—
Idaho fescue	X	X	X	—	—	—	X
King spikefescue	X	X	X	X	—	—	—
Pine reedgrass	—	—	X	X	X	X	—
Wheeler bluegrass	—	—	X	X	X	X	X
SEDGES							
Elk sedge	—	X	X	X	X	X	X
Ross sedge	X	X	—	X	X	X	X
FORBS							
Arrowleaf balsamroot	X	X	—	—	—	—	—
Aspen peavine	—	—	—	X	—	—	—
Baneberry	—	—	—	X	X	X	X
Bearberry	—	X	X	X	X	—	X
Bedstraw	—	X	X	X	—	X	—
Fireweed	—	—	—	X	X	X	—
Heartleaf arnica	X	X	X	X	X	X	X
Horsetail	—	—	—	X	—	X	—

(*continued*)

TABLE 10.2   (*Continued*)

	Limber Pine Woodland	Ponderosa Pine Forest	Douglas-fir Forest	Aspen Forest	Lodgepole Pine Forest	Spruce-fir Forest	Whitebark Pine Woodland
Longstock clover	—	—	—	X	—	—	—
Lousewort	—	—	—	X	X	X	—
Meadowrue	—	—	X	X	—	X	—
Oregongrape	—	X	X	X	X	—	—
Pyrola	—	—	—	X	X	X	—
Silvery lupine	—	X	—	X	X	X	—
Sticky geranium	—	—	—	X	X	X	—
Twinflower	—	—	—	—	X	X	—
Weedy milkvetch	X	X	X	X	X	X	X
Western coneflower	—	—	—	X	—	—	—

Note: — = absent or less common

*Only in northwestern Wyoming and the Northern Rocky Mountains

F I G. 10.4   Ponderosa pine forest in the Laramie Mountains south of Douglas. With the suppression of surface fires, tree density increases greatly (see fig. 10.5). The thick bark of mature ponderosa pine protects most trees from surface fires, though some would be scarred (see fig. 10.6). Elevation 2,037 m (6,700 ft).

F I G. 10.5   Tree density increases greatly following the suppression of fires in ponderosa pine forests, such as in this area near Laramie Peak south of Douglas. Such forests are highly susceptible to hot crown fires, which are difficult to control. Elevation 2,006 m (6,600 ft).

TABLE 10.3   Mammals, birds, and reptiles found in mountain landscapes

MAMMALS

Bear, Black	Marmot, Yellow-bellied
Bear, Grizzly[a]	Moose
Bobcat	Mouse, Deer
Chipmunk, Least	Pika[b]
Cougar	Pocket gopher, Northern
Coyote	Porcupine
Deer, Mule	Squirrel, Red
Elk	Vole, Southern red-backed
Hare, Snowshoe	Woodrat, Bushy-tailed

BIRDS

Bluebird, Mountain	Kinglet, Ruby-crowned
Chickadee, Black-capped	Nutcracker, Clark's
Chickadee, Mountain	Nuthatch, Red-breasted
Crossbill, Red	Nuthatch, White-breasted
Finch, Rosy	Owl, Boreal
Flicker, Northern	Owl, Great horned
Grosbeck, Pine	Pipit, American
Grouse, Blue	Ptarmigan, White-tailed[b]
Grouse, Ruffed	Raven, Common
Hawk, Red-tailed	Robin, American
Jay, Gray	Solitaire, Townsend's
Jay, Stellar's	Tanager, Western
Jay, Pinyon	Woodpecker, Downy
Junco, Dark-eyed	Woodpecker, Hairy

REPTILES

Garter snake, Wandering

[a] Now rare; restricted to the Greater Yellowstone Area in Wyoming.
[b] Alpine

establishment by reducing competition from grasses and forbs (Mandany and West 1983). Both fire suppression and livestock grazing are commonplace at lower elevations, and consequently it is difficult to find ponderosa pine forests representative of those that existed in presettlement times.

Some investigators believe that ponderosa pine seedling establishment is episodic, occurring when years of high seed production coincide with favorable conditions for seedling establishment and a sufficiently long fire-free period to enable the seedlings to develop a tolerance for the next surface fire (Peet 1988). Also, surface fires may burn unevenly through the understory. Consequently, the age structure of ponderosa stands is often patchy, with only a few age classes represented (fig. 10.7; Peet 1988).

In addition to fire, forests dominated by ponderosa pine are susceptible to bark beetles, in particular the mountain pine beetle. Outbreaks of these small insects have been especially common on the Laramie Mountains near Casper and Laramie Peak, and southward on the Front Range of Colorado. The trees are thought to become susceptible when they are unable to produce enough resin to defend themselves, as might occur when the trees are older or when climatic conditions lead to water stress (see chapter 11).

DOUGLAS-FIR FOREST

Douglas-fir and ponderosa pine sometimes coexist, and indeed they share several ecological characteristics in Wyoming. Specifically, mature Douglas-fir, like ponderosa pine, has a thick bark,

F I G. 10.6   Trees with thick bark, such as ponderosa pine, are often not killed by surface fires, even though the base of the tree is scarred when the heat kills a portion of the cambium. Such scars can be used to determine the year that a fire occurred.

which enables the tree to survive many surface fires. Also, both species occur in the lower montane zone, usually below 2,600 m (see fig. 10.1; Veblen and Lorenz 1986), and they are common pioneer species following fire. Usually, Douglas-fir forests are found at slightly higher elevations than ponderosa pine, or on somewhat more mesic sites (Daubenmire 1943; Veblen and Lorenz 1986), and typically Douglas-fir occurs on limestone or other sedimentary substrata (Patten 1963; Despain 1973; Loope and Gruell 1973; Lanier 1978; Weaver and Perry 1978). Both species may form even-aged stands. In Colorado, where the two species often occur together, Douglas-fir is more tolerant of the understory environment and may eventually replace the pine as succession proceeds (Goldblum and Veblen 1992).

In Wyoming, Douglas-fir is most common to-

ward the west, where it occurs below lodgepole pine forests, as, for example, in the Madison and Bridger ranges just outside Yellowstone National Park (Weaver and Perry 1978). Patten (1963) concluded that Douglas-fir was the climax tree on mesic sites at about 1,220–2,440 m elevation.

Fire history studies suggest that the mean fire-return interval in Douglas-fir forest is similar to that in ponderosa pine forests (Goldblum and Veblen 1992). The earliest study was done by Houston (1973) in northern Yellowstone National Park, where he estimated a mean fire-return interval of twenty to twenty-five years before effective fire suppression. With fire suppression and the elimination of native American fires, the trees have become more dense and sagebrush cover has increased. Similar observations have been made to the northwest in Montana, where McCune (1983) estimated a presettlement fire-return interval of sixty years and Arno and Gruell (1986) estimated thirty-five to forty years.[4] In addition to fire, outbreaks of western spruce budworm and the

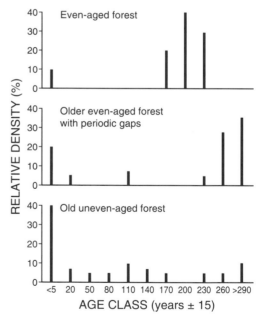

F I G. 10.7   The age structure of three different forest types. Note that an even-aged forest may have trees that became established during a ninety-year period. Most of the trees less than five years old will probably not survive. Lodgepole pine forests are often even-aged; spruce-fir forests typically are uneven-aged.

Douglas-fir beetle (another species of bark beetle) are likely as Douglas-fir forests age or the trees are damaged by fire (Amman and Ryan 1991).[5] Factors regulating the population sizes of these insects include the availability of host trees, climatic conditions, small mammals, insectivorous birds, and other arthropods (McCune 1983; Swetnam and Lynch 1989).

LODGEPOLE PINE FOREST

The most common tree in the mountains of northern Colorado, Wyoming, and much of the Northern Rockies is lodgepole pine (figs. 10.8, 10.9, 10.10). Forests dominated by this widespread species occur most commonly at middle elevations (1,800–3,200 m in northern Wyoming and 2,130–3,500 m in southern Wyoming; Green and Conner 1989). At higher elevations, lodgepole pine is usually found on drier sites and sometimes in areas that have been burned in the past century or two. Spruce and fir are dominant on the more mesic sites. Unlike Douglas-fir, Engelmann spruce, and subalpine fir, lodgepole pine is less common on limestone-derived soils (Despain 1973, 1990; Habeck 1987).

Traditionally, lodgepole pine has been viewed as a pioneer species because it does not appear to tolerate the forest understory environment. Indeed, other tree species are often common in the lodgepole understory, in particular Engelmann spruce, subalpine fir, and at lower elevations westward, Douglas-fir. Yet climax lodgepole pine forests have been identified on comparatively cool, dry, and nutrient-poor sites, where lodgepole pine seems to be the only tree that can survive (Moir 1969; Whipple and Dix 1979; Despain 1983; Lotan and Perry 1983; A. J. Parker 1986; Peet 1988). As gaps occur due to windstorms, parasites, and insects, new lodgepole pine seedlings grow into the canopy. Establishment is thus episodic and patchy, resulting in a forest with two or three age classes. The probability of lodgepole pine seedling establishment in the forest understory appears to increase with moisture availability, even without a canopy gap, but so does the chance of spruce or fir establishment.

Lodgepole pine is well adapted as a pioneer species following fire because of its ability to produce serotinous cones (figs. 10.11, 10.12). These cones are produced during most years, but they remain closed for many years until opened by intense heat, such as occurs during a fire. The large store of accumulated seed is thereby dispersed by wind when a mineral soil seedbed and low competition for resources help to ensure seedling establishment. More than a million seeds per hectare may be dispersed in a single year (Lotan 1975). Sometimes the seedlings are so dense that the developing stands are difficult to walk through, with fifteen thousand or more trees per hectare (fig. 10.13). Such forests are commonly referred to as "dog-hair" stands and could not develop without the storage of large amounts of seed in serotinous cones. The growth of individual trees is slow in such dense stands, but the trees still produce new seed and may survive for well beyond a century (though remaining less than 10 cm in diameter). Ecologically, seed production is a better measure of success than how large a plant grows.

The production of serotinous cones that store seed is a plant adaptation that facilitates the survival of the species. This adaptation is especially important for the thin-barked lodgepole pine, which is easily killed by trunk scorching, unlike ponderosa pine and Douglas-fir.[6] Notably, many serotinous cones with their enclosed seed may be burned during intense fires, with the result being fewer pine seedlings during early postfire succession. Pine seedling density was found to be spatially variable three years after the Yellowstone fires of 1988, with the lowest densities in the middle of very hot burns and the highest densities near the edges of burns where the fires were less intense and where live trees persisted with unburned cones (Anderson and Romme 1992; Ellis et al. 1993). Lodgepole pine density is apparently a function of seed mortality during fire in addition to the percentage of serotinous trees.

Some lodgepole pine do not produce serotinous cones (fig. 10.12), a trait that seems to be genetically determined. There are at least three recognized phenotypes: closed cones that require temperatures of 45°–60°C to open, intermediate cones that require cooler temperatures (35°–50°C), and nonserotinous cones that open at warm ambient temperatures of 25°–50°C (Perry

F I G. 10.8   A dense growth of bluejoint reedgrass and a variety of forbs (including fireweed) grow vigorously two years following fire in Yellowstone National Park. Elevation 2,310 m (7,600 ft).

F I G. 10.9   An even-aged stand of lodgepole pine in Yellowstone National Park. Tree-ring data suggest that this forest was initiated following a fire about ninety years ago. The wood "inherited" from the previous forest has been incorporated into the soil through decomposition. Elevation 2,432 m (8,000 ft).

and Lotan 1977). The trees can produce cones while still young (under ten years old), and there is evidence to suggest that the cones of some trees are nonserotinous until the trees reach an age of twenty to thirty years, after which the trees become serotinous (Lotan 1976). This plasticity can be adaptive, as good conditions for seedling establishment continue for more than a year after a fire, and the viability of dispersed seed often lasts for only a year or two. Moreover, the presence of some nonserotinous trees in every stand helps to explain why new seedlings emerge following disturbances other than fire (Muir and Lotan 1985a, 1985b) and why lodgepole pine can sometimes be observed invading sagebrush steppe or meadows even in the absence of fire (Patten 1969).

Recent studies suggest that the number of serotinous trees in a stand is highly variable. Brown (1975) and Lotan (1976) hypothesized that most

trees would be serotinous in stands where repeated, high-intensity fires occur and that the proportion of forest trees that are serotinous could be an index of fire history. In later studies, Muir and Lotan (1985a, 1985b) observed that the proportions of the trees that are serotinous and nonserotinous depend on the nature of the last disturbance: fires lead to more trees of the serotinous genotype because the heat opens the cones and an abundance of seed is dispersed when conditions for seedling establishment are ideal. Nonserotinous trees, however, are favored if the last disturbance was an insect epidemic or windstorm because most of the serotinous seed remains stored within cones. The primary seed source, then, is from nonserotinous cones. Muir and Lotan (1985a) suggested that having both cone types present in a stand in-

F I G. 10.10 After about 120–150 years, lodgepole pine can become susceptible to mountain pine beetle and other pathogens, such as in this area just north of Grand Teton National Park (elevation 2,432 m; 8,000 ft). As canopy trees die, an uneven-aged forest develops. Smaller trees growing in canopy gaps typically are Engelmann spruce, lodgepole pine (occasionally whitebark pine in Yellowstone National Park), and subalpine fir. After another 50–100 years of succession without a canopy fire, the forest is dominated by spruce and fir rather than lodgepole pine (see fig. 10.14).

creases the chance that there will be a seed source regardless of the type of disturbance. They suggested that management, in both the wilderness and timber production areas, should allow for both fire and nonfire disturbances so that both genotypes can be maintained, thereby maintaining biological diversity.

While studying cone serotiny in lodgepole pine, Muir and Lotan (1985b) also observed that considerably more seed is produced per year in nonserotinous cones than in serotinous cones, a pattern they attributed to the need for tough protective tissues in the serotinous cones to minimize seed predation by squirrels. In other words, trees with serotinous cones divert a larger proportion of available energy into cone tissues, leaving less energy available for seed production.

Lodgepole pine forests are often quite uniform structurally and even-aged (see fig. 10.7). Seedling establishment following a fire, however, occurs over a period of fifty years or more (Veblen 1986b; Veblen and Lorenz 1986; Peet 1988). As the can-

opy closes, lodgepole seedling establishment becomes less common. The mean fire-return interval is quite long in lodgepole pine forests, ranging from one hundred to three hundred years (Romme and Knight 1981; Romme 1982). Though most fires in this forest type do not burn more than a few hectares, infrequent fires burn thousands of hectares during dry years and have a major impact on landscape patterns (Romme and Despain 1989; Johnson and Larsen 1991). Comandra blister rust, mistletoe, mountain pine beetle, root rot, western gall rust, and windstorms can also affect landscape patterns (Krebill 1975; Amman 1977, 1978; see chapter 11).

The intensity of all forest disturbances varies greatly from place to place, creating great spatial variability in forest structure. Even-aged lodgepole pine forests, for example, can be found near uneven-aged stands. Moreover, even-aged doghair stands can occur adjacent to even-aged, low-density stands of the same age (Knight et al. 1981; Anderson and Romme 1991). The proportion of

F I G. 10.11   Serotinous cones on lodgepole pine re-main closed for many years, accumulating seed on the tree that is typically dispersed when the heat of a fire causes the cones to open. Notably, some lodgepole pine do not have serotinous cones (see fig. 10.12).

F I G. 10.12   The nonserotinous cones on this lodgepole pine have opened without the heat of a fire.

serotinous trees also varies from place to place (Tinker et al. 1994). Yet understanding succession and patchiness in lodgepole pine forests, or any coniferous forest, is not just a matter of cone ser-otiny and disturbance intensity. Of all the individ-ual plants observed one year after fire during a study in Yellowstone National Park, 67 percent originated by vegetative reproduction, 29 percent originated from seed stored in the canopy (lodgepole pine), about 1.5 percent originated from the soil seed bank, and about 2.5 percent from off-site or unknown sources (Anderson and Romme 1991). The small contribution of the seed bank is surprising, but this finding is consistent with other seed bank studies. Only 2 seedlings grew from 78 soil samples collected from a forest soil in Grand Teton National Park (James Krumm, pers. comm.). Contrary to expectations, the number of species was highest in soil samples from the most severely burned study area. Many

variables are involved in determining the nature of secondary succession.

SPRUCE-FIR FOREST

Engelmann spruce and subalpine fir often domi-nate subalpine forests because they can tolerate the lower temperatures just below treeline and because they have relatively low water use efficien-cies (Knapp and Smith 1982; Kaufmann 1985; Smith and Knapp 1990).[7] This means that they require relatively large amounts of water for every gram of plant tissue produced by photosynthesis. Their low water use efficiencies, partly a result of a tendency for the stomata to remain open even when the air is dry and transpiration is rapid, restricts these trees to cooler, more mesic environ-ments, such as those found in ravines and at higher elevations.

In contrast, lodgepole pine have stomata that close readily with the initiation of dry conditions during the summer, thereby conserving water (Lopushinsky 1969; Running 1980). Photosynthetic rates are reduced as well when the stomata close, but lodgepole pine still produces more plant tissue per gram of water than do spruce and fir (Knapp and Smith 1981; Kaufmann 1985). Consequently, lodgepole pine can survive on drier sites at lower elevations, where water stress is likely to develop. Ponderosa pine and Douglas-fir also have stomata that readily close during drought conditions (Lopushinsky 1969), a factor allowing these trees to survive at lower elevations as well.

Engelmann spruce and subalpine fir are comparatively tolerant of the understory environment (shade-tolerant), and thus subalpine spruce-fir forests are usually uneven-aged (see figs. 10.7, 10.10). Because the tree species dominant in the

F I G. 10.13 This one-hundred-year-old dog-hair stand of lodgepole pine trees near Dry Park in the Medicine Bow Mountains is very dense, probably because the previous forest that burned had a large number of trees with serotinous cones and the climatic conditions just after the burn were ideal for seedling establishment. With about 15,000 trees per hectare, there is little understory growth. Unlike the forest shown in fig. 10.5, dense forests such as this one are not caused by fire suppression. Elevation 2,785 m (9,160 ft).

overstory are the same as those in the seedling and sapling size classes, these forests are textbook examples of a climax community. Disturbances do occur, however, and lodgepole pine and aspen are often pioneer species that persist as dominants or co-dominants with spruce and fir for a century or more (or until the next disturbance). This successional sequence is repeated at high elevations throughout the northern hemisphere (MacMahon and Andersen 1982).

Various successional pathways for the development of spruce-fir forest are possible, depending on the nature and intensity of the disturbance, species composition prior to disturbance, and soil and microclimatic characteristics (fig. 10.14). Following fire, spruce and lodgepole pine are usually the first dominants (Brown 1975; Fisher and Clayton 1983; Johnson and Fryer 1989; Veblen et al. 1991), either together or separately, depending on moisture availability. After a hundred years or more, subalpine fir becomes more common and lodgepole pine less common. The probability of rapid succession to spruce-fir forest increases with fire suppression and the relative moistness of the site (Day 1972; Romme and Knight 1981). Hanson (1940) found evidence in the pollen record to suggest that landscape dominance by lodgepole pine alternates with dominance by Engelmann spruce and subalpine fir through several cycles. These cycles may reflect climatic changes, or they may be successional. Spruce-fir forests have also been observed to develop on former meadows, sometimes following the initial invasion of aspen (Schimpf et al. 1980; Pearson et al. 1987).

The relative abundance of spruce and fir varies greatly from place to place. In general, the largest and oldest trees are Engelmann spruce, as this species may live five hundred years or longer (Oosting and Reed 1952; Shea 1985; Alexander 1987a; Veblen et al. 1991). In contrast, subalpine fir is usually more common, and the trees are smaller and younger (rarely more than 250 years old; Veblen 1986a, 1986b; Peet 1988). Fir may also have ten to twenty times more seedlings, as fir seedling establishment is much more frequent than spruce. Apparently, the new roots of fir seedlings are better able to penetrate the considerable litter that accumulates over the soil surface (the forest floor),

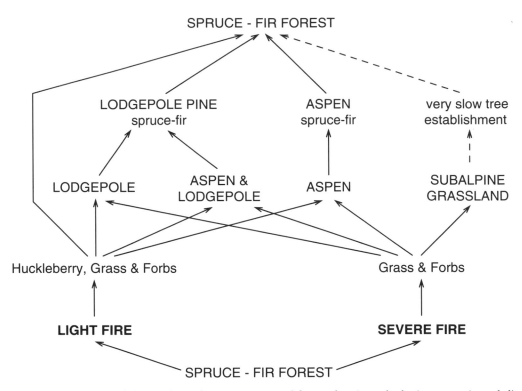

F I G. 10.14   Succession following fire in forests dominated by Engelmann spruce and subalpine fir can vary considerably. Severe fires at high elevations may lead to the establishment of a subalpine grassland that persists for centuries. Note also that the spruce and fir may function as both pioneer species and climax species. The young trees of aspen and lodgepole pine are less tolerant of the forest understory environment and become abundant only after disturbances such as fire. Adapted from Stahelin 1943.

whereas spruce seedlings are usually found where mineral soil has become exposed (such as around the tipped root system of fallen trees). Some investigators have also observed spruce seedlings on rotting logs at sites where the decomposing wood remains wet well into the summer (Lowdermilk 1925; McCullough 1948; Oosting and Reed 1952; Loope and Gruell 1973; Cui 1990). Spruce produces large amounts of seed every two to five years, and fir about every three years (Alexander 1987a).

Unlike the pines and Douglas-fir, spruce and fir are capable of vegetative reproduction when lower branches are pressed to the ground by snow. The branch often develops adventitious roots, after which the end of the branch begins to grow upright into a new tree. Eventually, the branch connection to the parent tree decomposes. This process, known as layering, produces clones of the original tree that are common locally at higher elevations. Clusters of subalpine fir in meadows can often be attributed to this cloning process (fig. 10.15).

That subalpine fir reproduces so much more effectively than spruce in the forest environment has led to speculation on how spruce persists as a codominant in the forest. Veblen (1986a) concluded that the explanation lies in the different life history "strategies" of the two species. Fir is shorter-lived but produces many seedlings, whereas Engelmann spruce compensates for poor reproductive success with increased longevity. Thus, the greater proportion of subalpine fir seedlings and saplings does not mean that it will eventually become the sole dominant.

It is tempting to view spruce-fir forests as being in a state of equilibrium. Continuing periodic disturbances, however, have important effects on these forests. Aplet et al. (1988) hypothesized that, following a large-scale crown fire, spruce and fir would together invade at the beginning of secondary succession (apparently without lodgepole pine in their Colorado study area). After one hundred to two hundred years, spruce could no longer reproduce in the forest understory, a period they labeled the "spruce exclusion phase." After another hundred years, many of the dominant fir and spruce would begin to die, initiating the "spruce reinitiation phase," when canopy gaps are created and spruce (and perhaps lodgepole pine) can again reproduce. The final phase they recognized was a "second generation spruce-fir forest," which would probably have considerable fuel accumulation and would be highly susceptible to a

F I G. 10.15 Occasionally subalpine fir and Engelmann spruce develop clones when lower branches, pressed to the ground by a heavy snowpack, produce roots (layering). A new tree then develops from the branch. Meadow plants include alpine bistort, alpine timothy, Idaho fescue, and sibbaldia, such as occurred here in the Medicine Bow Mountains (elevation 3,162 m; 10,400 ft).

stand-replacing crown fire during the next dry period. Although this is a plausible successional sequence, another crown fire could prevent the old-growth spruce-fir forest from developing. As Johnson and Fryer (1989) observed in Alberta, the fire-return interval may be short enough to ensure burning within the life span of the trees that invaded following the previous fire.

The stand development pattern described by Aplet et al. (1988) is undoubtedly influenced by insect outbreaks similar to those that occur in lodgepole pine forest. One common insect in spruce-fir forests is the spruce beetle, another bark beetle with a life cycle similar to the mountain pine beetle (Schmid and Hinds 1974; Alexander 1987a; Baker and Veblen 1990; Veblen et al. 1991a, 1991b). The spruce beetle attacks older stands with a high proportion of Engelmann spruce in the overstory. Usually, the susceptible stands have slow tree growth, suggesting that the trees are of low vigor. An abundance of downed trees, whether from logging or windthrow, may provide the energy base for the initial development of the outbreak. The infestation opens the canopy, greatly accelerating the growth of the abundant, nonhost subalpine fir and spruce that are too small to be susceptible to attack. The growth of other plants is stimulated as well (for example, aspen, lodgepole pine, and numerous shrubs and forbs). Fuel continuity is increased and fuel accumulation is accelerated, thereby hastening the time when a fire will occur.

ASPEN FOREST

Aspen forests are found from the foothills to the subalpine zone. They typically occur in depressions, ravines, and valley bottoms, or on the lee sides of ridges, where snow accumulates (fig. 10.16). Because of this distribution pattern, many ecologists believe that aspen requires more water than other trees. Kaufmann (1985), however, observed that the adult aspen uses considerably less water per unit leaf area than lodgepole pine, subalpine fir, and Engelmann spruce. The general restriction of aspen to moist areas is probably due to the intolerance of aspen seedlings to drought, not the intolerance of mature trees.

Aspen usually occurs in small groves, but ex-

F I G. 10.16   Aspen forests typically have an understory of subalpine fir, or in this area on the west slope of the Sierra Madre, lodgepole pine. Fire suppression leads to smaller areas dominated by aspen and larger areas dominated by conifers. Elevation 2,584 m (8,500 ft).

tensive forests of this species can be found in a few areas, as, for example, on the west slope of the Sierra Madre (Severson 1963). As discussed in chapter 3, the Sierra Madre aspen forests may be a result of the influence of the Arizona monsoon, which brings more reliable summer precipitation to the southern part of the state than elsewhere. Southward into Colorado and New Mexico, aspen becomes increasingly common, perhaps because of more reliable precipitation. Peet (1988) suggests that aspen occupies the same topographic positions in the Southern Rocky Mountains as lodgepole pine does in the Central and Northern Rockies.

The tendency for aspen to occur in small groves in Wyoming can be explained in part by its clonal nature, with new trees developing from root sprouts (or "suckers"; Schier et al. 1985).

Seedling establishment is rare, not because there is a lack of seed but because the conditions required for establishment are rarely encountered (Williams and Johnston 1984; McDonough 1985; Kay 1993). Possibly, the seedbed is too dry in Wyoming when it is warm enough for germination, or perhaps the seedbed dries out too quickly in undisturbed forests because of high evapotranspiration rates. Each grove thus develops as a genetically uniform clone. All the trees are actually part of one or a few plants. Although individual trees or shoots typically die after about a hundred years, the plant itself may live for thousands of years, perhaps dating back to Pleistocene times. Genetic variability thus exists between clones rather than within, and this is sometimes apparent where, for example, two clones are adjacent and one turns yellow earlier in the fall than the other. Great genetic variation undoubtedly explains why this species is found over such wide environmental gradients in Wyoming and, in fact, why it is the most widespread tree in North America.

Because of its ability to sprout, aspen persists in some coniferous forests until the next disturbance. Sprouting greatly facilitates secondary succession because the sprouts have more stored energy available to them than do seedlings. More than thirty thousand sprouts per hectare may emerge, especially following hot fires. Yet many of the young sprouts do not survive (Brown and De-Byle 1989; Bartos et al. 1991), and those that do are overtopped by conifers about seventy-five to a hundred years later—most commonly by subalpine fir, because its seedlings can become established on the aspen forest floor. Lodgepole pine and Engelmann spruce may replace the aspen under some conditions.

Where other species do not become established, aspen persists as a climax community, with sprouts replacing the older stems as they senesce (Reed 1971; Peet 1988). Generally, most aspen sprouts do not survive in the understory shade of an aspen forest, an interesting anomaly since it would seem that the small trees could tap the energy reserves made available by the larger trees, whose leaves are fully exposed to sunlight. Plant hormones produced by the dominant shoots apparently suppress the growth of the smaller ones below—a case of apical dominance.

Resource managers are concerned that fire suppression has led to a deterioration in the vigor of some aspen groves, especially if the capacity for sprouting is diminished or the level of elk browsing or beaver cutting is high (Houston 1973; Loope and Gruell 1973; Beetle 1974b; Boyce 1989). Aspen shoots can be killed by various pathogens, including root rot and canker diseases (Walters et al. 1982), especially if the bark is damaged. There seems little need to be concerned about aspen extinction, but it could become less abundant in some areas as aspen forests are slowly converted to coniferous forests. As aspen groves become less common, elk and deer in need of browse concentrate on those that remain, which could lead to their deterioration (see chapter 9). DeByle et al. (1987) recommended prescribed burning in some areas after calculating that there are now fewer fires than required to maintain the current abundance of aspen forests.

Aspen is unique because it is the only upland deciduous tree in Wyoming that grows in an environment that would seem to favor evergreen plants. The loss of all leaves each fall is not particularly efficient with regard to nutrient conservation, and this may partially explain why aspen seems restricted to depressions and other sites where nutrient availability is not a problem. As with conifers, aspen twigs are probably capable of reabsorbing nutrients from the leaves, another mechanism for nutrient conservation. And although aspen loses its leaves in the fall, there is chlorophyll in the bark. This adaptation has been found to contribute substantially to carbon uptake for the plant, even when ambient temperatures are near freezing or below (Pearson and Lawrence 1957, 1958; Strain and Johnson 1963; Covington 1975). The plant is effectively evergreen despite having deciduous leaves.

Overall, forests cover about 22 percent of Wyoming and are found primarily in the mountains where temperature, moisture, and nutrient conditions are sufficiently favorable to enable tree seedling establishment and growth. Although only six tree species are common, their adaptations vary considerably, and the influences of fires, insects, windstorms, and abrupt topographic changes create an interesting mosaic that includes meadows, shrublands, and mountain lakes. Individually, each forest appears quite simple, with fewer species of plants and animals than are found in some warmer and moister climates. The apparent simplicity of these ecosystems, however, is not indicative of the complex interactions that have enabled these forests to survive without management for thousands of years.

C H A P T E R   1 1

The Forest Ecosystem

Forests are characterized by trees, but 90 percent of the plant species are in the understory. They include such shrubs as buffaloberry, dwarf huckleberry, and ground juniper; forbs such as heartleaf arnica, lousewort, and pyrola; and various species of grasses and sedges (see table 10.2). Many animals and legions of microorganisms coexist with the plants. Clearly, forest ecosystems cannot be understood by studying the trees alone.

The density of trees and understory plants in forests varies considerably from place to place.[1] When tree density is high, the amount of understory biomass is comparatively low because of competition for light, water, and nutrients; but when the number of trees is reduced by some disturbance, the amount of understory vegetation often increases dramatically.[2] Disturbances benefit herbivores that depend on shrubs and herbaceous plants for food (Pase and Hurd 1957; Conway 1982; Woods et al. 1982; Brown and De-Byle 1989; DeByle et al. 1989). All forests are characterized by a canopy cover that oscillates over periods of decades or centuries between dense and sparse, and the understory biomass pulsates accordingly. This pulsating is not uniform throughout the forest, and canopy openings and flushes of understory growth occur at different times in different places. Though appearing to be static, forests change on the time scale of months as well as centuries.

Animals can have significant effects on forest ecosystems. Deer and elk, for example, can reduce

the rate of aspen regeneration, and seed-caching birds influence the distribution of limber and whitebark pine, just as in the foothills (see chapter 9). In lodgepole pine forests, red squirrels are important seed predators and dispersal agents (fig. 11.1).[3] Pine seed is an important source of food for the squirrels, but the pine has evolved adaptations to minimize the amount of seed that the squirrels consume (Smith 1970, 1975; Elliott 1974). For example, the cones are resinous and have sharp spines. Moreover, serotinous cones, which provide an especially rich food source, are sessile and are oriented on the branches in a way that makes them less accessible to squirrels. The squirrels have to work harder to extract seed, and consequently their jaw muscles have become enlarged (compared to squirrels in Douglas-fir forest; Smith 1970, 1975; Benkman et al. 1984).

Insects have also coexisted with conifers for millions of years, and periodically, when their population density is high, they have important effects on the forest ecosystem. These effects include creating openings in the forest canopy, with the result that understory plants grow faster. The rate of nutrient cycling may also increase because the insects digest tough leaf and bud tissues, creating detritus (including their own waste products and carcasses) that is more easily decomposed on the forest floor. Some ecologists have speculated that insects and other herbivores increase the level of NPP by increasing nutrient availability or by releasing younger trees from suppression (Matt-

F I G. 11.1   Red squirrels have coexisted with
lodgepole pine for a million years or more. Their
strong jaw muscles enable them to extract nutritious
seeds from closed pine cones. Photo by Robert Dorn
and Jane Dorn.

son and Addy 1975; Romme et al. 1986; Veblen et
al. 1991a, 1991b).

All organisms are subject to parasitism. Two
parasites are especially important in Rocky Mountain forests: dwarf mistletoe and comandra blister
rust. Dwarf mistletoe is a flowering plant that is an
obligate parasite on pines (fig. 11.2). The pale
green leaves and stems are capable of some photosynthesis, but the plants extend their root system into the host tree's sapwood and inner bark
(phloem), where they obtain water, nutrients, and
carbohydrates (Hawksworth and Johnson 1989).
The resulting tree deformation, slowed tree growth,
and occasionally tree death over extensive areas
have led silviculturists to label this parasite their
single most important problem in lodgepole pine
forests. Still, mistletoe is a native plant that has
coexisted with pine for millions of years.

Because of its economic importance, dwarf
mistletoe has been the subject of considerable research (Hawksworth and Scharpf 1984; Hawksworth and Johnson 1989). Mistletoe seeds are
ejected from the fruits by hydrostatic pressure,
often traveling 9–12 m and landing on the leaves

of another tree. With their muscilaginous coating,
the seeds slide to the leaf base, where they germinate in the spring. Infection occurs when the root
successfully invades the tree's phloem and xylem,
which usually occurs on one- to three-year-old
twigs. The mistletoe plant is hardly visible for several years, and new seeds are not produced for five
to six years. Both healthy and stressed pine trees
are susceptible to invasion. Clearcutting infested
stands is the only feasible control at present, since
attempts to kill mistletoe over large areas with
herbicides have been unsuccessful, and the selective harvesting of obviously infected trees invariably leaves trees where the mistletoe is not yet visible. Hawksworth (1975) speculated that periodic
fire in presettlement times helped to keep mistletoe populations at low levels and that the current policy of suppressing most forest fires has
allowed it to become more widespread.

In the case of comandra blister rust *(Cronartium comandrae),* three organisms are involved in

F I G. 11.2   Dwarf mistletoe is a parasitic flowering
plant common on lodgepole pine and limber pine.

addition to the fungus—lodgepole pine, typically big sagebrush, and a small herbaceous plant known as comandra or bastardtoadflax. Comandra is commonly found growing in dry mountain meadows and is itself an obligate root parasite (on sagebrush roots and various other plants). The disease therefore usually develops only where sagebrush, comandra, and lodgepole pine occur near one another. This arrangement is common and the disease is widespread (Geils and Jacobi 1984, 1987, 1990; Jacobi et al. 1993). Tree death cannot usually be attributed to the rust directly, but the treetops die and often a forked trunk develops. The disease is a problem, as it causes a reduction in the amount of harvestable wood, but as with mistletoe, it is part of a community that has existed for millennia.

With regard to insects, rapid population increases lead to major disturbances that trigger or hasten secondary succession (Romme et al. 1986; Veblen et al. 1991a, 1991b). The native mountain pine beetle and western spruce budworm are most often mentioned in this regard. Insect epidemics sometimes appear cyclic in nature (Brown 1975; Geiszler et al. 1980; Holland 1986; Gara et al. 1985; Baker and Veblen 1990), with the insects reaching peaks in old-growth forest and declining after the inevitable fire that follows.

Both lodgepole pine and ponderosa pine forests are subject to outbreaks of the mountain pine beetle (fig. 11.3) (Raffa and Berryman 1982; Amman and Cole 1983; Lotan and Perry 1983; McGregor and Cole 1985; Christiansen et al. 1987; Knight 1987; Amman 1989; Bartos and Amman 1989). In most years the beetle population is low, surviving in weakened or recently fallen trees in the forest (Christensen et al. 1987). Periodically, however, the population increases rapidly and many trees are killed (fig. 11.4). Understanding the life cycle of the beetle and the causes of its population explosions is fundamental to understanding the dynamics of lodgepole and ponderosa pine forests that occur below 2,750 m elevation. Above this elevation, the winters are too severe for the beetle larvae.

In late July and August, the female beetles, about 5 mm long, emerge from the bark and fly toward other trees. Research has suggested that

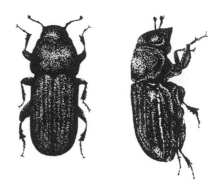

F I G. 11.3   Bark beetles, such as the mountain pine beetle, are native insects that have coexisted with pine trees for millions of years. The adults are about 5 mm long.

they are attracted most often to the larger trees (>20 cm in diameter), an attraction that seems adaptive because trees of this size would provide the beetle larvae with more food (phloem, cambium, and sapwood). In addition to visual cues, the beetles are apparently attracted by the odors emitted by weakened trees.[4]

After landing, the female beetle bores a hole through the bark and then begins to cut galleries for egg laying in the phloem and outer sapwood. If the tree is vigorous, however, resin is soon produced in such quantities that the beetle is "pitched out" before any damage to the tree occurs. Such trees are easily recognized by globs of resin (2–3 cm in diameter) adhering to the external bark surface (in which a dead beetle can often be found). Raffa and Berryman (1982) suggested that, in addition to providing a physical barrier, the resin is toxic to the beetle larvae and to pathogens introduced by the beetle. Less vigorous trees are unable to produce enough resin for defense, in which case fine sawdust is found around the holes made by the beetles and at the bases of trees. With the successful invasion of a tree, the female emits an aromatic chemical, a pheromone, that attracts other mountain pine beetles to the tree.

If successful, the female beetle lays eggs in the galleries. A fungus, known as the blue-stain fungus because of the bluish color it gives to the wood,[5] is also introduced by the beetle. The

F I G. 11.4 Numerous yellow spots of resin and saw-dust indicate that this lodgepole pine has been invaded successfully by mountain pine beetles (elevation 2,538 m; 8,350 ft). Each spot represents the entrance of one female beetle. Within a year the leaves will turn reddish-brown and the tree will be dead. Insect invasions such as this can accelerate succession and the rate of fuel accumulation. See fig. 11.5.

fungus prevents water and nutrient flow through the sapwood and thereby kills the tree within a year (Christiansen et al. 1987).

In late summer the eggs hatch and the larvae begin feeding on the inner bark. They move horizontally around the tree, cutting through the phloem, and thereby prevent the movement of carbohydrates from the leaves to the roots. Although this "girdling" surely contributes to the death of the tree, trees stressed only by girdling can live for four to five years. Clearly, the blue-stain fungi are the primary factor causing death. If the winter is not too cold, the larvae survive, and after further feeding and molting in the spring and early summer, new adults emerge to attack other live trees. The epidemic may continue for several years, spreading to adjacent forests with susceptible trees. It ends only after a cold winter kills many of the larvae or when enough susceptible host trees are no longer available—perhaps because the conditions causing host-tree stress have been ameliorated in the area. Beetle parasites and beetle predation by birds may also be factors in curtailing the outbreak, as observed by McCambridge and Knight (1972) while studying the spruce beetle.

Although beetles are always present in the forest, their populations increase explosively in some years. The cause of these population explosions is not well known but could hinge on tree vigor (Waring and Pittman 1985; Christensen et al. 1987). Pine forests older than about eighty years have often developed the maximum amount of leaf area and biomass that can be supported by the resources of a site. The trees thus come under increased competition for the resources they need. Defensive chemicals are thought to have low priority in the allocation of limited carbon compounds by trees (Waring and Schlesinger 1985) compared to the maintenance of leaf area and young roots, with the result that eventually less resin is produced. With less resin, the trees lose their ability to ward off beetle invasions. Damaged trees, whether from lightning strikes, fire, or logging activity, may also lose their ability to resist bark beetle attacks.

Tree age is just one factor that leads to bark beetle population increases. Defoliation or water stress caused by drought can also diminish the supply of energy available for resin production. Moreover, some scientists have suggested that the microclimate in closed, old-growth forests is a contributing factor in the population explosion of the beetle (Bartos and Amman 1989). According to this hypothesis, closed canopies minimize air movement, causing beetle-attracting odors to concentrate. Beetles are then attracted to such stands from a larger area, and with high beetle densities, even the more vigorous trees are susceptible to invasion (Amman 1989). Under such conditions, beetle populations increase rapidly within one year. Bartos and Amman reasoned that opening up the forest through selective logging increases air movement, with the "odor plume" being dispersed and less effective.

An additional dimension for understanding bark beetle epidemics was added by the research of Gara et al. (1985). They found that fire-damaged lodgepole pine is susceptible to infection by certain fungi, especially root pathogens (Tkacz and Schmidt 1986).[6] These damaged and infected trees then attract bark beetles, which persist in the trees until conditions are again favorable for the rapid expansion of the insects to neighboring trees. With extensive tree mortality caused by the insects, fuel accumulation is accelerated, which increases the probability of the next fire. Some trees survive the fire, however, and the cycle is repeated (fig. 11.5).

An important feature of the disturbances caused by bark beetles is that most of the mortality occurs in the larger trees. Smaller trees and saplings usually survive, and with increased light, water, and nutrient availability, their growth is often more rapid. New seedling establishment is possible also. The typically homogeneous, even-aged structure of the pine forest is thereby converted to a more heterogeneous, uneven-aged forest (fig. 11.6), and succession is accelerated (Brown 1975; Amman 1977; Romme et al. 1986). The production of herbaceous and shrubby vegetation in the forest understory also increases, providing additional forage for some animals until the maximum canopy leaf area is restored.

Productivity and Energy Flow

There are various pathways by which energy flows through ecosystems (see fig. 1.6). Only a small portion of the sun's radiant energy is used for photosynthesis (1–3 percent), with most of the energy being used to evaporate water and heat the rock, soil, and biomass on the earth's surface. The amount of energy available for photosynthesis, evaporation, and heat depends on the angle of the sun as well as warm or cold air masses directed toward an area by the jet stream. In mountains, the temperature typically decreases with increasing elevation because of heat radiation back into space at night (caused by a comparatively thin atmosphere and low air density). Because of this heat loss, mountain temperatures are usually lower and the growing season shorter. Still, water is readily available in mountains, and many plants have evolved adaptations for the short, cool growing season.

Tree growth rates are low in the Rocky Mountains compared to the Pacific Northwest and the Southeastern Coastal Plain of North America (fig. 11.7). This is apparent from the low site indices calculated for many forests in the region (Alexander 1987), with more than fifty years often required for trees to reach a height of only 15 m.[7] Pearson et al. (1987) estimated total NPP in the lodgepole pine forests of the Medicine Bow Mountains, including roots, at 2.5–3.2 Mg/ha/yr, a value that is on the low end of temperate forests worldwide.[8] Spruce-fir forests are probably similar. Forests at lower elevations or on more fertile floodplain soils, such as ponderosa pine forests in the Black Hills, undoubtedly have higher rates of production.

After live trees and soil organic matter, the largest component of the forest biomass is detritus. Composed almost entirely of dead plant material (boles, branches, and leaves), detritus gradually accumulates after fires. This occurs be-

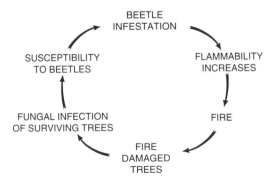

F I G. 11.5   A diagram illustrating the cyclic interaction believed to occur between bark beetles, fire, and certain fungi. Based on Gara et al. 1985.

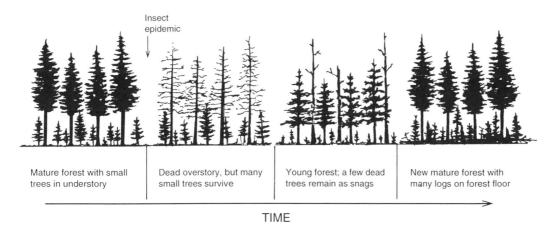

Insect epidemic

Mature forest with small trees in understory

Dead overstory, but many small trees survive

Young forest; a few dead trees remain as snags

New mature forest with many logs on forest floor

TIME

F I G. 11.6 Forest structure and succession can be altered greatly by epidemics of such insects as the Douglas-fir beetle, mountain pine beetle, spruce beetle, and western spruce budworm.

cause the Rocky Mountain environment leads to the production of plant tissues that have a high lignin content and high carbon-nitrogen ratios, both of which slow decomposition by bacteria and fungi. Decomposition is also limited by the relatively cool and sometimes dry conditions that prevail. Notably, Fahey (1983) observed that as much decomposition occurred under the snow of lodgepole pine forests during the six- to eight-month winter as during the short, cool, summer when the forest floor frequently dries out. Decomposition under snow is undoubtedly slow on a daily basis, but the soils are often unfrozen and moist.

Because detrital biomass accumulates over time, thereby increasing flammability, fire is another important pathway for energy flow in most ecosystems. Moreover, fire is a decomposition process that, like bacteria and fungi, converts organic material into the inorganic nutrients required for plant growth.

The combined biomass of detritus, bacteria, and fungi is far greater than the total biomass of birds, mammals, and insects. It is not surprising, therefore, that various animals have evolved that depend on detritus for a significant part of their energy. Small mammals commonly eat the fruiting bodies of fungi, such as truffles, mushrooms, and puffballs (Fogel and Trappe 1978; Maser et al. 1978; McIntire 1984), and birds commonly feed on these small mammals or insects and other invertebrates that derive much of their energy from detritus. Moreover, there is evidence that these animals are important in dispersing the spores of fungi (Maser et al. 1978, 1988) that are important not only for decomposition but, in the case of mycorrhizal fungi, for plant establishment and growth.

The importance of mycorrhizal fungi is suggested by the fact that up to 15 percent of the NPP in a coniferous forest can be allocated to the maintenance of these fungi on roots (Vogt et al. 1982). All studies thus far suggest that, rather than being

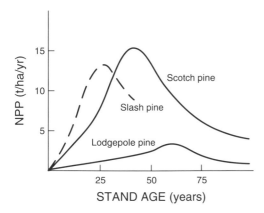

Scotch pine

Slash pine

Lodgepole pine

NPP (t/ha/yr)

STAND AGE (years)

F I G. 11.7. NPP changes significantly as a forest ages, increasing to a maximum and then declining. Note the low NPP of Wyoming lodgepole pine forests compared to slash pine in Florida and Scotch pine in northern Europe. Adapted from Knight 1991.

parasites, mycorrhizal fungi develop a mutualistic association in which both the vascular plant and fungus benefit (Allen 1991; Miller and Allen 1992). The fungi derive their energy from the plants, while the plants have increased access to water and nutrients. Vogt and co-workers estimated that the annual biomass increment in mycorrhizal fungi and fine roots accounted for about 45 percent of the annual NPP in a young coniferous forest in the Pacific Northwest and 75 percent in an older forest. They suggested that the decaying fine roots and fungi contributed more to nutrient availability than did decaying leaves, twigs, and branches above ground. Wyoming forests are probably similar, though comparable data are not available.

On average, little energy flows through animals in forest ecosystems. Yet they often influence the forest in subtle ways (such as seed planting by birds and the facilitation of nutrient cycling by insects; Mattson and Addy 1975). Moreover, the populations of herbivores and carnivores continually fluctuate, and their influences are greatly amplified when population sizes are high.[9] Ecosystems are dynamic, with all components changing over time, as does the relative amount of energy flowing through the grazing pathway.

Hydrology

Mountain landscapes are hydrologically distinct from the lowlands because, in general, they receive more water than can be evaporated during the year. In other words, the ratio between precipitation and evaporation (the *P-E* ratio) is greater than 1. Consequently, the mountains are the primary source of river water and groundwater for the whole region. Because 50–75 percent of the annual precipitation in the mountains occurs as snow, which accumulates during much of the winter, streamflow is characterized by a pronounced flood peak in the spring (fig. 11.8). Summer rains normally contribute very little water to streamflow because most rains are barely sufficient to wet the forest floor, to say nothing of thoroughly wetting the soil. Usually, and especially above the ponderosa pine and Douglas-fir zones, soil moisture reaches a peak just after snowmelt and then dries more or less steadily until fall (Knight et al.

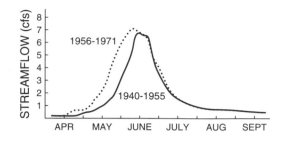

FIG. 11.8   A typical hydrograph (solid line) for a Rocky Mountain creek, showing how streamflow increases dramatically in the spring during snowmelt. More streamflow typically results following timber harvesting in small patches (dotted line). (Multiply cubic feet per second [cfs] by 101.9 to express streamflow in cubic meters per hour.) Adapted from Alexander 1987a.

1985). With the exception of wetlands and areas of prolonged snow duration, these conditions prevail in the mountains regardless of whether the vegetation is a forest, meadow, or alpine tundra. The amount of surface water draining from a Rocky Mountain landscape depends on at least six factors: the water content of the snowpack; the nature of the vegetation; the water-holding capacity of the soil and sapwood; climatic characteristics; the proportion of water that percolates into groundwater; and the patchiness of the landscape mosaic, including the potential for snowdrifting.

WATER EQUIVALENT OF THE SNOWPACK

The amount of water in snow cannot be predicted from snow depth alone, because snow can be comparatively wet or dry. For example, a snowpack that is 1 m deep may have the equivalent of 18–40 cm of water.[10] Snowdrifting is typical in Wyoming mountains, and thus there can be considerable spatial variability in depth as well as water content. Because of the importance of snow as a source of water for irrigation and downstream reservoirs, the Soil Conservation Service and Bureau of Reclamation routinely monitor the water content of snow in the mountains. By combining data on snow-water content with other information about the watershed, these agencies are able to predict the amount of runoff that will be available downstream.

VEGETATION

Two characteristics of Rocky Mountain vegetation that strongly affect runoff are the total amount of leaf area and whether the leaves are evergreen or deciduous. The effect of vegetation on water outflow is least when the leaf area index is low and the leaves are deciduous, such as in a meadow. Thus, probably 80–90 percent of the snow water accumulating in meadows is contributed to streamflow or groundwater. Comparatively little evaporates from the soil or transpires from the plants.

In contrast, forests develop large amounts of leaf area that can be evergreen or deciduous. Figure 11.9 illustrates the hydrologic budget of a typical lodgepole pine forest in the Medicine Bow Mountains, where about 35 percent of the annual precipitation is evapotranspired during a typical year and 65 percent flows to the stream or groundwater. Evaporation and transpiration each account for about half of the evapotranspiration. Evaporation occurs primarily as the evaporation of rainwater intercepted by the forest canopy and

the forest floor (or litter)—a pathway known as interception. Interception varies according to (1) the amount of plant surface area exposed above ground and (2) whether rainfall occurs as a slow drizzle or quick thunderstorm (Knight et al. 1985; Fahey et al. 1988).

Predictably, leaf area and other stand characteristics vary considerably from one forest to another. Knight et al. (1985) studied eight contrasting stands of lodgepole pine forest over a three-year period. Actual evapotranspiration for the period from early spring to late fall (when probably 95 percent or more of the annual evapotranspiration occurs) ranged from 21 to 53 cm, which was 33–95 percent (mean = 75 percent) of the total annual precipitation. For all stands and years, transpiration accounted for 50–61 percent of evapotranspiration, 9–44 percent of which occurred during the spring drainage period while snow still covered the ground (vernal transpiration). Estimated vernal transpiration and outflow varied considerably among the stands, with vernal transpiration accounting for 4–20 percent of the

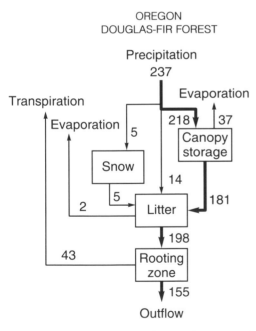

OREGON
DOUGLAS-FIR FOREST

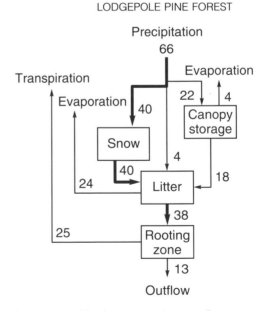

WYOMING
LODGEPOLE PINE FOREST

F I G. 11.9   Annual hydrologic budgets for stands of rain-dominated Douglas-fir in Oregon and drier, snow-dominated lodgepole pine in Wyoming. The boxes represent major storage compartments; the arrows are processes affecting water movement through

the ecosystem. Numbers are centimeters of water (e.g., 25 cm is the volume of water required to cover an area 25 cm deep). The figure for Douglas-fir is adapted from Sollins et al. (1980); data for lodgepole pine are from Knight et al. (1985).

snow water. The outflow beyond the rooting zone, that is, to the groundwater or streamflow, occurred only during snowmelt and accounted for 0–80 percent of the snow water, depending on climatic conditions during the year and on such stand characteristics as the leaf area index and soil storage capacity. Zero outflow was predicted for years of relatively low snow-water equivalents in stands with large amounts of evergreen leaf area capable of transpiration in the early spring.

Aspen forests present a different situation because the leaves are deciduous and absent in most of the snowmelt period, thereby precluding most vernal transpiration. Moreover, the leaves are shed in the fall, thereby precluding the possibility of transpiration during subsequent warm days. Some land managers have noticed that the flow of mountain streams increases after aspen leaf fall. The net effect of forest domination by deciduous trees is that less water is transpired and a larger portion of the water inputs is available for outflow. Aspen forests often change to evergreen coniferous forests as succession proceeds; streamflow may decrease as this shift occurs (Jaynes 1978; Harper et al. 1980; Knight et al. 1985).

Significantly, there are differences between the hydrology of lodgepole pine forests and spruce-fir forests (Lopushinsky 1969; Kaufmann 1985; Kaufmann et al. 1987). Kaufmann found that spruce-fir forests in the Rocky Mountains have considerably more leaf area than do lodgepole pine forests, and furthermore, that spruce and fir use water at higher rates than either pine or aspen. Consequently, streamflow should be less when spruce-fir forests are more common on a watershed.

Leaf area seems to be in balance with soil water availability in drought-prone forests, increasing until the amount of soil water cannot support additional leaves (Grier and Running 1977). Knight et al. (1985) studied two adjacent stands of lodgepole pine forest, both dating back to the same fire. One was a dense dog-hair stand with 15,000 trees per hectare, and the other was a more open stand with 2,000 trees per hectare. Despite large differences in tree density and total biomass, both stands had the same leaf area and about the same transpiration rates.

Of course, reductions in leaf area due to tree cutting, insect epidemics, fire, and windstorms will reduce evapotranspiration and increase streamflow (Gary and Troendle 1982; Troendle 1983, 1987; Knight et al. 1985; Alexander 1987; Troendle and Kaufmann 1987). Streamflow may continue at an elevated level for twenty-five years or more, declining to its predisturbance volume only after the original amount of leaf area has been restored.

WATER-HOLDING CAPACITY OF THE SOIL AND SAPWOOD

Another variable affecting the amount of surface runoff is the amount of storage available for additional water resulting from spring snowmelt or subsequent precipitation. Storage capacity depends on the amount of water already in the soil, soil depth, and soil texture, with deeper and finer textured soils holding considerably more water than shallow, coarse-textured soils. Plants create storage through water uptake and transpiration, which lead to soil drying. The amount of available storage generally increases during the summer, a function of the drying rate, the time since the last significant wetting event, rooting depth, soil depth, and texture. Soil moisture under lodgepole pine forests may be depleted at depths of 2 m or more, as that species has taproots reaching that depth (although 80 percent or more of the root biomass is in the top 20 cm; Pearson et al. 1984).

Another water storage compartment in forests is the sapwood. Normally thought of as a conduit for water transport from the roots to leaves, sapwood can store a large amount of water at times when transpiration is not occurring—for example, at night and in winter. The amount of water that can be stored in this way is a function of tree size and density, and it may be sufficient to sustain transpiration and photosynthesis for several days (Waring and Running 1978).

CLIMATE

The timing and amount of drainage (runoff) from mountain watersheds are also affected by climatic characteristics. Gradual warming causes a prolonged snowmelt, which provides additional time for vernal transpiration, thereby reducing the amount of drainage below the rooting zone. In contrast, some springs are cold until quite late, becoming warm when day lengths are long. Under such conditions, snow melts rapidly and there is little time for vernal transpiration. Moreover, the rate of snowmelt occasionally exceeds the maximum infiltration rate, with the result being overland flow. Overland flow, however, is not common unless the soil is already saturated or frozen. Water flowing overland is not available to the trees in the stand, and the potential for erosion increases greatly.

Fall climatic characteristics may be as important as those in the spring. For example, heavy fall rains could saturate the soil just before the onset of winter, when evapotranspiration occurs at slow rates (if at all). Little storage capacity then exists during the following spring, leading to the potential for greater spring runoff. Another climatic factor that leads to higher spring runoff is extremely cold autumn temperatures prior to snowfall, causing the surface soil to freeze. If frozen to a sufficient depth, the soil may not thaw until the following spring. The result is much less infiltration and more overland flow. Usually, mountain soils under considerable snowpack are unfrozen for most of the winter, but climatic conditions that cause freezing would increase the amount of surface runoff.

Climatic factors affecting runoff change abruptly across landscapes as well as from year to year. For example, north slopes in the Northern Hemisphere are more likely to accumulate snow because of less direct sunlight. In contrast, the snow on south-facing slopes typically melts between snowstorms, with a considerable amount of the soil moisture evaporating before additional snowfall occurs. Such periodic melting makes the snowfall events analogous to a summer shower, with each event producing too little moisture to cause substantial runoff. Thus, north slopes contribute more to runoff per unit area than do south slopes.

Similarly, higher elevations contribute more to runoff than lower slopes do because more snowfall commonly occurs there and the colder, high-elevation temperatures prevent substantial thawing before spring, even on south-facing slopes. Rocky Mountain streams sometimes have two flood peaks, the first caused by rapid snowmelt in the lower montane zone and the second by subalpine snowmelt. As might be expected, comparatively low mountain ranges, such as the Black Hills, have an earlier runoff than do higher mountains.

PERCOLATION TO GROUNDWATER

Surface runoff in the form of streamflow is greatly affected by the geologic substrate and whether percolation into aquifers is possible. Granites and other igneous and metamorphic rocks are usually impervious, which would lead to most of the runoff water leaving the watershed as streamflow. In contrast, watersheds underlain by sedimentary rock may have a substantial portion of the water draining into aquifers, where it contributes to the groundwater of intermountain basins. Geologic substrates, which vary markedly among watersheds and mountain ranges, clearly influence both the amount of water leaving an area as streamflow and the amount percolating to groundwater aquifers.

PATCHINESS OF THE LANDSCAPE MOSAIC

Heterogeneity or patchiness in landscapes is caused by disturbances and abrupt changes in various environmental factors (such as changes in topography, soil characteristics, and geologic substrate). Abrupt topographic change creates differences in the timing and periodicity of snowmelt and runoff, such as on north and south slopes. Patchiness in vegetation can also have significant effects. For example, a landscape where forests are frequently interrupted by small meadows probably contributes more water to streamflow than an area of continuous forest, because wind causes snow accumulation in small openings (Gary and Troendle 1982). Furthermore, snowmelt is more rapid in such openings, and the potential for transpiration is much lower. Consequently, more snow water flows beyond the rooting zone than it does from areas uniformly covered by forest. Watershed managers now recommend numerous small patch cuts to increase the water yield from a watershed. Alexander (1987), in a review of research for the Central Rocky Mountains, found that the largest increases in water yield (5 cm or more) were reached when 30–40 percent of a watershed was harvested in patches no larger than 1.2–2 ha. Large openings can have quite a different effect because more of the snow is blown into the forest, where the snowmelt is slower and there is a higher potential for evapotranspiration. Moreover, wind-scoured clearcuts often have inadequate snow to protect young trees from winter desiccation, thereby slowing the rate of tree establishment.

In general, abrupt changes between forests of different ages, or between forests and nonforests, have significant effects on snow distribution and melting. To predict the outflow from watersheds, the effect of patchiness on soils, topography, and vegetation must be considered in addition to the snow-water equivalent, climatic characteristics, soil storage capacity, characteristics of the vegetation, and geologic substrate. With the exception of geologic substrate, all these factors are subject to rapid change, whether from succession or disturbances. As with energy, various pathways for water flow are possible, and there is a continuously varying array of complex interactions that must be understood in order to make accurate predictions.

Nutrient Cycling

Unlike water and energy, nutrients can cycle within an area for indefinite periods. In other words, the same nutrient ion may be used several times in the plant, animal, and microbial biomass of a specific ecosystem before being lost by leaching, erosion, animal emigration, or volatilization. Nutrient cycling in Rocky Mountain coniferous forests is now fairly well understood, especially for nitrogen, a nutrient that limits plant growth in some mountain ecosystems (along with cool, short growing seasons and occasional late-summer droughts).

The potential for nutrient leaching is of special interest in mountain landscapes because so much of the water washes through the soils during the three- to six-week snowmelt period. How are nutrients retained within the ecosystem in the face of this "spring flush," especially considering that a large portion of the annual decomposition of detritus occurs during the winter and just before the

spring thaw? Do annual inputs from the atmosphere and rock weathering balance annual losses? What are the effects of disturbances and successional changes on the accumulation or losses of limiting nutrients? How does the chemical composition of streamwater depend on biotic processes in the forests and meadows of the watershed? Such questions have stimulated nutrient cycling research in Wyoming and elsewhere. All the intricacies of nutrient cycling in montane ecosystems cannot be examined here, but two topics, the mechanisms of nutrient conservation and the nitrogen cycle, are of special interest.

Inputs or additions of mineral nutrients to an ecosystem occur as the result of dryfall (dust), wetfall (rain and snow), rock weathering, animal immigration, surface or subsurface run-on, and in the case of nitrogen, the fixation (conversion) of atmospheric nitrogen by microbes into a form that can be used by plants (ammonium or nitrate). Losses occur through various processes, including erosion, leaching, animal emigration, and for nitrogen, the microbial process of denitrification. Although measuring all the inputs and outputs is a substantial challenge that has been only partially met, research in lodgepole pine forests suggests that nitrogen, and sometimes potassium and phosphorus, accumulate in undisturbed lodgepole pine forests even during years of heavy snowpack and large volumes of water outflow (Fahey et al. 1985; Knight et al. 1985; Fahey and Knight 1986). Nitrogen accumulation is probably due to uptake by both vascular plants and soil microorganisms. Fahey (1983) found that the nitrogen concentration of decaying wood increased by 80 percent, probably because of nitrogen uptake by fungi and bacteria, before beginning a slow decline (fig. 11.10). The microbial immobilization of phosphorus and calcium was also observed. Uptake by vascular plants is especially effective for nutrient immobilization when the biomass is accumulating and additional nutrients are being incorporated into new tissues (Gorham et al. 1979).

About 90 percent of the forest's nitrogen is in soil organic matter, with only 6 percent in aboveground detritus and 4 percent in living biomass (fig. 11.11). Annual nitrogen inputs from fixation and precipitation appear to be low in Rocky

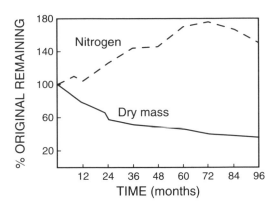

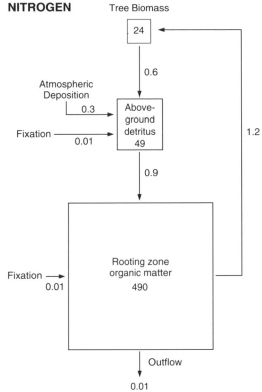

F I G. 11.10 Decomposing wood in lodgepole pine forests increases in nitrogen content before it begins to decline, slowly releasing this limiting element for the growth of other plants. Adapted from Fahey 1983.

Mountain coniferous forests, as are decomposition rates. Fixation rates may be higher in aspen forests (McNulty 1969; Mead 1982). Fahey et al. (1985) hypothesized that some of the nitrogen available for microbial growth in the detritus was translocated through fungal hyphae from the mineral soil to forest floor detritus. They also suggested, as have others (Harmon et al. 1986), that nutrient-enriched decaying logs may be an important source of nitrogen for sustained site productivity. Dead wood resulting from the last disturbance, usually a fire, is a major nutrient storage compartment. Silvicultural practices that remove excessive amounts of wood may be detrimental to long-term site productivity.

Once within the coniferous forest biomass, limiting elements are retained tenaciously (Gosz 1980; Fahey 1983; Knight 1991). Prescott et al. (1989a, 1989b) studied montane coniferous forests in Alberta—which are also dominated by lodgepole pine, Engelmann spruce, and subalpine fir—and found that 40–50 percent of the nitrogen and 50–80 percent of the phosphorus were reabsorbed from senescing tree foliage. More reabsorption occurred in pine than in spruce and fir. As a result of reabsorption, leaves falling to the ground are deficient in nutrients that may be limiting to decomposers. In general, the nitrogen concentration of detritus in coniferous forests is very low (<1 percent), a factor that undoubtedly contributes to the slow decomposition rates that

F I G. 11.11 Nitrogen distribution in a lodgepole pine forest. The arrows indicate annual nitrogen flows between ecosystem components (boxes); the numbers are g/m²/yr and g/m², respectively. Much of the nitrogen in the rooting zone compartment is in organic matter that apparently is not readily decomposable. Note (1) that nitrogen inputs to the ecosystem are larger than nitrogen losses, suggesting that nitrogen is accumulating, probably because it is a limiting factor for plants and microbial organisms; and (2) that the tree uptake estimate is larger than the sum of the input estimates to the rooting zone, which suggests that the soil nitrogen pool is gradually being depleted in this forest as biomass accumulates. Replenishment of the soil nitrogen pool may occur as the forest ages further (Fahey and Knight 1986) or following disturbances such as fire. Based on data in Fahey et al. 1985.

have been reported. Depending on site conditions, complete mineralization of lodgepole pine leaves requires twelve to twenty-two years (Fahey 1983). Boles, branches, and woody roots decompose slowly as well (Yavitt and Fahey 1982; Fahey 1983), with a 30-cm tree bole requiring about one hundred years for complete mineralization.

F I G. 11.12   One effect of forest fires is to mineralize small fuels such as leaves, twigs, and small branches. However, the organic matter and nutrients of roots, tree boles, and larger branches are eventually added to the soil. This photo was taken one year after a crown fire in the Medicine Bow Mountains; elevation 2,918 m (9,600 ft).

F I G. 11.13   In contrast to fire, timber harvesting removes the larger wood. The leaves, twigs, and small branches are typically left behind as slash, but the amount of organic matter added to the soil is less. Medicine Bow Mountains; elevation 2,888 m (9,500 ft).

The Effects of Disturbances on Ecosystem Processes

FIRE AND INSECTS

The effect of fire on ecosystem processes depends heavily on the intensity of the heat produced and the amount of vegetation burned (Hungerford et al. 1991). Surface fires are typical in foothills (as in ponderosa pine or Douglas-fir forests), usually burning only detritus, young trees, and some of the understory vegetation. Most understory plants respond quickly by sprouting, depending on rooting depth and fire intensity (ibid.). Nutrients contained within the detritus are transformed to ash, but the leaf area is reduced little if at all. The net effect of such fires is to enhance nutrient availability temporarily for the ecosystem and reduce the amount of fine fuels (branches and leaves) (Stark 1977);[11] little erosion occurs following the fires (ibid.). Trees that are damaged or fire-scarred may become susceptible to certain fungi and insects, but the immediate effects on water and nutrient fluxes are subtle.

In contrast, crown fires kill most of the plant biomass in coniferous forests and greatly reduce the amount of leaf area (figs. 11.12, 11.13). As a result, evapotranspiration is greatly reduced, and a much larger portion of the annual precipitation leaves the watershed as streamflow. Flood peaks could be higher than before the fire. With more water moving through the soil profile when nutrient uptake (immobilization) is low, there is a high probability of increased nutrient losses from the soil—and subsequent nutrient gains to streams

(Knight et al. 1991). Further, mineral soil is exposed when the forest floor is burned, thereby increasing the potential for erosion (Hungerford et al. 1991). How much erosion occurs depends on various factors, including slope, fire intensity, the amount of forest floor burned, the rate at which understory plant cover develops following the fire, and whether the burn was patchy or uniform over the landscape (Knight and Wallace 1989).

With regard to nitrogen specifically, some is volatilized by fire. This loss, however, is probably minor compared to losses from leaching. Nitrification typically accelerates after disturbances in coniferous forests, converting the relatively immobile ammonium cation to the highly leachable nitrate anion (Vitousek and Melillo 1979; Hart et al. 1981; Fisher and Gosz 1986; Knight et al. 1991; Parsons et al. 1994). As a result, the nitrogen pool in the soil may decline (Fahey et al. 1985; Fahey and Knight 1986). Notably, the loss of nitrogen after fire is probably not a problem, because the nitrogen requirements of early postfire vegetation are small and the lost nitrogen is replaced by continued atmospheric deposition, fixation, and the decomposition of soil organic matter—the largest pool of nitrogen in the forest ecosystem (see fig. 11.11). The greatest losses may occur on steep slopes where significant erosion is possible. As new biomass accumulates during succession, nitrogen again becomes limiting and is conserved within the ecosystem.

The effects of crown fires on streamwater quantity and quality will vary greatly from one watershed to another. Streams that drain watersheds where large areas have burned and that have a comparatively small portion of the watershed in riparian zones capable of filtering sediments and nutrients will usually be affected most dramatically (fig. 11.14). The patchiness of the burn may be important also.

Animal abundance and distribution patterns are also affected by fires. Abundant dead and dying trees create a habitat that is very different from the unburned forest. Herbaceous or shrubby vegetation accounts for most of the photosynthesis in areas that have been subjected to crown fires, with the effect that forage (including berries and other fleshy fruits) is more common. The habitat is

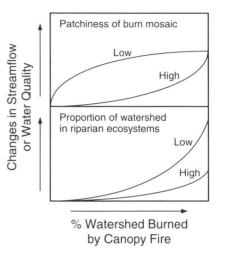

F I G. 11.14 Hypothetical relationships between changes in streamwater quality or quantity and the percentage of a watershed burned by canopy fire, the proportion of the watershed in riparian vegetation, and the patchiness of the burn mosaic. Adapted from Knight and Wallace 1989.

more open (such as between burned and unburned forest), and an additional edge favorable for some wildlife is created. Prey animals (such as small rodents) are more visible, and not surprisingly, raptors and some insectivorous birds are attracted to the burned forest (Taylor and Barmore 1980). Fires in extensive lodgepole pine and spruce-fir forests typically burn in a pulsating manner, creating a mosaic of hot burns, light burns, and unburned forests.

Burned patches are typically fringed by a "halo" of trees with brown needles (fig. 11.15), the result of singeing where fire intensity has been inadequate for complete combustion. Some of these trees die because the amount of singed needles is too high or because too much of the cambium is burned (Ryan and Reinhardt 1988). Others survive, often with fire scars. Such scarred trees can serve as epicenters for outbreaks of such insects as mountain pine beetles (Geiszler et al. 1980; Gara et al. 1985). Insect outbreaks seem to contribute to the development of a fuel complex that makes fires more probable in future drought

F I G. 11.15   Aerial view of a burned patch over Yellowstone National Park showing the typical fringe of trees that have been damaged by heat. Some of the trees in the fringe will survive, but they may be fire scarred and more susceptible to pathogens and insects.

years (Knight 1987). In this way, disturbances of one kind have an influence on the spread of disturbances of other kinds—a primary theme for landscape ecologists (Turner 1987).

Insect outbreaks in the singed tree halos, however, may be kept in check by insectivorous birds. Woodpeckers are known to be more common in burned areas (Taylor and Barmore 1980), and they undoubtedly forage also among the weakened trees along edges where some insects might be more common. Large mammals may also adjust their feeding patterns. Just as bison have been observed to track grassland fires in search of more nutritious forage, insectivorous birds, deer, and elk may track forest fires because of more abundant herbaceous plants (Knight and Wallace 1989; Singer et al. 1989).

Irregular burning could have significant effects on nutrient transfer by animals and the productivity of different areas. For example, elk spend more time near forest edges and commonly select habitats with both forage and cover nearby. New edges created by fire might be more attractive than old edges because of the vigorous growth of herbaceous plants. Nutrient availability for the plants in such sites is enhanced by reduced competition from trees and by nutrient deposition in urine and feces. Herbivory is known to affect the rate and direction of nutrient movement (Merrill 1978; Schimel et al. 1986; Ruess and McNaughton

1988). Furthermore, forage produced following fires may be more nutritious because of increased nutrient availability and a higher ratio of succulent green leaves to less palatable stems and standing dead biomass (Wood 1988; McNaughton 1979).

Insects may also change forest structure and the rate of plant growth. For example, as already discussed, some ecologists think that older stands with low NPP become more susceptible to the mountain pine beetle or western spruce budworm, insects that create canopy gaps where understory plants grow more rapidly, thereby elevating the rate of plant growth. Romme et al. (1986) observed that annual wood production per hectare usually returned to preoutbreak levels or higher within ten to fifteen years. Because of the rapid recovery of wood production, they suggested that the effects of the beetles could be considered generally neutral or even beneficial in some situations.[12]

Could bark beetle outbreaks increase forest flammability? Although flammability may increase in the first year or two after an infestation because of dead leaves still attached to the trees, the risk of destructive fire in years two to twenty may be lower because (1) the leaf and twig biomass declines, reducing fuel continuity, and (2) the addition of fine fuels to the forest floor does not increase the fine fuels significantly in forests old enough to be susceptible to a beetle outbreak. Accelerated growth in understory trees may increase fuel continuity and the risk of fire after about twenty years.

TIMBER HARVESTING COMPARED WITH FIRES

Increased water and nutrient outflow and changes in animal abundance and distribution can result after timber harvesting just as they do after a fire. As might be expected, the magnitude of the effects depends on whether the harvesting was done by clearcutting or selective thinning. Clearcutting has been observed to increase water and nutrient outflow (Vitousek and Melillo 1979; Stottlemyer 1987; Troendle 1987; Troendle and Nilles 1987; Prescott et al. 1989a; Knight et al. 1991), apparently for the same reasons that losses are thought to increase after crown fires: higher volumes of water outflow and less biotic demand for the nutrients.

Both clearcutting and crown fires change the forest ecosystem considerably, and it is tempting to equate the two disturbances. Still, there are important differences (see figs. 11.12, 11.13). During clearcutting, most of the bole wood is removed, and the slash (branches and leaves) is left behind. Indeed, the forest floor usually increases in biomass because of the slash. Little mineral soil is exposed. Erosion is slow, except on steep slopes or along roads because of the high infiltration rates that are possible with the abundant organic matter covering the mineral soil. There is also little shade available after a clearcut.

In contrast, following a crown fire, most of the boles remain. Harvesting essentially removes bole wood,[13] whereas burning removes mostly leaves and branches. Burning also consumes much of the forest floor, thereby exposing the mineral soil and facilitating seedling establishment (especially for lodgepole pine and Engelmann spruce). Furthermore, the amount of shade cast by dead standing trees may be considerable, simulating a shelterwood harvest rather than a clearcut. After a fire, many of the nutrients formerly tied up in the biomass (living or dead) exist as ash, which is easily dissolved in water or blown about by wind. More nutrient loss can occur after a fire than after a clearcut because the ash and exposed mineral soil are more susceptible to erosion, though increased leaching is probable after both disturbances, and harvesting does remove some nutrients.

Erosion may be greater following some fires, but large amounts of wood are left behind that eventually become incorporated into the soil (Harmon et al. 1986; Maser 1988; Harmon and Hua 1991). The incorporation of wood into the soil is thought to be important in sustaining the long-term productivity of forest ecosystems. The forest soils that exist today are partially the result of organic-matter additions from wood and leaves over a period of ten thousand years or more. A continuing supply of wood to the soil is as important as supplying wood to lumberyards, but the amount of wood and other detritus required for soil sustainability has yet to be determined.

It is disconcerting that forest fires—a natural phenomenon—might sometimes lead to more erosion and nutrient loss than clearcutting, which is often viewed as an artificial, human-caused disturbance. Nevertheless, erosion and nutrient loss must be placed in the context of present conditions. Erosion from mountain forests before road construction and certain other human activities was probably less common than it is today. Furthermore, the addition of some sediments and nutrients to streams could have sustained a higher level of aquatic productivity and possibly higher biological diversity. Various studies have hypothesized the beneficial effects of erosion and nutrient additions to aquatic ecosystems after fires, even to the point of suggesting that pulses in aquatic productivity might be tied to periodic fires that burn a substantial portion of the watershed (Romme and Knight 1982; Minshall et al. 1989). Such a relationship may still exist in large wilderness areas.

Fires may be acceptable in wilderness areas, but can the same be said of those parts of our national forests where wood production is an important goal? Typically, forest managers recommend the suppression of fires in such areas. As a result, fuels have accumulated, aspen forests have changed to coniferous forests, dwarf mistletoe may have become more abundant and widespread, and epidemics of some insects may have become more frequent. These trends all lead to conditions where fires are inevitable and cannot be suppressed.

Timber harvesting in one form or another has been proposed to minimize the problems created by fire suppression, but usually that solution requires costly roads that can increase erosion if not carefully designed. Even when harvesting is done with great care, some feel that recreational and wildlife values are diminished in the process. Such concerns are often valid and are an enigma for the many forest managers who have the same concerns. Furthermore, inevitable fires can cause some of the same losses. Although prescribed burning is often viewed as a compromise and might be feasible, it is also subject to debate, because many see fires as something to avoid entirely (perhaps because of towns or summer homes built within the forest). Because both prescribed fires and timber harvesting are controversial and because stopping disturbances in forests creates other, equally significant problems, forest managers face a trying predicament.

Perhaps one of the most acute examples of this problem is in ponderosa pine forests managed as roadless areas, as, for example, on Casper Mountain and near Laramie Peak. In such areas, fires have been suppressed, and timber harvesting has often not been permitted. Consequently, former savannas have been converted to more dense forests, and as the trees age or become more water- and nutrient-stressed because of increased competition, they become susceptible to outbreaks of mountain pine beetle. Moreover, fuels have increased to the point that prescribed fires are now difficult to control and expensive (a result of the need to contract additional expertise). Timber harvesting to reduce the fuel complex before prescribed burning is also expensive—and inappropriate in a wilderness area. The only solution may be a large group of laborers (possibly volunteers) willing to reduce by hand the fuel complex long distances from roads, probably by piling and burning excess fuels. If done properly, easy-to-control surface fires—the kind that must have characterized such forests in presettlement times —might be prescribed to maintain the system in a more natural and aesthetically pleasing state.

Selective harvesting is often a feasible option for forest management in the Rocky Mountain region. Clearcutting is not the only alternative (Alexander 1986b, 1987a; Smith 1987a, 1987b). Selective cutting leaves many of the trees in place (fig. 11.16) and is comparable to the thinning that is sometimes caused by certain insects (such as the mountain pine beetle). Moderate levels of evapotranspiration are still possible, less water is available for nutrient leaching, and the root systems of the remaining trees continue to take up nutrients and minimize erosion. One study in a lodgepole pine forest found that killing 60 percent of the trees reduced leaf area by 43 percent and led to a 92 percent increase in water outflow, compared to clearcutting, which increased the water outflow by 277 percent (Knight et al. 1991). Yet there was no significant effect on nitrate and total nitrogen outflow with the selective cutting. In contrast, clearcutting increased nitrate concentrations in the soil solution by ten to forty times, and total nitrogen outflow by six times. The marked differences in

FIG. 11.16   Selective harvesting, or thinning, preserves much of the forest structure in this lodgepole pine forest in the Medicine Bow Mountains (elevation 2,949 m [9,700 ft]). Small lodgepole pine and subalpine fir are present in the forest understory.

nitrogen outflow from the thinned and clearcut stands suggest that surviving trees in thinned stands are important for nutrient immobilization (Parsons et al. 1994).

Regardless of whether the forest is subjected to partial or total tree death—whether by burning, cutting, or insect epidemics—the nutrients lost are probably a small portion of the total nutrient capital in most Rocky Mountain ecosystems. The leaching of nutrients is elevated for perhaps ten years or even more, but the disturbances do not retard the continuation of nutrient input processes. Most calculations suggest that the lost nutrients are replaced within a few decades at the most (Knight et al. 1991). Soil erosion is probably a more serious concern. Still, with proper precautions it should be possible to ensure the continuation of soil development in the Rocky Mountains. Though having significant impact at the time they occur, both crown fires and clearcuts are infrequent events, unlike the annual cultivation of agricultural lands.

ECOSYSTEM RECOVERY

Succession is slow in Rocky Mountain coniferous forests because of the short, cool growing season. Understory plants, however, often grow vigorously within a year or two of a disturbance, accumulating nutrients in their biomass, reducing the potential for erosion, and restoring some of the leaf area. Sometimes understory recovery is slower than might be expected because the understory plant cover was sparse in the forest before the disturbance. Intense soil heating may also retard recovery because of the death of more roots with the potential for sprouting, but this effect is most likely to occur in small patches where fuel accumulation is high. Tree seedling establishment is usually conspicuous within a few years.

If lodgepole pine becomes established after a major disturbance, maximum total biomass accumulation rates of 2.5–3.2 metric tons/ha/yr are reached forty to sixty years later (see fig. 11.7; Pearson et al. 1987). Biomass accumulation occurs primarily in living vegetation throughout stand development, except for brief episodes of dead wood increases associated with the mortality of large trees. It is difficult to say when biomass accumulation rates approach zero (that is, when NPP will be balanced by heterotrophic respiration), and indeed the stand may burn again before this happens. Given sufficient time, fuels will develop to the point where fires are inevitable, whether the ignition source is lightning or humans.

With regard to nutrient accumulation, Pearson et al. (1987) found that the forest floor (including dead wood) immobilizes the largest amounts of nitrogen, phosphorus, calcium, and magnesium, at least during the first forty to eighty years of stand development. Living biomass appears to be the second most important factor in nutrient immobilization, especially after sixty to eighty years. Nutrient increment rates estimated by Pearson and co-workers remained positive even in very old stands, suggesting that episodes of nutrient loss are associated with large-scale disturbances rather than old age. Nutrients lost after burning and timber harvesting are replaced during forest recovery (Knight et al. 1991).

The nature of old-growth forests varies with elevation and species composition. Old-growth ponderosa pine and Douglas-fir forests, with periodic surface fires, would be quite open or savanna-like. Occasional snags, and dead trees on the forest floor, would be expected, and several age classes of trees would be represented. The younger cohorts would probably be found in patches. In contrast, old-growth forests dominated by lodgepole pine, spruce, fir, or aspen would be more dense, there would be more snags, and the detritus would include many fallen trees in various stages of decomposition. The species diversity of both kinds of old-growth forest might be high but would depend on the land area occupied by the old-growth ecosystems. Patches of old-growth that are small and isolated by expansive tracts of adjacent younger forests, whether created by fires or timber harvesting, may not be biologically unique. The fragmentation of extensive forests, by whatever means, may be a more severe disturbance for some organisms and processes than the loss of a specific forest stand (Franklin and Forman 1987). Determining the extent of fragmentation that can be tolerated without undesirable effects is a major challenge for both forest managers and scientists.

Whatever the forest type, all disturbances reduce leaf area to some extent, thereby enabling more sunlight penetration to the understory vegetation and probably reducing the amount of water withdrawn from the soil by transpiration. Streamflow often increases (Eaton 1971; Hart and Lomas 1979; Megahan 1983; Alexander 1987a; Troendle 1987; Troendle and King 1987). Furthermore, the demand for limiting nutrients by plants and microbes may be reduced, which enables more nutrients to flow into streamwater. Such changes are temporary, however, as ecosystem development over the ensuing decades leads to the restoration of predisturbance rates of transpiration and nutrient immobilization. Succession following disturbances may be viewed as a recovery process, but forest managers must confront various challenges that are not easily resolved by either intensive management or "preservation." One of the most interesting challenges is to determine which management practices are compatible with the emerging national goals of sustainable land use and the preservation of biological diversity—and which are most appropriate for a Rocky Mountain state with a short, cool growing season and rapidly growing human populations in the nearby metropolitan areas of Denver and Salt Lake City (see chapter 17).

CHAPTER 12

Mountain Meadows and Snowglades

Sometimes referred to as parks, mountain meadows lend aesthetically pleasing diversity to the landscape (fig. 12.1). At lower elevations, the meadows are similar to foothill grasslands (see chapter 9); bluebunch wheatgrass is a characteristic species. At higher elevations, Idaho fescue and tufted hairgrass become common, along with numerous other sedges and forbs. Occasionally, forbs are dominant (for example, on the moist slopes of the Teton Range). Wet meadows are especially common along streams and in depressions, where melting snow provides a reliable source of water throughout the summer (Ellison 1954; Starr 1974; Windell et al. 1986). Mountain silver sagebrush, shrubby cinquefoil, and various willows are common shrubs in such areas. As in forests, species composition and productivity vary with elevation, moisture availability, soil depth, and geologic substrate (table 12.1; Hurd 1961; Johnson 1962; Smith and Johnson 1965; Wilson 1969; Pond and Smith 1971; Mutel 1973; Weaver 1974, 1979; Potkin 1991).

A frequently asked question about mountain meadows is why they have no trees. The answer depends on the place. Meadows often occur on fine-textured soils, such as those that develop on shale, or where alluvium and colluvium accumulate in drainageways or near the bases of slopes (fig. 12.2; Dunnewald 1930; Daubenmire 1943a; Jackson 1957; Patten 1963; Wilson 1969; Despain

1973; Peet 1988). The absence of trees has been attributed to the soils remaining too wet for too much of the summer, which is thought to be unfavorable for conifer seedlings. It is also possible that the fine-textured soils of meadows are too dry for trees, primarily because infiltration is slower. Trees can be found on fine-textured soils, but usually on north slopes where more snow accumulates, where there is less direct solar radiation, and where water stress is slower to develop (Jackson 1957; Despain 1973). Another factor thought to contribute to the absence of trees on fine-textured soils is too much competition for tree seedlings from established herbaceous plants (Daubenmire 1943b). Cold-air drainage or frost pockets may also restrict tree establishment.

Elsewhere, meadows are found where the soils are too shallow and too dry for tree growth (Jackson 1957), such as on the warmer south and west slopes or on ridge tops, knolls, and high plateaus, where winter winds remove the snow. Transitions between the meadows and forests in upland settings are sometimes the most difficult to explain, as edaphic differences are typically not obvious (Behan 1957; Cary 1966; Miles and Singleton 1975). Some evidence suggests that the forests occur on soils with higher phosphorus concentrations (Behan 1957; Cary 1966), but other investigators have reported the opposite (Dunnewald 1930; Patten 1963). Nutrient deficiencies and eas-

F I G. 12.1   A mountain meadow or "park" in the Wyoming Range west of Big Piney. Mountain big sagebrush and Idaho fescue are common, with short willow shrublands in the riparian zone. Arrowleaf bal-samroot and lupine are abundant in the foreground. The forests at this elevation (2,614 m; 8,600 ft) are dominated by Engelmann spruce, lodgepole pine, and subalpine fir.

ily eroded soils could also prevent tree seedling establishment, such as on soils that develop from shales (J. D. Love, pers. comm.).

Perhaps the best-studied meadow in the Rocky Mountain region is Cinnabar Park in the Medicine Bow Mountains (fig. 12.3). Some have concluded that this dry meadow is slowly moving downwind, with trees invading on the upwind side (where more snow accumulates) and older trees dying on the downwind edge (Miles and Singleton 1975; Vale 1978). Doering and Reider (1992), however, concluded that the park was stable. They suggested, after a detailed analysis of soil profiles, that trees were absent in the meadow because of a 12-cm accumulation of fine-textured surface soil that provided a better environment for grasses and forbs. Surface-soil texture in the adjacent forest is coarser. The presence of a distinct stone line under the fine-textured surface soil in the meadow (fig. 12.4) and its absence in

the forest led Doering and Reider to conclude that the meadow originated after a large-scale disturbance (perhaps a fire). Subsequent wind erosion removed much of the fine material but left a layer of stones. The park area was subsequently covered by new deposits of fine, windblown material. The dry nature of the park was attributed to the difficulty of water percolation beyond the shallow stone line, with most water being held in the fine-textured surface soil, where it would evaporate rapidly, and to the strong winter winds that greatly reduced snow accumulation on the meadow.

Although some meadows have abrupt boundaries and appear to be influenced primarily by edaphic characteristics and topography, others are being invaded by lodgepole pine and other trees. The origin of these meadows is often difficult to explain, but tree invasion suggests a change of one or more environmental factors over time. Those factors mentioned most often are less frequent

TABLE 12.1   Characteristic plants of subalpine meadows in Wyoming

	Sagebrush Meadows	Dry Subalpine Meadows	Mesic Subalpine Meadows	Wet Meadows and Sedge Bogs
GRASSES				
Alpine bluegrass	—	—	X	—
Alpine timothy	—	—	X	X
Canby bluegrass	X	X	—	—
Cusick bluegrass	—	—	X	—
Danthonia	—	X	—	—
Idaho fescue	X	X	X	—
Junegrass	X	X	—	—
Mutton bluegrass	X	—	—	—
Nodding bluegrass	—	—	X	—
Patterson bluegrass	—	—	X	—
Sheep fescue	—	—	X	—
Slender wheatgrass	—	X	—	—
Thickspike wheatgrass	—	X	—	—
Timberline bluegrass	—	—	X	—
Trisetum	—	—	X	—
Tufted hairgrass	—	—	X	X
SEDGES				
Beaked sedge	—	—	—	X
Dunhead sedge	—	X	—	—
Ebony sedge	—	—	X	—
Hood sedge	—	X	X	—
Needleleaf sedge	X	X	—	—
Obtuse sedge	—	X	—	—
Rock sedge	X	X	—	—
Sheep sedge	—	—	X	X
Water sedge	—	—	—	X
FORBS				
Alpine avens	—	X	X	—
Alpine bistort	—	X	X	X
Alpine sagewort	—	X	X	—
Arrowleaf balsamroot	—	—	X	—
Ballhead sandwort	X	X	—	—
Elephanthead lousewort	—	—	—	X
Fireweed	—	—	X	—
Fringed sagewort	X	X	—	—
Lambert locoweed	X	X	—	—
Marsh marigold	—	—	—	X
Mountain bluebells	X	—	X	—
Mouse-eared chickweed	X	X	—	—
Mulesear wyethia	—	—	X	—
Pale agoseris	—	X	X	—
Sibbaldia	—	X	X	—
Silky locoweed	X	X	—	—
Silvery lupine	X	X	X	—
Sulfurflower wildbuckwheat	X	—	—	—

(continued)

TABLE 12.1   (*Continued*)

	Sagebrush Meadows	Dry Subalpine Meadows	Mesic Subalpine Meadows	Wet Meadows and Sedge Bogs
Tufted fleabane	X	—	—	—
Yarrow	—	X	X	—
SHRUBS				
Drummond willow	—	—	—	X
Grayleaf willow	—	—	—	X
Planeleaf willow	—	—	—	X
Mountain big sagebrush	X	—	—	—
Mountain silver sagebrush	—	—	—	X
Shrubby cinquefoil	—	—	—	X

Note: — = absent or less common

FIG. 12.2   An extensive mountain meadow in the Bighorn Mountains, located where a shale in the Gros Ventre formation is exposed. Shales lead to the development of fine-textured soils that are not favorable for tree seedling establishment (elevation 2,250 m; 7,400 ft). Common plants in the meadow include fringed sagewort, hoary balsamroot, Idaho fescue, junegrass, prairie smoke, silvery lupine, starry cerastium, yarrow, and wild geranium. Aspen groves occur in the ravines. Forests dominated by lodgepole pine and limber pine are found on the adjacent, coarse-textured soils derived from sandstones in the Flathead formation. Similar meadows exist wherever fine-textured sedimentary rocks are exposed on the surface, e.g., on the east slopes of the Wind River Mountains.

F I G. 12.3 Cinnabar Park, a dry mountain meadow in the Medicine Bow Mountains. It is located on a high plateau (2,918 m; 9,600 ft), and the snow typically drifts into the downwind forest on the left, creating a snowglade without trees where the snow persists until August. The band of lodgepole pine trees ("ribbon forest") functions like a snow fence. Canby bluegrass, Idaho fescue, junegrass, pussytoes, sheep fescue, and starry cerastium are common in the park; tufted hairgrass and other wet meadow species are found in the snowglade. See fig. 12.4.

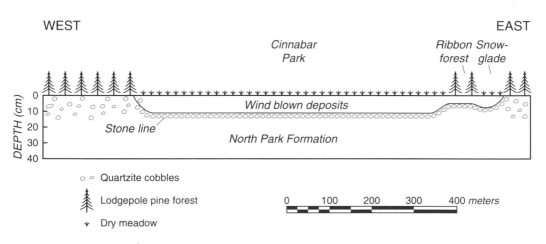

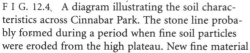

F I G. 12.4 A diagram illustrating the soil characteristics across Cinnabar Park. The stone line probably formed during a period when fine soil particles were eroded from the high plateau. New fine material has gradually accumulated to a depth of approximately 15 cm, but the establishment of tree seedlings in the meadow is still difficult. Adapted from Doering and Reider 1992; see fig. 12.3.

fires, climate change, and grazing by domestic livestock (which can reduce competition from herbaceous plants for tree seedlings; Dunwiddie 1977). There is also the possibility that tree invasion occurs slowly following some disturbance in the distant past, as, for example, on parts of Libby Flats in the Medicine Bow Mountains (Billings 1969).[1]

Ribbon Forests and Snowglades

Drifting snow is another possible cause of meadows. Trees simply cannot become established where snow persists until late in the summer, in part because the soil remains too wet for too long. Buckner (1977), in a detailed study of snowglades in Colorado, listed the following five effects of snow accumulation: a shortened growing season, more pocket gophers, late seed germination or germination failure, the breakage and bending of those few trees that do become established, and a higher frequency of parasitic snow molds. The growing season is short in such areas, and various pathogens (such as the blackfelt snowmold, or *Herpotrichia nigra*) can kill lower branches buried in snow until midsummer (fig. 12.5; Cooke 1955).

The ribbon forest–snowglade pattern takes at least two forms. Commonly, there is just one band of trees and one band of meadow on the lee side of a more extensive meadow, such as in Cinnabar Park, where the late-spring snowdrift may be 15 m deep or more. The origin of the band of trees and single snowglade on the lee side of a meadow is difficult to explain. Billings (1969) suggested that a "forest fire near timberline . . . changes the snowdrift pattern enough that trees in an unburned area to the lee of the burn are killed by late-lying snow during the summer. The dead trees are replaced by a wet type of snowglade meadow." Support for Billings's suggestion comes from the observation that huge snowdrifts do form on the lee side of some clearcuts, usually 3–6 m into the uncut forest downwind, and the effect of the drift is to reduce tree vigor. Dead standing trees can be found even though the upwind clearcut is only ten to twenty years old.

Elsewhere, usually near upper treeline, many bands of forest and meadow may occur in sequence (fig. 12.6; Arno and Hammerly 1984).

F I G. 12.5   Tree branches that remain covered with snow late into summer, such as in snowglades, are often killed by blackfelt snowmolds. This photo shows an infected lodgepole pine sapling.

Again, the role of wind and snowdrifting seems obvious because the bands are perpendicular to the prevailing westerly winds and because the deepest snow occurs in the glades. Buckner (1977) concluded that the bands of trees were a result of bands of microsites favorable to tree establishment. Although tree establishment has usually occurred on slightly elevated sites, as young trees grew, they presumably caused drifting to occur downwind, which greatly reduced the chances of tree invasion. Thus, snowdrifting accentuated the pattern. Buckner hypothesized that the pattern should be especially obvious in areas of higher snowfall, such as on the western side of mountain ranges. He found no evidence that his study area had once had a continuous forest cover, and thus disturbance was not a critical factor in the formation of the multiple-band snowglades he studied.

A similar phenomenon has been observed where the trees occur in doughnut- or U-shaped

F I G. 12.6   At higher elevations with strong winds, such as in this area north of Medicine Bow Peak, parallel ribbon forests and snowglades may develop. The trees are predominately Engelmann spruce and subalpine fir; elevation 3,344 m (11,000 ft). Photo by William K. Smith.

groves (see fig. 9.13). These groves, sometimes referred to as timber atolls, may be fringed by aspen, such as on the northern end of the Sierra Madre, or by Engelmann spruce and subalpine fir at higher elevations in northwestern Wyoming (Griggs 1938). These unusual groves are typically surrounded by sagebrush steppe or subalpine meadows, and they probably developed because of the snow-fence effect of the few trees that did become established. As the trees develop, the only sites where new seedlings or sprouts grow is around the edges of the snowbank created by the original trees. Eventually, trees become established around all or most of the snowbank. The persistence of the groves is favored because of the more mesic conditions created by snow that drifts into the center. Although trees cannot survive in the center because the snow persists too long, the snowdrift provides a more dependable supply of moisture for the trees on the perimeter.

The forest atoll or ribbon forest–snowglade pattern may characterize a small portion of the mountain landscape, but it is a pattern commonly observed from the air. The atolls and snowglades can be of considerable significance economically as a naturally developed high-water-yield ecosystem in which most of the snow accumulates where there is comparatively little transpiration. A common practice of watershed hydrologists to increase streamflow is to create small openings in the forest where snow can accumulate (see chapter 11).

Livestock Grazing on Mountain Meadows

Mountain meadows have long been an important component of the summer range for elk and deer, and during the past century many ranchers have come to depend on such meadows as a way of resting their rangelands. The diversity of the landscape, with rangelands at different elevations, increases the carrying capacity for both wildlife and livestock (Coughenour 1991a, 1991b; Coughenour and Singer 1991). Essentially, there is a wave of new plant growth that the animals can follow from the lowlands in the spring to high mountain meadows later in the summer. Because of topographic diversity, there is always a new supply of nutritious, green forage.

Livestock managers have learned that the grazing season is an important consideration in mountain meadows (Paulsen 1975; Turner and Paulsen 1976; Thilenius 1979; Thilenius and Smith 1985). Ranchers prefer to rest their low-elevation rangelands in the spring, when plants are actively growing and most susceptible to damage by grazing. For this reason, there has traditionally been a desire to move livestock onto the mountain meadows as soon as possible. Unfortunately, late-persisting snow prevents those meadows from being accessible, and when they are available, often the soils are so wet that damage from trampling is possible. Moreover, the mountain plants are subject to damage if grazed excessively in the early stages of their growth, just as in the lowlands.

Today, the grazing of mountain meadows is restricted in many areas to a shorter summer period than it was in the early 1900s. Fewer animals are permitted on federal land, and often a shepherd must be employed to keep the animals moving (especially sheep) so that specific areas are not grazed excessively. As a result, some ranchers now find it uneconomical to move their livestock to the mountains for such a short time (essentially midsummer to early fall). Of course, the decline in livestock grazing in mountain meadows is not due to a single factor. The amount of land area in meadows has probably declined as well over the years, owing to fire suppression and the subsequent development of more dense forests. This trend might be especially apparent in mountain ranges that were subjected to considerable burning or tree cutting in the late 1800s and early 1900s. The best example in Wyoming may be the Sierra Madre, where extensive timber harvesting in conjunction with copper mining created large areas of rangeland. Much of this area has again been covered by trees.

Overall, mountain meadows are widely appreciated for the benefits they provide in terms of livestock forage, water yield, and aesthetically pleasing landscapes. As important, the edge between the forest and meadow provides habitat for many birds and mammals.[2] Some meadows are being invaded by trees, but only rarely are forests converted to meadows after fire or timber harvesting. Shifts in the location of forest-meadow boundaries may be an indication of the climatic changes possible within the lifetime of many plants.

CHAPTER 13

Upper Treeline and Alpine Tundra

The tree species commonly found at upper tree-line are Engelmann spruce, subalpine fir, limber pine, and in the northwest, whitebark pine. All are short at this elevation, and at their extreme upper limits their windswept, shrubby growth forms are known as krummholz. Sometimes the trees grow only as shrubs less than a meter tall. Others have a vertical trunk flagged with branches only on the leeward side, shaped by blowing ice and snow (fig. 13.1). Branches and leaves near the ground, in contrast, are dense and more vigorous because they are protected by drifting snow. The trees often occur in ribbons because of the interactive effect of wind, snow, and microtopography (see fig. 12.6), or they are widely spaced in a matrix of alpine or subalpine meadows (fig. 13.2).[1]

Though wind undoubtedly contributes to treeline formation, most investigators have concluded that heat deficiency is the primary cause of both arctic and alpine treelines (Griggs 1938, 1946; Daubenmire 1954; Wardle 1965, 1968, 1971; Tranquillini 1979; Arno and Hammerly 1984; Stevens and Fox 1991). By their nature, trees have woody tissue that must be maintained by carbon-rich compounds manufactured by photosynthesis in the leaves. Photosynthesis is a temperature-dependent process, and at some point temperatures are too low for too much of the growing season to allow the level of photosynthesis required by large plants. Hence, the trees become shorter and even shrubby, and at some point even the small seedlings cannot survive. Worldwide, trees are absent where the July mean temperature is lower than 10°C or the mean daily maximum temperature in July is lower than 11.1°C (Tranquillini 1979; Arno and Hammerly 1984). Perennial herbaceous plants and small shrubs are clearly better adapted for the cold alpine environment, probably because a larger portion of the plant biomass is capable of photosynthesis. Less photosynthesis is required to maintain smaller plants, and the roots and rhizomes store carbohydrates for the production of new leaves each spring (Mooney and Billings 1960).

Predictably, some trees are well adapted for the cold environment. For example, treeline conifers have clustered leaf arrangements that tend to maintain leaf temperatures up to 10°C warmer than the ambient air (Smith and Carter 1988). This temperature difference occurs because the leaf clusters minimize wind movement between the leaves and present a larger mass for the absorption of solar radiation. Furthermore, the flexibility of limber pine and whitebark pine branches helps to reduce breakage due to wind. Wardle (1965) observed that because of this flexibility, limber pine may develop into erect trees in the same location where Engelmann spruce and subalpine fir can survive only as prostrate krummholz.[2]

F I G. 13.1   Trees at upper treeline typically have the krummholz growth-form, with dense lower branches that are protected by snow during the winter. Upper branches develop only on the leeward side of the trunk because of abrasion from blowing ice and snow. Most krummnolz trees are Engelmann spruce or subalpine fir. Common plants in the adjacent alpine fellfield include alpine avens, alpine bistort, alpine sagewort, and spike trisetum. Elevation 3,283 m (10,800 ft).

Wind and cold are clearly important aspects of the alpine environment, but drought is another factor affecting plant survival at high elevations (Lindsay 1971; Tranquillini 1979; Wardle 1981; Hadley and Smith 1983, 1986). Because of lower atmospheric pressures, water vapor diffusion—and consequently evapotranspiration—is more rapid at higher elevations (Smith and Geller 1979), which increases the probability that water stress will develop.

The complexity of understanding treelines is illustrated further by the hypothesis that plants in alpine environments are predisposed to winter abrasion because the summer is too short. According to this hypothesis, the maturation of leaves is not possible before the onset of winter (Tranquillini 1979). Although plausible, Hadley and Smith (1986, 1987, 1989) concluded that this hypothesis does not apply in the Medicine Bow Mountains. They did find winter abrasion of the leaf cuticle and epidermis, but they observed no evidence to suggest that the explanation for tree absence was a short growing season rather than the force of winter winds.

Other factors that may be important in determining treeline elevation are more intense ultraviolet radiation (Caldwell et al. 1980), snow depth (Hermes 1955), and the excessive reflection of solar radiation from snow onto leaves—all of which can interfere with physiological processes, including photosynthesis. Moreover, warm days alternating with freezing nights during much of the growing season leads to soil particle movement owing to the shrinking and expanding of ice crystals. This process, known as cryoturbation or congeliturbation, often prevents tree seedling establishment. Raup (1951) proposed that the alpine or arctic treeline could be considered a transition from relatively stable to relatively unstable soils, and that the control of climate on tree establish-

F I G. 13.2   Upper treeline in the Bighorn Mountains. The trees are Engelmann spruce and subalpine
fir. Common plants in the alpine turf are alpine
avens, alpine bistort, alpine forget-me-not, alpine
sagewort, blackroot sedge, selaginella, spike trisetum,
and stonecrop. Elevation 2,918 m (9,600 ft).

ment was indirect through its influence on congeliturbation. Tree distribution is patchy, perhaps
because congeliturbation is patchy.

The activities of the Clark's nutcracker and
Stellar's jay can also be an important factor affecting tree distribution (see chapter 9). Although less
common in the alpine zone, these species typically
cache the seeds of limber pine and whitebark pine
on southern slopes that are often snow-free during winter. Doing so helps to ensure that the seed
can be found again. Notably, whitebark pine and
limber pine are often more common on southern
slopes (Bamberg 1971; Steward 1977).

Timberline trees often occur in clusters or islands that have developed through layering, as in
the case of spruce and fir. Tree clusters develop
because the successful establishment of one tree
creates a microenvironment favorable for the establishment of others. Snowdrifting around es-

tablished trees provides protection and more water for young seedlings, and more drifting might
occur as more tree biomass develops. The snow
provides protection from abrasion and a better
supply of moisture. Other plants benefit from this
microenvironment as well, such as dwarf huckleberry, mountain gooseberry, and other species
typically associated with the less severe environment of subalpine forests and meadows.

Although snow accumulation is commonly
thought to be advantageous for tree survival in
winter, snowbanks that persist into early summer
provide favorable environments for snow-mold
fungi that can kill the leaves and twigs of conifers,
most frequently pine (see fig. 12.5). These fungi
grow in the snow and completely enmesh twigs in
a feltlike mass of hyphae. Seedlings and young
trees with all their foliage below the snow may be
killed, but when a substantial portion of the tree is
above the snow, only the buried branches are affected.

A common question about both arctic and alpine treelines is whether the trees are advancing
into the tundra or retreating, a question that pertains to considerations of climatic change and the
possible influence of global warming. In an early
study in the Teton Range, Griggs (1938) concluded that the alpine treelines were essentially
static, though he did suggest that livestock grazing
may facilitate the tree invasion of alpine meadows.
Billings (1969) found evidence of fire at treeline
in the Medicine Bow Mountains, which had the
effect of lowering the treeline because new tree
establishment was so slow. Recent studies by
Carrera et al. (1991) suggest that Rocky Mountain treelines have advanced and retreated with
changes in climatic conditions during the Holocene. Carrera and co-workers calculated that the
alpine treeline would rise 80-140 m with less than
a 1°C increase in the mean summer temperature.

Subalpine forests and krummholz in Wyoming give way to the treeless alpine tundra at elevations ranging from about 3,500 m (11,480 ft)
in the Medicine Bow Mountains to the south to
about 3,000 m (9,840 ft) in the Beartooth Mountains to the north (fig 13.3; Peet 1988). Daubenmire (1954) suggested that the alpine treeline is
about 110 m lower per degree of latitude between
30° and 60° north latitude. Aspect is important as

F I G. 13.3   Alpine tundra in the Beartooth Mountains of northwestern Wyoming (elevation 3,300 m; 10,825 ft). The bedrock in this area is Precambrian granite and gneiss.

well, with the treeline often occurring at higher elevations on the warmer south slopes. Moreover, longer snow persistence shortens the growing season, with the result that the treeline is lower on north and west slopes (Daubenmire 1954; Hermes 1955; Komarkova and Webber 1977). Deeper snow accumulations often occur on west slopes because of heavier snowfall, and snow persists longer on north slopes because of less direct solar radiation.

The Alpine Mosaic

The vegetation above treeline is commonly referred to as alpine tundra. Many alpine plants are also found in the arctic, as the climate in both areas is characterized by cool, short growing seasons.[3] There are, however, prominent differences (Billings and Mooney 1968; Billings 1973, 1974b). In particular, the arctic has long summer days

with little diurnal temperature fluctuation and low ultraviolet and visible light intensities. In sharp contrast, alpine zones have high ultraviolet and visible light intensities, shorter summer photoperiods, and great diurnal temperature extremes in the summer—often from more than 20°C during the day to below freezing at night. Also, permanently frozen ground (permafrost) is rare in Wyoming alpine zones (Péwé 1983).

In the alpine, the combination of extreme temperature fluctuations during a twenty-four-hour period, the rapid development of drought in summer (away from melting snowdrifts), rapid rates of evapotranspiration, soil mixing caused by frequent freezing and thawing (cryoturbation), and a cool, short growing season leads to perhaps the most severe conditions for plant growth to be found anywhere. Still, thousands of plant species worldwide have become adapted to this environment (Daubenmire 1943b; Billings 1974a).

The predominant growth forms of alpine plants are perennial herbs, including grasses, sedges, and low woody or semiwoody shrubs. Most of the biomass is below ground (Thilenius 1975, 1979). As seedling establishment is difficult in the alpine environment, annuals are rare.[4] Lichens are common, but they are not a dominant feature of alpine vegetation (unlike in the arctic tundra, which can have a dense cover of the lichen known as reindeer moss). The aboveground net primary productivity of alpine tundra has been estimated at 40–128 g/m²/yr, which is high, considering that the growing season may be only thirty to seventy-five days long (Bliss 1962; Terjung et al. 1969; Tieszen and Detling 1983).

Some unusual adaptations have evolved to allow plant growth in the alpine environment (Bliss 1962; Billings 1974; Bell and Bliss 1979). For example, all alpine plants are capable of photosynthesis and growth under cold conditions, even when temperatures are near freezing or below (Scott and Billings 1964; Sakai and Otsuka 1970), and many are drought-tolerant. Evergreen and wintergreen leaves are common, enabling growth early in the spring, when moisture is available; some plants are capable of initiating growth even while still covered with 50 cm of snow (Billings and Bliss 1959; Curl et al. 1972; Kimball et al. 1973; Hamerlynck 1992). The rapid translocation of stored carbohydrates, even under cold conditions (Wallace and Harrison 1978), allows for rapid growth in the spring. Rochow (1970) studied alpine pennycress at high and low elevations in the Medicine Bow Mountains, finding that the higher populations broke dormancy and completed their growth more rapidly than populations at lower elevations.

An interesting distribution pattern was detected by Bell and Bliss (1979) in alpinesedge, a sedgelike plant common in the Beartooth Mountains and Rocky Mountain National Park. As is true of some other montane plants, alpinesedge has green overwintering leaves that are photosynthetically active in winter. The leaves are also major sites for carbohydrate storage. A loss of the leaves thus represents an extraordinary loss of energy. For this reason, the plant is restricted to habitats that are comparatively well protected from the wind, so that leaf erosion by blowing ice and snow is not excessive. Also, the plants apparently require sites where they are not covered with snow for much of the winter. In addition to these special habitat requirements, which were also observed by Bamberg and Major (1968), alpinesedge is not very tolerant of grazing—again because so much of the carbohydrates is stored in the leaves. Because alpinesedge seedlings are so rare, Bell and Bliss wondered whether meadows dominated by this species might be irreplaceable if subjected to intense livestock grazing. Unlike mixed-grass prairie, alpine plants probably did not evolve with a long history of grazing by large ungulates.

Another plant adaptation is the cushion plant growth form, in which the branches and leaves grow in dense mats close to the soil surface. The soil surface is comparatively warm, and because the branches are very dense, they absorb more solar radiation and minimize cooling by wind. Dark color and epidermal hairs also help to maintain warmer leaf and bud temperatures (Spomer 1964).

Reproduction in alpine plants is largely vegetative because of the difficulties associated with seedling establishment in such a rigorous environment. Nevertheless, many plants have evolved special adaptations for seed production. Some plants have unusually large and showy flowers or inflorescences, presumably to attract scarce insect pollinators during short warm periods, when they are active. Flowers with large petals that serve as windbreaks are even thought to provide a warmer microenvironment for insects (fig. 13.4; Knutson 1981). Because pollination often does not occur, a high percentage of the plants are capable of producing seeds without fertilization *(apomixis).*

In addition to having one of the most stressful environments, alpine landscapes are also among the most variable (fig. 13.5, table 13.1). Much of the land area is exposed rock with little or no plant growth, such as on talus slopes, boulder fields, mountain peaks, and cliffs. Other extensive areas are covered by persistent snowbeds or glaciers. Where soils have developed and little snow accumulates, the alpine zone can be divided into fellfields, which are moderately dry habitats with rocky soil; alpine meadows, which are comparatively free of large rocks and may be dry, mesic, or wet; willow thickets; and sedge bogs.[5] Sphagnum

F I G. 13.4   The big flowers of big-flowered hymenoxys attract pollinating insects in alpine tundra during the short, cool summer.

bogs in Wyoming are rare and are known to exist in only one or two localities. Soil texture and depth, patterns of snowdrifting, soil moisture availability, and cryoturbation are important variables in determining which plants and animals will be found in a specific area.

Billings (1974b) suggested three terms that now are widely used in describing vegetation pattern in the alpine tundra: *macrogradients, microgradients,* and *mesogradients.* Macrogradients occur over distances of kilometers and reflect differences in climatic conditions. In contrast, microgradients refer to the vegetation changes that occur over distances of a few decimeters or meters. Such abrupt changes can be caused by the presence of boulders, which absorb and store considerable heat from the intense solar radiation that occurs at high elevations during the day. Snow accumulates on the lee side of the boulders, and rainwater or condensation on the boulders drains rapidly to the soil below, where it is less likely to evaporate. Because of these influences, the boulder provides a warmer, more mesic microenvironment, which

allows some plants to grow where they might not otherwise.

Intermediate between macrogradients and microgradients are mesogradients, which are determined largely by topography. As illustrated in fig. 13.5, windswept ridges or flats are usually fellfields, which are often dominated by cushion plants and other xerophytes that require little snow protection during the winter. The lee sides of ridges typically have drifted snowbanks that persist until midsummer and consequently have a low cover of plants because of the short growing season. The area just below an alpine snowbank has a dependable source of water for most of the growing season, as well as a longer growing season, and consequently the vegetation there has higher cover and productivity. Moreover, the dominant species are different. Abrupt changes in plant species composition occur along gradients around melting snowbanks, where plants that are uncovered later in the growing season produce flowers in a shorter time (Billings and Bliss 1959; Holloway and Ward 1963). Similar "catch-up" behavior of plants under late-lying snowbanks has been observed by others (Webber et al. 1976; Knight et al. 1977). Billings and Bliss found that annual production in the alpine zone is proportional to the number of snow-free days and that if the snow release comes after late July, little plant growth is possible (such as in snowglades).

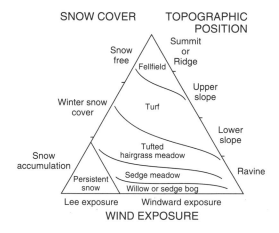

F I G. 13.5   The distribution of different kinds of alpine vegetation in relation to topographic position, wind exposure, and snow cover. Adapted from Johnson and Billings 1962.

TABLE 13.1 Characteristic plants of alpine vegetation types in Wyoming

	Fellfield	Alpine Turf	Wet Meadow	Willow Thicket
FORBS				
Alpine avens	X	X	—	—
Alpine bistort	—	—	X	X
Alpine forget-me-not	X	X	—	—
Alpine mountainsorrel	X	X	—	—
Alpine pussytoes	X	X	—	—
Alpine sagewort	—	X	—	—
Bigflowered hymenoxys	X	X	—	—
Blueleaf cinquefoil	X	—	—	—
Dwarf clover	X	X	—	—
Elephanthead lousewort	—	—	X	X
Leafybract aster	—	—	X	X
Lightpetal dryad	X	—	—	—
Marsh marigold	—	—	X	X
Moss silene	X	—	—	—
Parry clover	—	—	X	X
Parry lousewort	X	—	—	—
Rocky Mountain nailwort	X	—	—	—
Rosecrown stonecrop	—	—	X	X
Selaginella	X	—	—	—
Sibbaldia	X	X	—	—
Sticky polemonium	X	—	—	—
Tufted phlox	X	X	—	—
Yarrow	—	X	X	X
GRASSES				
Alpine bluegrass	X	—	—	—
Alpine timothy	—	X	—	—
Purple reedgrass	X	—	—	—
Scribner wheatgrass	X	—	—	—
Sheep fescue	X	—	—	—
Timberline bluegrass	X	X	—	—
Trisetum	X	X	—	—
Tufted hairgrass	—	—	X	X
SEDGES				
Alpinesedge	—	X	—	—
Blackroot sedge	—	—	X	X
Drummond sedge	—	—	X	X
Ebony sedge	—	—	X	X
Obtuse sedge	—	—	X	X
Rock sedge	—	—	X	X
Water sedge	—	—	X	X
SHRUBS				
Alpine kalmia	—	—	X	X
Arctic willow	X	—	—	—
Cascades willow	—	—	—	X
Grayleaf willow	—	—	—	X
Planeleaf willow	—	—	—	X
Snow willow	X	—	—	—
Bog birch	—	—	X	X

Note: — = absent or less common

207

CRYOTURBATION AND SUCCESSION

Another factor causing patterns in alpine landscapes is the diurnal as well as seasonal freezing and thawing of moist soils (Caine 1974). Soils expand when they freeze, pushing comparatively large objects (such as stones and boulders) in the direction of least resistance, usually toward the surface. After thousands of freeze-thaw cycles, the larger objects are sorted from smaller ones, and various kinds of "patterned ground" develop (such as frost boils, stone nets, and stone polygons; fig. 13.6). Known as cryopedogenesis, this process alters soil structure seasonally, if not daily.[6] Cryopedogenesis apparently damages root systems and creates new microhabitats. If moist soils freeze and thaw on a twenty-four-hour cycle, needle-ice may form, which can lead to the forma-

tion of frost boils. Pocket gopher digging may complicate the pattern created by freeze-thaw cycles.

Another frost-related geomorphic process is *solifluction*, in which frost and gravity combine to move wet soil downslope. Solifluction occurs most often on lee slopes, where snow accumulates and the soils become wet. Formed over long periods, solifluction lobes or terraces cause a distinctive pattern where they occur in alpine landscapes (fig. 13.7).

Such frost-related processes, combined with the burrowing of pocket gophers, are thought to be continual sources of disturbance to alpine vegetation. In fact, some ecologists question whether a stable climax tundra ecosystem ever develops (Churchill and Hansen 1958). Some species are favored by soil perturbations (Bamberg and Ma-

F I G. 13.6   Alternating freezing and thawing over long periods causes large stones and rocks to be separated from fine-textured soil particles, forming stone "polygons" or nets, such as in this area in the Beartooth Mountains. The alpine turf is dominated by alpine avens, alpine bistort, alpine sagewort, Drummond sedge, snow willow, tufted hairgrass, tufted phlox, and other alpine plants (Johnson and Billings 1962). Elevation 3,162 m (10,400 ft).

F I G. 13.7   Solifluction lobes or terraces develop in alpine tundra on slopes in the Beartooth Mountains, creating different environments for plant growth over short distances. Water seeping through the lobes provides an ideal environment for barrenground willow. Elevation 3,222 m (10,600 ft). See Johnson and Billings 1962.

jor 1968), whereas others are more characteristic of relatively stable sites. Marr (1961) suggested that the major factors that determine the species composition of alpine plant communities are cryoturbation, the burrowing of pocket gophers, and the depth and duration of snow cover.

Other scientists have observed clear successional trends in alpine ecosystems, despite periodic disturbances. For example, with gradual weathering and soil development, boulder fields develop into fellfields, fellfields into alpine turf, and wet meadows dominated by water sedge develop into mesic meadows dominated by tufted hairgrass (Murdock 1951; Lewis 1970; May 1976). Spence

(1985), however, studied a chronosequence of stands on glacial moraines in the Teton Range and found that compositional differences were not attributable to succession. Continual disturbances, mostly frost-related, seemed to occur. He concluded that competition, one of the processes thought to be important in the later stages of succession, was a minor factor in determining the species composition of alpine tundra. Spence did note that fairly stable alpine communities could develop where there was little soil water and consequently little cryoturbation (such as in boulder fields and some fellfields). Where cryoturbation is common, Billings and Mooney (1959) found that areas of a few square meters or less changed from a frost hummock to a sorted polygon and back to another frost hummock—an example of cyclic succession.

Another example of cyclic succession was suggested by Phillips (1982) for the stone polygons of Rocky Mountain National Park in Colorado. Phillips observed that typical polygons in Colorado are 2–3 m across and have a soil depth of 0.75–1.5 m. The polygons are outlined by large rocks, with fine-textured soil in the centers. Plants characteristically found on the rocky edges are different from those in the centers. The cycle starts with a level, well-vegetated turf of alpinesedge and rock sedge on a well-developed soil in the center of a polygon. The high leaf area of the turf allows for high rates of evapotranspiration, which stabilizes the site because less water remains in the soil. Drier soils are less likely to experience cryoturbation.

With continued soil development and the accumulation of organic matter in the turf, a convex surface is formed that is eventually subject to wind abrasion. With erosion, cushion plants invade where the turf had once grown, and with less snow accumulation on the more exposed surface, the site remains dry or becomes drier. Desiccation cracks form, and with continued erosion a concave surface develops with a thin surface horizon and sparse vegetation. Snow, however, then accumulates in the depression and the site becomes wetter, which leads to more cryoturbation. Other plants invade and begin to trap fine soil particles. With time, the surface horizon becomes thicker, and a new turf develops on a level surface, completing the cycle.

AVALANCHES

Snow avalanche tracks are a conspicuous feature of landscapes where large volumes of drifting snow accumulate, such as on the lee side of high ridges and where the topography is steep enough that snowslides (avalanches) are common (fig. 13.8). Such areas are particularly common in Colorado, northwestern Wyoming, and parts of the Northern Rocky Mountains. Usually, the snow accumulates in the alpine zone and periodically slides along a track that extends down into the subalpine forest, ending in the "runout zone." Depending on the amount of snow and other conditions, avalanches can be gentle slides over short distances, or dramatic, thunderous events that break large trees in the valley bottom. Sometimes several avalanches may occur on the same track in a single winter.

The vegetation of avalanche tracks is different from the alpine tundra or forest on either side. In the tundra, the track is typically sparsely vegetated because the catchment area usually has snow that persists until midsummer, thereby shortening the growing season. Damage to small alpine plants is minimal because snow movement typically does not occur at the soil-snow interface. Most physical damage to plants occurs below the treeline, where tall, woody plants are more common. Even there, small trees and flexible shrubs are not damaged. Trees with diameters greater than 10 cm, however, usually break because they provide too much resistance to the sliding snowpack (Potter 1969; Mears 1975; Johnson 1987; Patten 1987). As a result, the trees in avalanche tracks (if they are present) are usually small, especially where snowslides occur every year or two. Flexible shrubs dominate the vegetation where avalanches occur frequently (Eversman 1968; Stauffer 1976; Cushman 1981; Malanson and Butler 1984; Butler 1985; Johnson 1987; Patten 1987).[7]

In addition to the conspicuous contrast of avalanche tracks and adjacent vegetation, there is an interesting pattern within the tracks that is associated with elevation. Most avalanches involve low volumes of snow, extend only short distances down from the catchment area, and occur almost

FIG. 13.8   An avalanche track in Cascade Canyon, Grand Teton National Park. Tree damage is less frequent at the bottom of the track, enabling the trees to grow taller there. As the trees grow larger, they offer more resistance to the avalanche and are more likely to break or be uprooted. The trees are subalpine fir and Engelmann spruce. The more flexible shrubs higher on the slope include bearberry honeysuckle, mountainash, Rocky Mountain maple, rusty menziesia, and serviceberry. Elevation at the bottom is 2,200 m (7,216 ft).

every year. In contrast, large avalanches extending down to the runout zone may occur only once every fifty to a hundred years. Others are intermediate in extent and frequency. Thus there is a vertical gradient in the vegetation that is correlated to avalanche frequency: small plants and flexible shrubs dominate at higher elevations, with trees becoming more common lower on the track because of longer periods of undisturbed growth. Predictably, infrequent large avalanches create the most disturbance because trees are broken and plant and animal species composition changes significantly (Patten 1987).

The disturbances that occur in alpine tundra are distinctly different from those occurring in ecosystems at lower elevations, and nowhere in the region is the environment more rigorous. Still, through natural selection occurring over a million years or more, plants and animals have evolved adaptations for surviving in the alpine environment. Indeed, they are not found elsewhere. In one sense, the tundra is a highly stress-tolerant ecosystem. Yet ecosystem recovery or succession after disturbances is slow, and for this reason some people refer to the tundra as fragile. To be sure, humans can cause undesirable changes that will persist for many decades if not centuries.

V

Landscapes of
Special Interest

CHAPTER 14

The Yellowstone Plateau

Early explorers along the headwaters of the Missouri and Columbia rivers returned with tales of abundant fish and wildlife, geysers, boiling mud, steam rising from holes in the earth, hot springs, spectacular water falls, and a deep canyon with walls of yellow stone. So incredible were their stories that few took them seriously. Some newspapers hesitated to print their reports. Cornelius Hedges, a member of the 1870 expedition led by Gen. Henry D. Washburn, was probably trying to preserve his reputation as a keen observer when he commented, "I think a more confirmed set of skeptics never went out into the wilderness than those who composed our party, and never was a party more completely surprised and captivated with the wonders of nature" *(Helena Daily Herald,* 9 Nov. 1870). Washburn himself complained of the wind and cold in August but was filled with "too much and too great a satisfaction to relate," after walking along the rim of the Grand Canyon of the Yellowstone River. A year later, Ferdinand V. Hayden led an expedition sponsored by the United States Congress. Among his entourage were the artist Thomas Moran and the photographer William Henry Jackson. With their documentation and the urging of many, Yellowstone National Park (YNP) was established in 1872—the first national park in the world.

The accolades continued. In 1875 Gen. W. E. Strong wrote, "Grand, glorious, and magnificent was the scene as we looked upon it from Washburn's summit [Mount Washburn]. No pen can write it—no language describe it." Today the wonders of Yellowstone still entice and excite many visitors, even in an age when an earthrise on the moon is more likely to elicit superlatives comparable to those of explorers in the 1800s. Sometimes the park is tranquil with elk, moose, and bison at rest in expansive meadows. The rivers are clean, and the thermal features are fascinating curiosities that relax hikers with time to spare on a summer day. Such tranquility, however, belies the long winters that plants and animals must endure, the forest fires that rage across the landscape at intervals of two or three centuries, and the molten rock that exists close to the earth's surface. In the Eocene, 35–55 million years ago, a series of volcanic eruptions literally buried adjacent mountain ranges over thousands of square kilometers (Keefer 1971; Schreier 1983; see chapter 2). These eruptions occurred intermittently, with new forests developing after each one (fig. 14.1; Dorf 1964). The last major eruptions occurred about 660,000 years ago, and like the fires of 1988, they could occur again.

A prominent geologic feature of the Yellowstone landscape began to develop about 2 million years ago, when pressure generated by molten rock created an oval dome that was 48 by 64 km. Eventually the dome collapsed into the void left by the volcanic ejection of gas, lava, and ash, creating a caldera (depression) in the area of Yellowstone Lake and westward that occupies about a third of the present-day park.[1] Another caldera had formed

F I G. 14.1   Forests were buried by volcanic deposits in Yellowstone National Park during the volcanic eruptions that occurred periodically in the Eocene, about 50 million years ago. The fossil trees include specimens of chestnut, magnolia, maple, oak, redwood, sycamore, and walnut. National Park Service photo by James Peaco. Elevation 2,554 m (8,400 ft).

a short time earlier in the Island Park area of eastern Idaho, just west of Yellowstone. These oval or circular depressions were partially filled by lava that continued to flow from cracks in the earth's crust, forming a rolling plateau composed of a rock known as rhyolite. The walls of the caldera are still plainly visible in some areas (such as the cliffs north of the Madison River). Mount Washburn, itself the remnant of a volcano, lies on the northeastern edge of the caldera, Mount Sheridan is on the southern edge, and Canyon Village is in the bottom of the caldera.

With regional uplifting and continental cooling, glaciers began to form in the Rocky Mountains and covered most of Yellowstone during at least three separate advances. The most recent glaciation, known as the Pinedale, began about twenty-five thousand years ago and covered 90

percent of the park.[2] Most of Yellowstone Plateau was glaciated. In some places the ice is believed to have been more than 800 m thick. Like the volcanic ash before, ice covered many mountains in the area (including Mount Washburn).

With melting, the ice disappeared over most of the area by about twelve thousand years ago. Glacial moraines dammed the Yellowstone River near the Upper Falls, causing an increase in the size of Yellowstone Lake. The shoreline, however, gradually retreated southward because of continued land shifting. Lake-bottom sediments were exposed that now are covered by the extensive grasslands and shrublands of Hayden and Pelican valleys. The trees of the region do not grow well on fine-textured soils.

Based on fossil pollen records from lake sediments, Whitlock (1993) determined that tundra was prevalent in Yellowstone at the time of glacier retreat about 14,000 years ago. By about 11,500 years ago, Whitlock concluded, Engelmann spruce had formed a spruce parkland on the comparatively more fertile substrates developed from andesitic rocks. Whitebark pine and subalpine fir joined the spruce several centuries later (11,000–9,500 years ago), but the less fertile rhyolitic soils of the Central Plateau remained unforested during this period. The climate gradually became warmer, but it was still cooler and more humid than today. Continued warming led to the expansion of lodgepole pine throughout the region. Limber pine was present at low elevations, Whitlock determined, and Douglas-fir expanded to higher elevations between 9,500 and 4,500 years ago. In the past 4,500 years, the climate has become cooler and wetter at high elevations, and as a result, spruce and fir are more common there. Douglas-fir savannas are now common at lower elevations.

Yellowstone National Park occupies an area of 9,259 km[2] (89 × 97 km). The highest mountain peaks occur in the Absaroka Mountains on the eastern boundary, rising to slightly more than 3,333 m; the lowest elevation is at Gardiner where the Yellowstone River leaves the park (1,620 m). The average elevation in the park's central plateau is about 2,377 m, considerably higher than Jackson Hole to the south. More than 150 lakes constitute 5 percent of the park, but all are small except for the Yellowstone, Shoshone, Lewis and Heart lakes (which account for 94 percent of the lake surface area; Gresswell 1984). Most soils have developed on glacial till. There are two kinds of volcanic rock: rhyolite and the comparatively nutrient-rich andesite. The highest mean annual precipitation (about 60–70 cm) occurs on the high southwestern plateaus and the Absaroka Mountains to the east, with the interior and lower elevations of the park receiving much less (28 cm at Gardiner; Despain 1990). The town of West

Yellowstone receives a mean of 58 cm. There is great year-to-year variability in precipitation, but most comes as snow and spring rain. More summer precipitation occurs in the center of the park (Despain 1990).

THE LANDSCAPE MOSAIC

Today, 80 percent of YNP is forested (fig. 14.2).[3] Like other parts of the Northern Rockies, Douglas-fir forests predominate at lower elevations (1,820–

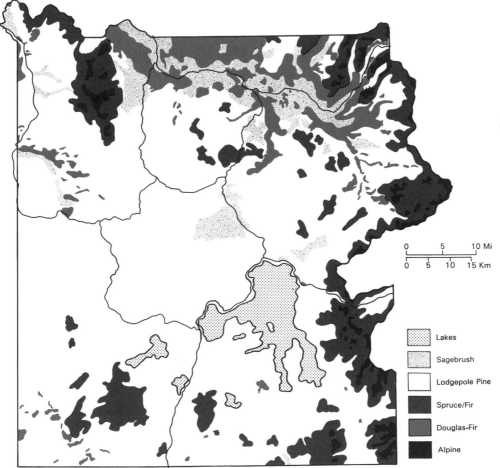

© Linda Marston, 1993

F I G. 14.2   Major vegetation types, roads, and large lakes in Yellowstone National Park. The Northern Range occurs at the lowest elevations on the north edge of the park. Based on a map prepared by Despain 1990. The Absaroka Mountains on the eastern border and the Gallatin Mountains in the northwestern corner rise to 3,344 m (about 11,000 ft); the

lowest elevation is 1,601 m (5,265 ft), where the Yellowstone River flows northward out of the park. The park covers an area of 8,982 km² (3,468 mi²). Annual precipitation ranges from 25 cm in the Northern Range to about 180 cm on the southwestern plateau.

2,500 m), such as in the Yellowstone and Lamar valleys to the north. Lodgepole pine forests occur at midelevations (2,300–2,700 m), and a mix of Engelmann spruce and subalpine fir dominates at higher elevations and in other mesic locations, such as on north-facing slopes (Despain 1990). Ponderosa pine is not found in the park, probably because the growing season is too cool and dry. Aspen is comparatively rare (<10 percent of the land area), occurring primarily in small groves of a few hectares at lower elevations to the north in association with Douglas-fir forests and sagebrush steppe.[4] Interspersed throughout the forests are shrublands, grasslands, meadows, geyser basins, wetlands, lakes, and alpine tundra (above 3,000 m).

The forests of YNP appear uniform to many visitors. Yet Despain (1990) identified twenty-nine different forest environments (habitat types) based on estimated climax vegetation. Lodgepole pine now predominates over large areas, partially because of fires during the past few centuries. If fires did not occur, Despain concluded that subalpine fir and Engelmann spruce would become the dominants over most of the park after several hundred years of succession. Like others, he did observe some forests where lodgepole pine appeared to be the climax species, with several age classes represented, especially on the drier, less fertile soils derived from rhyolite (Despain 1983, 1990). Andesitic soils derived from older Absaroka volcanic rocks are more fertile, and succession to spruce and fir following fire is thought to occur more rapidly. Whitebark pine is more common at high elevations, where andesitic rocks are also more common (fig. 14.3). Tree-dominated riparian woodlands are not common in the park, but where they do occur, the principal trees are aspen, balsam poplar, narrowleaf cottonwood, and sometimes Douglas-fir, Engelmann spruce, and lodgepole pine (Chadde et al. 1988).

Fire has played a significant role in all of the upland forest types, with the Douglas-fir forests experiencing surface fires every twenty to twenty-five years (Houston 1973). These frequent burns maintained open forests with low tree density and an abundant growth of understory plants that are important for elk, mule deer, and bison—especially in winter. With fire suppression, the

FIG. 14.3   Whitebark pine is the predominant tree-line species in northwestern Wyoming and the Northern Rocky Mountains, as shown here near the top of Mount Sheridan in Yellowstone National Park. Elevation 3,134 m (10,308 ft) at the summit.

forests have become more dense (Houston 1973; Arno and Gruell 1983). Big sagebrush density in adjacent shrublands also increased, probably for the same reason. Big sagebrush is sometimes found in the understory of Douglas-fir, an indication that the trees have expanded into the shrubland. At higher elevations, the mean return interval for fires is a century or more, and stand-replacing crown fires are more typical (Romme 1982; Romme and Despain 1989).

Mountain pine beetles have been another significant factor affecting lodgepole pine forests, especially from 1970 to 1983 at lower elevations on the western side of the park (Despain 1990). The cause of beetle epidemics during this time is not well known, but it could have been the result of fire suppression in the adjacent Targhee National Forest (see chapter 11). The epidemic did not spread far into the park, perhaps because of the colder winters at higher elevations.

Nonforested vegetation types are also common in YNP (Patten 1959; Despain 1990). Foothill grasslands dominated by bluebunch wheatgrass and Sandberg bluegrass are found at low elevations near Gardiner (<38 cm annual precipitation). Also found in this area are desert shrublands dominated by fringed sage, Gardner saltbush, greasewood, pricklypear cactus, and winterfat. Meadows at low elevations along the Lamar River have Kentucky bluegrass (near the site of the old Buffalo Ranch), Wheeler bluegrass, sheep sedge, and bearded wheatgrass, among other species.

At higher elevations, Idaho fescue, junegrass, thickspike wheatgrass, and bearded wheatgrass become dominant meadow species. Big sagebrush is found on mesic sites, especially in the Northern Range, and silver sagebrush and shrubby cinquefoil are found in meadows, where the soils stay wetter for a longer time (such as in the Hayden and Pelican valleys). Interspersed throughout the subalpine forests are meadows dominated by tufted hairgrass, sedges, and a variety of forbs. Several species of sedge are especially common in wet areas (water sedge, mud sedge, Buxbaum sedge, beaked sedge, and fewflowered spikerush; Mattson 1984). In the riparian zone there can be both meadows and shrublands (fig. 14.4). Riparian meadows typically have bluejoint reedgrass, tufted hairgrass, and other grasses, sedges, and forbs. Shrublands include alder, water birch, shrubby cinquefoil, silver sagebrush, and various species of willows (Geyer willow, tealeaf willow, sage willow, Wolf willow, and others; Chadde et al. 1988; Chadde and Kay 1991). Dwarf willows (<10 cm tall) can be found above treeline, along with other alpine plants common to the region (see chapter 13).

Several unique plant species are found around hot springs and geyser basins, including Ross bentgrass—a plant that is endemic to Yellowstone. The mineral-rich water of these thermal features creates a saline environment, and consequently the characteristic plants include halophytes such as seaside arrowgrass, alkali cordgrass, meadow barley, and baltic rush (Despain 1990). The chemical composition of the water varies from one geyser basin to another, with the result that species composition varies too (both terrestrial and aquatic; Sheppard 1971). For example, the rock formed at Mammoth Hot Springs

F I G. 14.4   Mountain meadows add diversity to the landscapes of Yellowstone National Park, such as in Hayden Valley along the Yellowstone River. The meadows in this area often have mountain silver sagebrush and shrubby cinquefoil. Idaho fescue and Wyoming big sagebrush are common where the soil is slightly drier; water sedge and beaked sedge are common where the soil is flooded during much of the summer. The trees in the foreground are lodgepole pine. Elevation 2,250 m (7,400 ft).

is known as travertine and is rich in calcium carbonate, but the water of other springs is rich in silica and forms sinter (geyserite) as it evaporates. Much of the color variation between geyser basins is caused by the different species of bacteria and algae able to survive at near-boiling temperatures (Brock 1978). New geothermal springs develop periodically, sometimes killing the trees over large areas (fig. 14.5).

Rivers are a popular feature of the Yellowstone landscape because they are big and clean, a result of large watersheds that are still dominated by plants instead of humans. Pulses of sediment-rich water do occur following natural erosion events, especially in the spring and after heavy thunderstorms over recently burned forests, but most of the time the rivers run deep and clear. They

F I G. 14.5   The lodgepole pine in this area died because of the saturated soil resulting from new hot springs that have developed during the lifetime of the trees. The forest is being converted to a meadow by geothermal activity.

also offer some of the finest trout fishing in the world. Canyons, point bars, cutbanks, terraces, and broad floodplains are dispersed throughout the park, sometimes forming a complex and dynamic riparian mosaic (see chapter 4). Rapids and waterfalls are common.

The Grand Canyon of the Yellowstone River is especially famous. About 457 m deep, the canyon is cut in rhyolite softened by hydrothermal activity. It begins at Lower Falls, with a drop of 94 m, and continues for about 32 km to Tower Falls.

Land Management Issues

The management of YNP has stimulated controversy almost from the beginning. Some early settlers had visions of damming the rivers, marketing elk meat, and building hotels with hot spring pools (Haines 1977). They feared that designating the area as a national park would restrict economic development. No dams were constructed, but a few hotels were permitted, including the Old Faithful Inn. A more recent disagreement arose when park managers decided that bears should no longer take food from tourists, nor should they have access to garbage dumps. Some doubted that the grizzly could survive without this food source; others complained they would no longer see the bears along highways. The most recent controversies have been over fire management, wolf reintroduction, and whether the elk population is too large.

Over the years, management priorities in Yellowstone have changed. For example, predators such as the cougar and wolf were initially shot at every opportunity because the hunting of elk, bison, and antelope was an attraction for many visitors. These animals were fed hay in winter to increase their numbers (Singer et al. 1989). All hunting in the park is now forbidden. Moreover, winter feeding has stopped, and there is strong sentiment nationally to reintroduce the wolf.

Managing natural areas such as YNP for the permanent protection of natural phenomena and as a recreational destination for people has not been an easy assignment. But it is one of the most noble challenges of modern society. Currently, three interrelated issues are of special ecological interest: large mammals, fire management, and the so-called natural regulation policy.

ELK, BISON, WOLVES, AND GRIZZLY BEARS

Reading a list of the names of Yellowstone's large mammals—elk, bison, wolves, grizzly bears, mule deer, pronghorn antelope, mountain goats, and bighorn sheep—is akin to contemplating tyranosaurus rex, apatosaurus, and pterodactylus coexisting in a landscape with roads, campgrounds, and mysterious plumes of steam and water emanating from the earth. Little wonder that the early explorers referred to Yellowstone as a wonderland. Today there are only a few places where a comparable assemblage of large mammals can be found. Each species has its own story.

The Greater Yellowstone Area provides habitat for the largest herd of elk in the world (about 31,000 summer in the area, mostly in YNP itself).

F I G. 14.6   Mountain sagebrush steppe and foothill grasslands are widespread in the Northern Range of Yellowstone National Park. The forests and woodlands at this elevation (2,128 m; 7,000 ft) are dominated by Douglas-fir.

A large number even winter there, especially in the Lamar and Yellowstone River valleys (the Northern Range), where more than sixteen thousand were counted in December 1992 (figs. 14.6, 14.7). Others winter to the south in Jackson Hole, to the east in the Sunlight Basin and along the Greybull River, north into Montana, and west into Idaho.

Animals moving out of the park in the fall are subjected to predation by the many hunters that contribute significantly to the regional economy. Because of this hunting pressure and the absence of wolf predation, elk exist in a different environment than that of presettlement times. Many wildlife biologists believe, however, that population numbers are determined primarily by food availability during the winter (Houston 1982; Despain et al. 1986). Commonly, a third of the Northern Yellowstone herd dies (about five thousand animals) when the winter range is subjected to deep snow cover and extended cold periods, especially if the preceding summer was characterized by drought (Singer et al. 1989). Despite

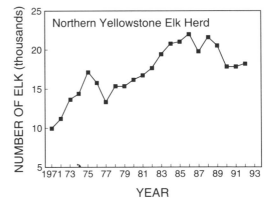

F I G. 14.7   Elk populations on the Northern Range fluctuate considerably but have been increasing gradually since the early 1970s. Some evidence suggests that elk were as numerous in the early 1800s as they have been in the late 1900s (Schullery and Whittlesey 1992).

such a high winter mortality, elk populations recover quickly (fig. 14.7). Only about 5–7 percent
of the herd is harvested by hunters annually—
usually less than seven hundred animals.

Significantly, YNP game populations may have
been low in the late 1800s because of hunting
pressure. Strong wrote in 1875 that "the terrible
slaughter which has been going on since the fall of
1871 has thinned out the great bands of big game,
until it is a rare thing now to see an elk, deer, or
mountain sheep. . . . No attempt has been made
to enforce the act in the Park" (Strong 1968). Indeed, enforcing the 1872 act that created YNP was
impossible in the beginning. Even the first superintendent, Nathaniel Langford, was a volunteer
(Haines 1977). He had to make his living elsewhere and entered the park only twice during his
five-year term. No funding was provided for park
protection. Finally realizing that natural area
management required wardens, the United States
sent its cavalry to provide fire control and protection in 1886—fourteen years after park establishment (the National Park Service was not
created until 1916). Hunting was allowed but restricted. Animals were fed, and some fences were
put up to keep them in the park. By 1912, the
park's elk population in the summer was estimated, no doubt crudely, at more than thirty
thousand (Houston 1982).

Some observers believe that elk are now having
an adverse effect on their winter range. One of
the first to express this concern was Theodore
Roosevelt, an avid sportsman and conservationist
(Haines 1977). Convinced that indeed this might
be a problem, managers authorized the shooting
of elk in the park (1,976 animals in 1956; Houston
1982). The American public and Congress, however, began to oppose this form of active management in the 1960s. It was stopped after Senate
hearings in 1968.

There were some ecologists who argued that the
park's rangelands, in fact, were not overgrazed.
Native plants seemed to be surviving, and there was
a strong sentiment to determine whether "natural
regulation," instead of hunting, would keep the elk
population at acceptable levels. Houston (1982),
after ten years of research, concluded that, "As of
September 1980, the data available do not indicate
that reductions of elk . . . are necessary in the

park." That conclusion was probably correct, but
the elk population is considerably larger now than
at any time since annual censuses began (about
19,000 elk and 2,600 bison in 1988, and 16,000
and 2,500, respectively, in 1992; fig. 14.7). Such
high numbers have been attributed to a series of
mild winters that have allowed more of the animals to stay in the park and thereby avoid being
hunted.

Several studies are now nearing completion on
the effects of elk and bison on the winter range. At
a recent conference on research in YNP, no evidence was presented that would suggest excessive
grazing (Coughenour et al. 1993; Wallace 1993).
Despite the consumption of 45 percent of the biomass by elk, Frank and McNaughton (1993)
found that NPP on the Northern Range was 36–85
percent higher than in exclosures, suggesting
compensatory growth (see chapter 5). Variations
in climatic conditions appear to be more important than grazing intensity in causing the forage
fluctuations that were observed.

Perhaps the reason that high levels of grazing
are not having an adverse impact on herbaceous
vegetation is that much of the foraging occurs
when the plants are dormant during the fall and
winter. In the spring, elk and bison disperse to
higher elevations as new plant biomass is produced. By moving up the mountain to where nutritious forage is available, the herbivores extend
their foraging period and minimize grazing time
in a particular area (Coughenour 1991a, 1991b;
Coughenour and Singer 1991)). Grazing intensity
is thereby comparatively low in the summer, when
plants store energy in root systems for the following year. The seasonal migrations of large herbivores must be considered in evaluating carrying
capacity.

In contrast to the herbaceous plants of grasslands, aspen and willows have been significantly
affected by large ungulates (figs. 14.8, 14.9; Kay
1990; Chadde and Kay 1991; and see chapter 9).
The buds and twigs of these plants are an important source of food, especially for elk in late winter
and spring, and because of the heavy browsing in
some areas, aspen trees appear to be restricted to
exclosures. Yet small root sprouts often persist.
Despain (1990) maintained that aspen are usually
not killed by browsing and that root sprouts con-

tinue to be a source of food for the animals (see chapter 9). Aspen bark is also eaten by elk in winter, leading to the formation of conspicuous scar tissue (fig. 14.10). Despain suggested that this scar formation, combined with root sprouting and the production of chemicals that reduce herbivory, is an important mechanism that enables aspen to tolerate browsing.

Kay (1990, 1993) argued that the cause of heavy or excessive browsing on woody plants is the large number of elk that now winter in YNP. Comparing historic and modern photographs, he noted an abundance of willow and a lack of scar tissue on aspen bark in the 1800s. Such comparisons, he concluded, falsify the assumption that the current level of elk influences on the winter range is natural. Kay also maintained that the winter concentrations of elk on the Northern Range have been caused in the past century by human pressures on prehistoric winter ranges outside the park. If significant adverse effects on winter range plants could be documented, Kay would advocate herd reduction or the acquisition of more winter range. It is important to determine the extent of aspen and willow deterioration in relation to succession, fire, and climatic trends, as well as the current level of herbivory (as discussed in chapter 9).

Notably, Schullery and Whittlesey (1992), after a thorough review of observations made by explorers before 1882, indicated that the record is not sufficiently detailed to conclude that in any given year the elk population equaled, exceeded, or was less than at present. The authors were convinced, however, that the park was a winter range for thousands of elk before 1882.

Bison management is controversial in much the same way. Bison were nearly extinct in 1901, when saving them became a national goal (Blair 1987; Crowe 1990). The Buffalo Ranch was established in the Lamar River valley of YNP, with land plowed and planted for hayland. Today, the population is at about 2,600—the highest in recorded history. Is the population too high for the winter range, which they share with elk? Bison are a significant feature of the park, and they probably influence their rangelands in much the same way as the original Yellowstone herds did.

There is an added dimension to bison manage-

ment in Yellowstone, as they frequently emigrate to Montana. Because the animals are known to carry the bacterial disease brucellosis and because some believe that bison can transmit the disease to cattle, Montana has passed laws authorizing the shooting of bison that leave the park (Keiter and Froelicher 1993). The transmission of brucellosis between free-roaming bison and cattle has yet to be proved, but this concern is one more example of the difficulties associated with park management when the boundaries do not encompass all ecological processes.

With growing support for letting populations fluctuate without human intervention in such large natural areas as Yellowstone, wolf reintroduction has become a popular cause. Many have wanted the return of this predator so that the ecosystem will be "complete." Now, with hunting forbidden in the park and continuing concerns over inordinate grazing, wolf advocates have another motive: predation by wolves might reduce and disperse the winter herds that some feel are too large.

Wolf reintroduction is still controversial, mostly because of potential livestock losses outside the park and the fear that control efforts in the future, if needed, will be difficult because of the endangered-species status of the wolf. Also, some critics believe that the wolf will adversely affect the grizzly bear (another endangered species) through competition for food. Regardless of the decision on reintroduction, wolves are dispersing southward from Glacier National Park, and they might soon arrive in Yellowstone—if they are not already there. Numerous observations indicate that the wolf is native to the area (Schullery and Whittlesey 1992).

In contrast to wolves, bears have a more diverse diet that includes berries, whitebark pine seeds, succulent grasses and forbs, fleshy roots, insect larvae, carrion, trout, and the young and old of large mammals.[5] Following hibernation, the grizzly bear ranges widely during the summer in search of food. Black bears feed mostly in the forest, but the grizzly prefers open areas. Wildlife biologists have observed that the animals gain weight primarily in years when berries or whitebark pine seeds are abundant. At such times, the animals' behavior has been described as a "feeding

F I G. 14.8 An 1893 photograph of Yancy's Hole in the Northern Range of Yellowstone National Park (looking eastward; elevation 2,280 m; 7,500 ft). Note the tall-willow shrublands along the creeks, and the open woodland on the ridge, which appears to have been recently burned. Photo by F. Jay Haynes (H-3020), reproduced with permission of the Haynes Foundation Collection, Montana Historical Society, Helena.

frenzy" (F. Hammond, pers. comm.). Remote areas with fish, open meadows, carrion, berry-producing shrubs, and whitebark pine are sure to be prime grizzly bear habitat and have been mapped as such by the Interagency Grizzly Bear Study Team.

Now rare and restricted to remote areas, the grizzly was once common throughout the mountains and plains. Its current range is probably less than 1 percent of what it was in the 1800s (Strickland 1986). With fewer than 1,000 animals in Montana, Wyoming, Idaho, and Washington combined, the grizzly was listed by the United States Fish and Wildlife Service as a threatened species in 1975. In the Greater Yellowstone Area the population was at least 200 in 1987. Most reports suggest that the grizzly population has increased, possibly to around 300, but many wildlife biologists are still concerned about its survival in the region.

Of recent concern to grizzly bear biologists is the demise of many whitebark pine stands in the Northern Rockies owing to blister rust, succession to subalpine fir and Engelmann spruce, and the 1988 fires (Arno and Hoff 1989; Schmidt and McDonald 1990). Whitebark pine seeds are thought to be important to the building of the fat reserves necessary for winter survival. Because individual trees produce large amounts of seed only every three to five years, it is important that the bears have large numbers of trees over large areas so that some seed is available each year. The bears may, however, be able to develop fat reserves with other foods.

F I G. 14.9   Compare this 1988 photograph of Yancy's Hole to fig. 14.8. The trees on the ridge have become more dense because of postfire succession, and the tall-willow shrublands have essentially disap-peared, probably a result of heavy browsing by elk. Photo by Charles E. Kay (3051-11), Utah State University.

FIRE MANAGEMENT

During most of the park's history, the fire management policy—if not practice—has been to suppress fires (Schullery 1989). This policy began with the United States cavalry in 1886 when it was sent to stop illegal hunting and in general to protect the park. Fire control, however, would have been difficult over such a large wilderness area. Fires that burned slowly enough to be put out by the cavalry would often have burned out by themselves before consuming more than a few hectares, as they do today (Schullery 1989). When ignitions occurred during a drought in areas with abundant fuel, the fire would probably have burned out of control before fire fighters arrived, which is also the case today. Fire spread might have been cur-tailed in some cases. Burning, however, was and continues to be a significant component of the landscape mosaic, which is appropriate in a national park, where fire is widely recognized to be a natural environmental factor.

The effective suppression of most fires became possible only after World War II with the availability of paratroopers and modern equipment. This suppression, and possibly that of the cavalry, caused increased tree and shrub density in the lower elevation Douglas-fir forests (figs. 14.11, 14.12; Houston 1973; see chapter 10), but research suggests that the higher forests were not influenced in a significant way (Romme and Despain 1989; Knight 1991). Fuel accumulation was slower there, and only about twenty years passed from the time of effective suppression until the Na-

F I G. 14.10   The scars on these aspen were caused by elk feeding on the bark and are commonly seen near winter ranges.

tional Park Service adopted a policy in 1972 that allowed some fires to burn:

1. Permit as many lightning-caused fires as possible to burn in designated areas.
2. Prevent wildfire from destroying human life, property, historic and cultural sites, special natural features, or threatened and endangered species.
3. Suppress all human-caused fires (and any lightning-caused fires whose suppression is deemed necessary) in as safe, cost-effective, and environmentally sensitive way as possible.
4. Resort to prescribed burning when and where necessary and practical to reduce hazardous fuels.

This policy was revised after the wildfires of 1988. Now, some fires will still be allowed to burn, but only under more carefully defined conditions. Most fires capable of burning large areas will probably be suppressed at the outset if possible.

Much was learned about fire management in the summer of 1988, which was the driest and windiest on record in the Yellowstone region.

Of course, large-scale burning is common in western North America and Alaska, where often-flammable coniferous forests are a predominant feature of the landscape. In 1988, however, there were fears that YNP, along with a critical tourism industry, was being destroyed. Such fears were fostered inadvertently by news editors approving overly dramatic headlines, public leaders making premature pronouncements, managers and scientists with insufficient information, and by smoke in the air for miles around that made life unpleasant. Newspaper headlines and public pronouncements were more moderate after the 1988 fires were extinguished. Although approximately 36 percent of YNP (<10 percent of the Greater Yellowstone Area) had been subjected to some burning, only about 15–20 percent of the park was subjected to crown fires, which give the impression of devastation (fig. 14.13; Despain et al. 1989). The geologic features that attract so many tourists are apparently unscathed, and the large mammal populations appear to thrive (Singer et al. 1989). Moreover, park visitation since 1988 has been higher than ever before (3 million in 1992; YNP data).

Controversy about fire in Yellowstone continues, but there are three points on which there is agreement. First, no one doubts that over the years fire has played a natural role in creating the mosaic of different vegetation types (Taylor 1973, 1974; Houston 1973, 1982; Lotan 1975; Arno 1980; Romme 1982; Romme and Knight 1982; Pyne 1989; Romme and Despain 1989; Despain 1990). Second, no one disputes that fires can occur in the absence of humans; lightning ignitions are common. Finally, everyone agrees that the suppression of every fire leads to fuel accumulations of such magnitude that suppression becomes impossible. There remains, however, some uncertainty about the importance of fire suppression in creating unnaturally high flammability in subalpine forests, and questions remain about the desirability and potential of prescribed burning.

The Importance of Suppression. Romme and Despain (1989) concluded that past fire suppression may have had an influence on the spread and severity of fires in 1988, but that it was not an important factor in creating the large scale of the fires. Apparently, extensive fires are best explained by the extremely dry and windy climatic conditions combined with adequate fuel accumulation that had developed through natural succession over the preceding 250–300 years (when the last comparatively large fires occurred). The investigators learned that the large area that burned in the late 1600s and early 1700s was the result of several large fires in a period of two decades or more (figure 14.14), and that in their study area over the last three hundred years, never had so much burned in one year as in 1988. Only time will tell whether the same fire pattern will occur in the late 1990s. Additional large fires could occur in the near future.

Two other factors are relevant in considering the effect of fire suppression. Fires on adjacent lands often burn into the park, and although suppression in the back country of Yellowstone may have been difficult before the 1940s because of inaccessibility and limited resources, fire suppression on adjacent, better-roaded national forests might have been more effective. Suppression outside park boundaries therefore essentially means fire suppression within. Also, fires started by native Americans have been eliminated in the Greater Yellowstone Area. There is evidence that fire frequency was higher in areas of Indian occupation than elsewhere (Barrett and Arno 1982; Arno 1985; Gruell 1985), and some have argued that the elimination of Indian fires created "unnatural" fuel accumulations in YNP since fire frequency caused by lightning alone would have been lower (Chase 1986; Bonnicksen 1989).

Other factors must be considered, however, in deciding on the effect of native American fires: First, little evidence is available on the effect of Indian fires in the Greater Yellowstone Area. The study by Barrett and Arno was done in western Montana at lower elevations. Arno (1985) and Gruell (1985) concluded that Indian-caused fires were probably less important at higher elevations, though fires could have spread into the higher mountains from the foothills. Second, fires that affect a large area probably occur when fuel and climatic conditions are right; the source of

ignition is comparatively unimportant. Yellowstone has many lightning ignitions each year, the vast majority of which go out without suppression.

With its abundance of fish and game, the Yellowstone region must have been used by native Americans. They surely started some fires, accidentally if not intentionally. These fires might have maintained the forests in a more open, less flammable condition than is common today, especially at lower elevations. Yet early explorers of Yellowstone wrote about extremely dense forests. For example, General Strong noted, "The trees have been falling here for centuries, and such a network of limbs, trunks, and stumps has been formed, that to face it with a horse . . . is enough to appall the stoutest heart . . . and add to this a perfect labyrinth of fallen timber, with limbs and branches entwined in every imaginable shape." Other explorers also commented on the dense forests (Haines 1977).

Thus, the large-scale 1988 fires could have been a "natural" event, despite the large area burned in one year, or a "near-natural" event, as concluded by Romme and Despain (1989) after considering the possibility of effective fire suppression. Such distinctions are fraught with ambiguity.

Using Prescribed Fires to Avoid Wildfire. Prescribed fires are traditionally those ignited by managers during climatic conditions that create moderate or low fire hazard. Usually the goal is to reduce fuel loads, improve wildlife habitat, and improve conditions for the growth of certain trees. Although prescribed burning is comparatively simple in Douglas-fir and ponderosa pine savannas at lower elevations, where many of the trees are not killed by surface fires (see chapter 10), the logistic problems are substantial in the more dense subalpine forests. Lodgepole pine forests are hard to ignite under most weather conditions, but when sufficiently dry, crown fires develop easily. Such fires are difficult to contain within designated boundaries, especially if unexpected winds occur. Moreover, forests with heavy fuel accumulations may continue smoldering until weather changes allow the fires to continue (J. K. Brown, pers. comm.). Some prescribed fires may thereby end up burning under weather con-

F I G. 14.11   A photograph taken by J. P. Iddings in 1885 of the Northern Range in Yellowstone National Park (Tower Junction area; elevation 1,920 m; 6,298 ft). Dead trees (probably Douglas-fir and limber pine) and the general lack of big sagebrush in the foothill grassland suggest that this area was recently burned. Two small stands of aspen are visible. From the J. P. Iddings Collection of the U.S. Geological Survey (USGS 152).

ditions more hazardous than when a prescribed ignition was thought to be safe.

Human-ignited prescribed fires are used only in circumstances where managers are confident of containing the fire within designated boundaries. Restricting the fires to the burning of surface fuels is often preferred, though canopy fires are sometimes prescribed (J. K. Brown, pers. comm.). In Rocky Mountain subalpine forests, to restrict prescribed burning to easily controlled surface fires—if that were possible—would be to change dramatically the nature of the fire disturbances that most likely prevailed in presettlement times. That is the dilemma that makes prescribed fire difficult to implement in western coniferous forests: to preserve fire as a natural process, some unpopular and possibly expensive wildfires must occur.

Though permitted by park policy, human-caused prescribed fires have rarely been used in YNP, because of a desire to minimize human influence. It has seemed better to let meteorologic and fuel conditions determine where fires should occur rather than depend on human decisions. The Park Service is committed to this policy, even though it has frequently led to difficult questions about the difference between fires caused by humans and those caused by lightning. Although not different physically, there could be differences in fire location and the extent of fire spread. For example, humans with drip torches can elevate temperatures to ignition points in several locations through prolonged exposure to flames, greatly increasing the chances that a fire will occur in an area that a lightning strike would fail to ignite. Moreover, some lightning ignitions will occur when human-caused prescribed fires would not be approved because of extreme fire hazard

F I G. 14.12   This photograph is of the same area shown in fig. 14.11 but was taken eighty-five years later (in 1970). Douglas-fir density and big sagebrush cover have increased considerably. Note also that the aspen stands have essentially disappeared, possibly be- cause of invasion by Douglas-fir with fire suppression. The historic fire-return interval in this area was 20– 25 years (Houston 1973, 1982). Idaho fescue and needle-and-thread grass are common. Photo by Douglas B. Houston.

conditions. As more is learned, it may be possible to mimic lightning fires more faithfully through prescribed burns.

Right or wrong, the goal of the Park Service has been to maintain a landscape mosaic, away from visitor centers, that is affected primarily by lightning-caused fires. Although park managers have had difficulty in addressing challenges about whether the mosaic is indeed "natural," there seems little need to modify the policy until substantial evidence is presented that the existing mosaic is artificial—owing to fire suppression or because human-caused prescribed fires were necessary to protect human life and property.

For some observers, the need for human-caused prescribed fires was made eminently clear in 1988. Noting that gateway communities and visitor centers were threatened because adjacent forests were too close and too flammable, they argued that human-caused prescribed fires during moderate fire hazard conditions, especially those near human habitations, could have minimized the anxiety created by the 1988 fires. Brown (1991) substantiated this claim, suggesting that the cost of fuel reduction near developments would have been reasonable compared to the cost of fighting the 1988 wildfires. Some fuel reduction could have helped to protect developed areas, despite the fact that some fires can be started by embers blown in from more than a mile away (called *spotting*).

Brown also concluded that the scale of the 1988 fires would not have been reduced significantly with a program of human-caused prescribed burns, given the traditional resources and attitudes of both the Park Service and Forest Ser-

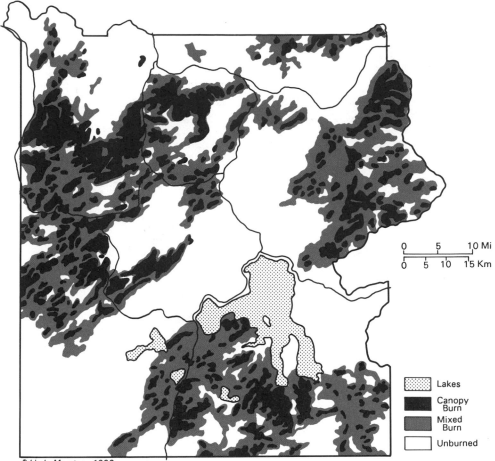

© Linda Marston, 1993

Lakes

Canopy Burn

Mixed Burn

Unburned

F I G. 14.13   Approximately 36 percent of Yellowstone National Park burned in 1988, but only 15–20 percent of the park had crown fires that gave the impression of devastation. Mixed burns were less intense or continuous and often consisted of surface fires that did not kill many trees. Adapted from Despain et al. 1989.

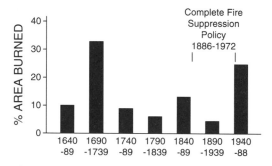

F I G. 14.14   The percentage of a 15,000 km² (5,790 mi²) area in Yellowstone National Park burned over fifty-year periods. Adapted from Romme and Despain 1989.

vice. Large areas of young forests, with comparatively little fuel, were burned in 1988 as well as older forests with abundant fuel (Romme and Despain 1989). Human-caused prescribed fires would have been set only under moderate fire conditions, with the result that only small areas would have burned. The resources needed to contain such fires if weather conditions and fire intensity had changed quickly would not have been sufficient to deal with more than a few fires each year. And not every year would have been acceptable climatically. Thus, assuming that human-caused prescribed fires had been used after 1972, when fire was officially accepted as a necessary process in the Greater Yellowstone Area, Brown concluded that too little of the forest away from human habitations would have been burned to prevent the large scale of the 1988 fires.

The current fire management policy does not preclude human-caused prescribed fire for the purpose of reducing hazardous fuels, and the manual removal of fuels with saws has been accomplished in some areas (primarily during the 1988 fires). Some may not like the more parklike surroundings thereby created, but there is no other choice (short of removing visitor centers altogether) if easier protection from fire is desired. Human habitations in coniferous forest face the same uncertain future as those on a floodplain or ocean beach.

THE NATURAL REGULATION POLICY

Many have questioned why fuel reduction programs in Yellowstone were not initiated around human habitations when they were allowed by policy. There are two answers: First, experience in the park before 1988 indicated that lightning-caused fires burned only a few thousand acres annually, and then only in dry years. Second, since YNP was different from most national parks because of its large size, there were compelling reasons to manage the park with a minimum of human intervention—a natural regulation policy. Human-caused prescribed fires seemed unnecessary, and preventing them was one way of maintaining a more natural state, which was similarly being maintained by allowing elk winter kill rather than hunting in the park.

The natural regulation policy has been sharply criticized by Chase (1986), Bonnicksen (1989), and Kay (1990). They argue that Yellowstone has already been subjected to so many adverse human impacts that allowing natural disturbances to occur without anticipatory management can only lead to further declines in park resources. Commonly mentioned adverse impacts include the loss of the gray wolf, declines in beaver and aspen, the elimination of native American fires, an alleged excessive fuel accumulation due to fire suppression, and a perception that there are too many large ungulates. Aldo Leopold recognized the dilemma of managing parks in 1927, when in a letter to the superintendent of Glacier National Park,

he wrote that "the balance of nature in any strict sense has been upset long ago. . . . The only option we have is to create a new balance objectively determined for each area in accordance with the intended use of that area." His son, Starker Leopold, later chaired a committee that recommended the restoration of parks to something close to the conditions existing before European settlement (Leopold et al. 1963). In general, critics of the Park Service want proactive management plans with specific goals and a scientifically sound program to assure that progress toward goals is being achieved.

Such recommendations seem highly appropriate for many national parks and natural areas, especially those of small size and low ecological diversity. In fact, a recent workshop that involved National Park Service employees also recommended improved management goals and more careful monitoring (Agee and Johnson 1988). Still, wildlands like those in and around Yellowstone may be large enough and natural enough for the kind of semipassive management now in effect (Houston 1971). Because of Yellowstone's large size and complexity, as well as inadequate funding for research, it has been difficult to establish goals that will withstand the criticism of being either too human-centered or not human-centered enough. Moreover, human values and goals change over time.

When possible, there is merit in taking a passive or semipassive approach to wildland management. To exercise active management consistently with precise goals preempts the opportunity to learn about ecological phenomena that have occurred for millennia. This in itself is one of the major scientific values of wildlands that should be protected. Preserving such values does not preclude large parks from being "for the enjoyment of the people" most of the time. The costs of park management may be especially high in some years (such as in 1988), but the economic benefits of preserving the ecological processes and diversity of places like Yellowstone are substantial year after year.

Given the certainty of climatic change during our tenure as Yellowstone stewards and the complex interrelations that characterize this ecosystem, it is difficult to know what management actions should be adopted.[6] Yellowstone is big, but is it big enough to preserve grizzly bears? Or wolves? What are the natural fluctuations and events that we should learn to expect as normal? What is the effect of human activities outside the park on important features within the park, and vice versa? What level of management is necessary? These questions merit our most careful attention. As Yellowstone scientists have warned us, however,

"We would be well advised to retain enough humility to know that nature will not always be controlled despite our best, most carefully planned management." They might have added that the Yellowstone Plateau will probably bulge and collapse again. Fortunately, Yellowstone has nearly all its original species and ecological processes intact. Furthermore, the area as a whole still offers a powerful sense of wildness and exceptional opportunities for research.

CHAPTER 15

Jackson Hole and
the Tetons

Nowhere in the Rocky Mountains is there such an abrupt rise in elevation as in Grand Teton National Park (fig. 15.1). From the intermountain basin known as Jackson Hole at an elevation of 1,981 m, the Teton Range rises 2,215 m to the top of Grand Teton at 4,186 m. Seven peaks are above 3,658 m. The Teton Range is the result of uplift along the Teton Fault, an 80-km-long crack in the earth's crust, where slippage continues to occur (fig. 15.2; Love and Reed 1971). While the mountains are gradually rising higher, in some places at the rate of about 30 cm every one hundred years, the valley floor east of the Tetons continues to sink.

Both the Yellowstone Plateau and the Tetons are extremely dynamic geologically, but in different ways. In the Yellowstone region, volcanism has created a landscape of solidified lava and consolidated ash. In contrast, the formation of the Tetons is more typical of the Rocky Mountains, where uplifting along a fault line and subsequent erosion have exposed some of the oldest rocks on earth (primarily Precambrian granite and gneiss). The present-day Tetons, however, are the youngest mountains in the region, having developed during the past 9 million years (Love and Reed 1971).

The processes that formed the Teton landscape are no less amazing than those that formed Yellowstone and the Absarokas. Before uplift, and some 40 million years after the formation of moun-

tains to the north, east, and south, the Teton–Jackson Hole region was flatter. The Precambrian rocks were buried under 6,000 m of sedimentary and volcanic rocks. With uplifting along the Teton Fault, much of the softer strata were eroded, exposing the harder rocks that now form the mountains. Today, all of the sedimentary rocks have been eroded from the steep eastern slopes and highest peaks, except for a telltale sandstone remnant on the top of Mount Moran. Geologists know that this sandstone is part of the same Flathead sandstone that now lies 10,000 m below the valley floor, a vertical displacement of nearly 11.2 km. The west slopes of the Tetons are gentler and are still covered with sedimentary strata.

Glaciation was another important factor in shaping the Tetons and Jackson Hole. The first episode, known as the Buffalo Glaciation, was the largest and began about 200,000 years ago. Ice covered most of Jackson Hole to a depth of 30–900 m, covering Signal Mountain, most of the Mount Leidy Highlands, and Blacktail, Miller, and East and West Gros Ventre buttes. The Bull Lake Glaciation was about half as large, occurring about 30,000 years ago, and was followed by the Pinedale Glaciation about 18,000 years ago. It was the Pinedale that created the terminal moraines at the foot of the Tetons that led to the formation of Jackson, Jenny, Leigh, Bradley, Taggart, and Phelps lakes (see fig. 2.6). The Pinedale Glaciation

F I G. 15.1   The Tetons rise abruptly above the Snake River and Jackson Hole to a maximum elevation on Grand Teton of 4,186 m (13,772 ft). Shrublands dominated by Mountain big sagebrush are characteristic of the flat glacial outwash plains (elevation 2,006 m; 6,600 ft), including the lower terrace carved during the Pleistocene epoch. Blue spruce, Engelmann spruce, narrowleaf cottonwood, river birch, silver buffaloberry, and various willows are common in the riparian woodlands; Douglas-fir occurs on the slopes to the left (south). See fig. 4.7 for a photograph taken about 90° to the right.

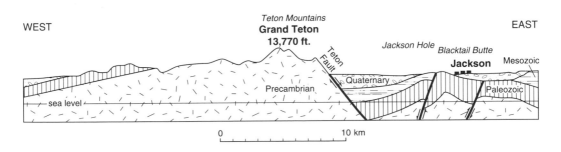

F I G. 15.2   A diagram illustrating the geologic structure of the Teton Range and Jackson Hole. Continuing displacement along the Teton fault has created a mountain range without foothills. Adapted from Love and Reed 1971.

F I G. 15.3   Melting glaciers created this kettle moraine topography when huge blocks of ice were buried and subsequently melted. The vegetation is mountain big sagebrush steppe. Photo by W. B. Hall and J. M. Hill, from the collection of J. D. Love.

also shaped the jagged peaks and U-shaped valleys of the Tetons and formed the potholes east of Jenny Lake. Also known as kettles, the potholes were formed when huge blocks of ice were left buried in a moraine. As the blocks melted, symmetrical depressions were created (fig. 15.3).

Human History

Anthropologists believe that native Americans were in Jackson Hole when the glaciers retreated about 9,000 years ago. Artifacts found on the northern end of Jackson Lake document the existence of hunting camps dating back about 2,500 years (Frison 1984; Love 1984). Considering the short, cool growing season of the valley, the region was probably used primarily for hunting in the summer.

The first Caucasian to enter Jackson Hole is believed to have been John Colter, who in 1807 requested a discharge from the Lewis and Clark Expedition so that he could hunt and explore on his own. He was followed by various trappers, including David E. Jackson, for whom the valley is named. Ferdinand V. Hayden led the first scientific expedition to Jackson Hole in 1872 (1872–78). One member of his entourage was Thomas Moran, who made sketches of the landscape while others collected information on natural resources. Many of the lakes and mountain peaks were named by Hayden, often after members of the expedition (Bradley, Taggart, St. John, Moran).

The settlement of Jackson Hole started in the 1880s, and the town of Jackson was formed in 1897. Homesteaders raised livestock and various crops (Righter 1982). Jackson Lake was enlarged with a dam (completed in 1906) for the purpose of storing more irrigation water for Idaho farmers (Palmer 1991). Perennial concerns about possible dam failure, triggered by the 1976 failure of the Teton Dam in Idaho, finally led to reinforcements that were completed in 1989.

A significant development occurred in Jackson Hole with the establishment of the National Elk Refuge in 1912. Located just north of Jackson, the refuge was established to provide a winter feeding ground for elk that had become a nuisance to ranchers. Winter feeding is a controversial topic among wildlife biologists, but it continues because the traditional winter range for these migratory animals is now occupied by more than ten thousand people. The refuge covers 9,840 ha and supports approximately 7,500 elk in the winter. It also provides habitat for 46 other species of mammals and 175 species of birds.

The abundant wildlife and spectacular scenery of Jackson Hole increased in popularity, and some began to think that the area should become a national park. Pierce Cunningham, a rancher near Spread Creek, circulated a petition in 1925 asking Wyoming or the federal government to set aside the valley "for the education and enjoyment of the Nation as a whole." Grand Teton National Park (GTNP) was established in 1929. John D. Rockefeller, Jr., took on the mission of buying as much of the land in Jackson Hole as possible, with the intent of enlarging the park through a land donation to the federal government (Schreier 1982;

Righter 1982). This was accomplished in 1950, though not without great criticism from many landowners. Before legislation could be passed enabling park enlargement, provisions were made for continued livestock grazing and the hunting of elk in certain areas. The booming economy of the valley eventually negated the animosity felt toward Rockefeller. In 1972, the land along the highway connecting Yellowstone and GTNP was given to the National Park Service by the United States Forest Service and named the J. D. Rockefeller, Jr., Memorial Parkway, thereby acknowledging his vision.

The Landscape Mosaic

Jackson Hole is surrounded by mountains that extend to above treeline. In general, the vegetation is typical of the region. With regard to GTNP alone, 58 percent is nonforested (alpine tundra, boulder fields, meadows, grassland, and shrublands), 28 percent is lodgepole pine forest, 7 percent Engelmann spruce–subalpine fir forest, 4 percent Douglas-fir woodland, 2 percent whitebark pine woodland, and 1 percent aspen groves (Greater Yellowstone Coordinating Committee 1987). Sagebrush steppe dominates the floor of Jackson Hole, except on glacial moraines, where lodgepole pine and spruce-fir forests are common (fig. 15.4). Rivers and streams are fringed by riparian woodlands or shrublands dominated by alder, balsam poplar, basin big sagebrush, blue spruce, common juniper, Engelmann spruce, lodgepole pine, narrowleaf cottonwood, russet buffaloberry, silver buffaloberry, willows, and other species (see chapter 4). Extensive willow shrublands occur near Jackson Lake (such as the Willow Flat area southwest of Jackson Lake Lodge).

Aspen groves are found in moist upland areas, especially at lower elevations where snow accumulates or where there is groundwater seepage to the surface. Loope and Gruell (1973) noted that aspen groves are frequently interspersed in sagebrush steppe and that Douglas-fir is a common associate. The mosaic of sagebrush steppe, aspen groves, and Douglas-fir woodlands is similar to that observed in the northern valleys of YNP. As in Yellowstone, fire and elk browsing are important environmental factors.

F I G. 15.4   This view of Jackson Hole from the slopes of the Tetons shows forests of lodgepole pine and subalpine fir on glacial moraines, and mountain big sagebrush steppe on the glacial outwash plains (elevation 2,006 m; 6,600 ft). Note that some trees are slowly invading the sagebrush steppe, perhaps because of fire suppression. The Gros Ventre Mountains rise to the east beyond the Snake River.

The drier lower slopes of the Tetons typically have mixed foothill shrublands (see chapter 9) or Douglas-fir forest (fig. 15.5). Five Douglas-fir habitat types were identified in the area by Steele et al. (1983), with the distinguishing understory species being ninebark and Rocky Mountain maple on comparatively moist sites, and mountain snowberry, pine reedgrass, and spiraea on the drier sites. Lodgepole pine is common on some mountain slopes, such as on Signal Mountain, the west slopes of the Tetons, and the mountains to the north and east of Jackson Hole, but it is infrequent on the east face of the Tetons. Engelmann spruce and subalpine fir forests occur on mesic north slopes in the Teton Range and in canyon bottoms; whitebark pine is common above 2,440 m.

Douglas-fir and lodgepole pine are the most common trees at lower elevations in the Jackson Hole area, typically occurring together on Black-

F I G. 15.5 West of the Snake River, patches of mountain big sagebrush are found dispersed in a shrubland dominated by low sagebrush. Lodgepole pine and subalpine fir dominate the forests on glacial moraines; Douglas-fir and limber pine occur on the drier slopes of the Tetons, and whitebark pine occurs near treeline. Forests of Engelmann spruce and subalpine fir are restricted to the cool north slopes of canyons and around some lakes at the mouth of canyons. Elevation 2,006 m (6,600 ft).

tail Butte and in the Gros Ventre Mountains to the east, where soils are deep and fertile enough to support tree growth. In some areas (for example, around Jackson Lake), Douglas-fir extends higher than lodgepole—a distribution pattern that is unusual for the Rocky Mountains. One possible explanation for this anomaly is that excessive water stress to lodgepole pine develops on the steep east face of the Tetons. In contrast, the soils of the glacial moraines on the valley floor, where lodgepole is abundant, have larger amounts of silt and clay (Oswald 1966; Love and Reed 1971) and probably provide a more mesic environment late in the summer. This explanation assumes that Douglas-fir in this area is more drought-tolerant than lodgepole pine. Another explanation hinges on temperature inversions caused by cold-air

drainage, which may cause temperatures in the valley bottom to be cool enough for lodgepole pine. Soil parent material may also be involved. Despain (1973) found that Douglas-fir was commonly associated with sedimentary rocks in the Bighorn Mountains. Farther south in Jackson Hole, near Hoback Junction, Douglas-fir is the most common tree at lower elevations, with lodgepole pine and aspen being abundant only at higher elevations. Lodgepole pine forest is common on Signal Mountain, with Engelmann spruce and subalpine fir common on north slopes at higher elevations (at least as saplings).

In general, the vegetation distribution is closely tied to the geologic substrate. With the recent completion of a detailed geology map for GTNP (Love et al. 1992), ecologists will be able to learn more about plant establishment and growth in relation to rock-weathering rates, nutrient supply, and the various soils that develop under diverse climatic and topographic conditions. In some areas, such as the Gros Ventre Mountains, sparse plant growth may be caused largely by the shallow, infertile soils that have developed from Cretaceous shales.

Predictably, fire and mountain pine beetles have had a significant effect on the landscape mosaic. Without fire, most of the aspen and lodgepole pine forests would change to forests dominated by subalpine fir and Engelmann spruce (Reed 1952; Oswald 1966; Steele et al. 1983). Loope and Gruell (1973) found stands of Engelmann spruce in moist habitats at the base of canyons and around some lakes, where fires have apparently been infrequent. With regard to mountain pine beetles, many dead trees are still visible from the beetle epidemic that occurred in the 1970s. The beetles killed many overstory trees, thereby stimulating the growth of understory plants such as spiraea, russet buffaloberry, pachistima, huckleberry, and trees that had previously been suppressed. Generally, the forests appear to be more flammable now, twenty years after the beetle outbreak.

At higher elevations on the Tetons, Douglas-fir forests are replaced by subalpine forests dominated by Engelmann spruce and subalpine fir, or by whitebark pine woodlands on drier sites that are subjected to more wind (fig. 15.6). Much of

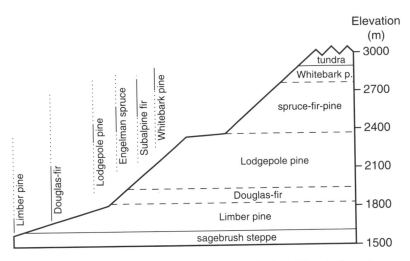

F I G. 15.6   The elevational distribution of major forest types near Grand Teton National Park. The solid vertical lines show the altitudinal ranges over which each tree species is important as a forest domi-nant; the dotted lines indicate the range over which the species can be found. Adapted from Baker 1986 and Whitlock 1993.

the Tetons, however, are above treeline. Alpine meadows and fellfields are common where soil has developed, such as around Solitude and Amphitheatre lakes, but the steeper slopes are essentially barren. Avalanche tracks extend from above treeline into the forested valleys and have clear vegetation patterns controlled by avalanche frequency (see chapter 13).

THE SAGEBRUSH MOSAIC

Much of Jackson Hole is an unusually flat sagebrush steppe, interrupted only by the Snake River valley and a few buttes and moraines. Geologists refer to the area as a glacial outwash plain because it was deposited by the meltwaters of the glaciers that developed 12,000–30,000 years ago. One of the primary effects of the meltwater was to wash away much of the silt and clay that would have greatly enhanced the nutrient and water storage capacity of the soil. Subsequently, the soils were enriched with loess deposits, blown in from the Snake River plains of Idaho.

Close examination of the sagebrush steppe indicates that at least three plant communities are represented. West of the Snake River and north of Jackson Lake Lodge, there is an intriguing mosaic of patches of low sagebrush interspersed in a ma-

trix of big sagebrush (see fig. 15.5). Low sagebrush is known to be an indicator of shallow soils, and possibly of soils with impeded drainage or lower fertility (Hironaka 1963). The second community occurs east of the Snake River, where there is a more complete cover of mountain big sagebrush, Idaho fescue, and other species. Low sagebrush is rare or absent, perhaps because it cannot compete in this environment. The abundance of big sagebrush suggests a deeper soil with a considerable water-holding capacity. The third sagebrush community is found toward the south of GTNP, near the Jackson airport and Blacktail Butte, and is characterized by the association of bitterbrush with mountain big sagebrush. The presence of bitterbrush suggests sand or gravel in the soil. The distribution of these three communities is probably a reflection of different glacial outwash patterns as well as differences in the rock types found in the mountains to the west and east of the valley (Sabinske and Knight 1978).

As discussed in chapter 6, there is evidence to suggest that sagebrush increases in cover with fire suppression and grazing pressure. The high cover of sagebrush in Jackson Hole could be attributable to these factors. Indeed, even some lodgepole pine trees are moving into the sagebrush steppe in a few areas. Big sagebrush, however, was a conspicuous part of the Jackson Hole landscape long before European settlers had moved into the valley. Osborne Russell noted that Jackson Hole was covered with sagebrush in 1835, and Walter W. De Lacy observed in 1863 that part of the valley east of the Snake River was covered with the "largest and thickest sagebrush" that he had ever seen (Dorn 1986).

THE GROS VENTRE LANDSLIDE

The mountain ranges to the east of Jackson Hole have diverse origins, with the volcanic Absarokas to the northeast, the gravel conglomerates of the Mount Leidy Highlands and the sedimentary strata of the Gros Ventre Mountains to the east, and the granite and gneiss of Jackson Peak to the southeast. All are much older than the Tetons. Fine-textured soils with high water-holding capacities have developed from the sedimentary rocks of the Gros Ventre Mountains, soils that are highly susceptible to slippage when saturated. Landslides and mudflows are common in the Gros Ventre Mountains (Blackwelder 1912; Voight 1978; Love and Love 1988), and the largest slide in United States history occurred there on 23 June 1925 (fig. 15.7). Despite a dense cover of coniferous forest, a block of earth 2.4 km long, 0.8 km wide, and 90 m deep in places slipped down the mountainside, crossed the Gros Ventre River, and created a dam 300 m long and 69 m high (Voight 1978; Lawrence and Lawrence 1984). All this occurred in about two minutes. Lower Slide Lake, now about 5 km long, was created, and ranch buildings were inundated.

The forests also slipped about 300 m in elevation, with many of the trees coming to rest at an angle. Lawrence and Lawrence (1984) observed that only the smaller trees survived, usually those less than forty years old. Smaller trees were probably flexible enough and light enough to avoid significant breakage. Seedlings of subalpine fir, limber pine, aspen, cottonwood, and other trees have since become established among the surviving Douglas-fir, Engelmann spruce, lodgepole pine, and aspen. In contrast, the scar left by the slide remains largely unvegetated and is visible from the main highway through GTNP, about 13 km to the east.

F I G. 15.7 Aerial view of the Gros Ventre landslide, which created Lower Slide Lake when it dammed the Gros Ventre River in 1925. The slide is more than 1.6 km (1 mi) long and 0.8 km wide. The forests are dominated by lodgepole pine at lower elevations and Engelmann spruce and subalpine fir at higher elevations. Photo (June 1955) by P. E. Millward, from the collection of J. D. Love. Elevation 2,128 m (7,000 ft).

Fears that the new dam would break discouraged tourism. Indeed, two years later the upper 18 m of the dam did fail, creating a flood of water, rock, and mud that claimed the lives of six people. The town of Kelly was swept away, and a 6-m wall of water hit Wilson two hours later, 40 km downstream. A new alluvial fan was created in the Kelly area, but the water level was back to normal the next day. Kelly has been rebuilt at the same location, and new shrublands and riparian vegetation have developed over the past sixty-five years. A trip to the Gros Ventre Slide provides a reminder of what could happen again.

Management Issues

Some of the current controversies over resource management in Jackson Hole and GTNP are essentially those of Yellowstone. A prominent issue for many years has been the adverse effects of elk and livestock on aspen groves. Especially near the National Elk Refuge, aspen are heavily browsed (see chapter 9). Browsing is a plausible explanation for the decline of some groves, but fire suppression might be part of the explanation. Subalpine fir and Douglas-fir are replacing the aspen through natural succession in some areas away from the refuge, thereby possibly increasing the browsing pressure on remaining groves. To maintain vigorous aspen groves, the Bridger-Teton National Forest is using prescribed burning as well as the cutting of conifers in aspen stands, but the debate continues (Boyce and Hayden-Wing 1979; Boyce 1989; Kay 1993). Are there too many elk? Considering the benefits of the refuge, is the loss of some aspen stands a small price to pay? Could the elk herd be managed so that aspen stands will have a chance to recover after a period of heavy browsing? Notably, only about 1 percent of GTNP proper has aspen cover, but elk hunting is allowed in the park because of concerns about too many animals in the valley.

Fire management is another issue that is essentially the same in GTNP and Yellowstone (see chapter 14). There is no doubt that fires have historically played an important role in shaping the vegetation mosaic. In 1897, T. S. Brandegee (1899) explored Jackson Hole and observed: "It is only occasionally that tracts of timber of merchantable size are found. . . . This condition appears to be due simply and solely to fires which have swept over the country so completely and persistently that scarcely any part has been entirely exempt from them. . . . Under present conditions the tree-bearing regions as a whole decrease, while the aspen areas increase at the expense of those now producing conifers." Two prominent artists, pho-

tographer W. W. Jackson and painter Thomas Moran, complained in 1878 and 1879, respectively, of the smoke that obscured the Tetons (Jackson and Driggs 1929; Fryxell 1932).[1] Gruell (1980) compiled a series of repeat photographs that documented the increases in tree and shrub density between 1899 and 1971. He attributed the changes to fire suppression. Loope and Gruell (1973) concluded that extensive fires had occurred between 1840 and 1879, with most stands originating after fires in 1856 and 1879, and that Douglas-fir stands had an average fire-return interval of fifty to a hundred years before settlement. They also found evidence of fires in the 1600s and 1700s.

Fires occurred on the sagebrush steppes of Jackson Hole as well as in the forests. Frank Bradley (1873), a member of the 1872 Hayden Expedition, wrote: "Large areas of sage had been burned off, and the grasses had grown up densely, forming fine pasturage and on these we again encountered antelope."

With the introduction of livestock, fires in sagebrush may have become less common because of consistently lower amounts of fine fuels in late summer or fall, when a fire is most likely. Of course, bison may have reduced the fine fuels as well, at least in some years. Historic fire-return intervals are difficult to determine in grasslands and shrublands, but the current high density of big sagebrush and the presence of some young trees in the shrublands suggest that fire frequency is lower now than it was in presettlement times.

Loope and Gruell (1973) were strong advocates of a prescribed burning program in the Jackson Hole area and were instrumental in developing a fire management plan for GTNP similar to the plan now in effect. Under the present policy, fires are allowed to burn if they are started by lightning in areas where human life or property do not appear to be in danger, and if adequate resources are available to extinguish the fires. Unlike YNP, the fires of 1988 did not burn a large area in GTNP (<1 percent, or about 24 ha). Other, larger fires have occurred in the past, such as the Waterfalls Canyon fire that burned 1,400 ha in a period of four months in 1974 on the western side of Jackson Lake. The biggest problem with that fire was for tourists, who had driven many miles to enjoy the Tetons, and who, like Jackson and Moran, often could not see the mountains because of smoke.[2]

Unique to GTNP are the issues of livestock grazing, the presence of an airport, elk hunting, and the management of Jackson Lake and the Snake River. About 9 percent of the park is open to livestock grazing (Boyce 1989). For some visitors, the mere presence of domestic livestock in a national park is controversial. A few national parks, however, do allow certain historic occupations to continue.

The most recent controversy pertaining to livestock has been the feared spread of brucellosis from bison and elk to cattle (Keiter and Froelicher 1993). With evidence that bison have been a natural part of the Jackson Hole ecosystem, GTNP biologists introduced twenty animals to the park in 1948. The herd now numbers about 160, and they coexist with about 8,500 elk, 1,500 mule deer, perhaps 500 moose, and about 1,600 cattle.[3] Ranchers with park grazing permits routinely vaccinate their cattle against brucellosis, and there has been no problem. Some neighboring ranchers, however, have suffered losses from the disease. Although elk and bison wander beyond park boundaries, currently there is little or no evidence that the cattle acquired the bacteria from these native ungulates.[4]

As with livestock grazing, the airport and Jackson Lake reservoir predate the establishment of GTNP and are important parts of the valley's economy. For many, the noise of jet passenger planes has no place in a national park, but attempts to eliminate the airport have been unsuccessful. The current controversy is whether the runways should be lengthened to increase safety. With regard to Jackson Lake, part of the debate has been resolved with the recent strengthening of the dam. Engineers are now more confident that the dam will withstand damage from earthquakes. Of greater concern to some are the unnatural beaches created when the reservoir is low (because of de-

mands for irrigation water or the need to create storage capacity for floodwaters). All reservoirs have this problem, but now that the dam has been strengthened, it should be possible—except in drought years—to maintain the reservoir at higher levels.

Streamflow regulation has a significant effect on riparian landscapes (as discussed in chapter 4), and the Snake River is no exception. Mills (1991) found evidence of a trend toward greater channel stability, fewer side channels or braids, increased tree cover, and reduced shrub cover along the 42-km section of the Snake River in GTNP below the dam. Higher flows are preferred for rafting, a form of recreation that contributes significantly to the economy of Jackson Hole, but reduced channel migration and streamflow fluctuations are creating a riparian landscape that probably did not exist before the construction of Jackson Lake dam. Short of removing the dam, lowering the lake 12 m to its original level, and removing homes from the floodplain below, nothing will prevent an increasingly less natural riparian landscape from developing.

Overall, the landscape mosaic of Jackson Hole and the Tetons is strongly influenced by the geologic substrate. Glacial moraines and outwash plains, steep mountain slopes, avalanche tracks, the Snake River and its numerous tributaries, and other geologic features all exert strong influences on plant and animal life. Human influences are also significant: a winter feeding program concentrates elk in unusually large numbers on the elk refuge, a reservoir has been created with disturbing water-level fluctuations, and streamflow regulation on the Snake River has altered the riparian ecosystem significantly. Still, Jackson Hole and the Tetons constitute one of the most inspiring parklands in North America. The area also provides exceptional opportunities for learning about the natural history and ecology of the Rocky Mountain region.

CHAPTER 16

The Black Hills, Bear Lodge Mountains, and Devil's Tower

The Black Hills and Bear Lodge Mountains of western South Dakota and northeastern Wyoming originated as an elliptical uplift (anticline) at about the same time as the Rocky Mountains. Fifty million years of erosion have now stripped thousands of feet of sedimentary material from the mountains, exposing the granitic core. The faces of Mount Rushmore are carved in this granite. Today the Black Hills rise from 900 to 1,200 m above the surrounding plains, with the highest point being Harney Peak at 2,207 m (7,239 ft) above sea level. Erosion occurred largely toward the east, the direction in which two major rivers flow (the Belle Fourche around the northern end of the Black Hills and the Cheyenne around the southern end) (fig.16.1). Erosion also exposed one of the best-known volcanic necks (an igneous intrusion) in North America: Devil's Tower (fig. 16.2).

The Sioux Indians occupied the Black Hills just before the arrival of the Europeans (Froiland 1990). Probably the first to arrive were the Verendrye brothers, who explored the area near Sturgis in 1773. Many more came in the 1870s following the confirmation of gold by the Custer Expedition. In July 1874, Lt. Col. George A. Custer left Fort Abraham Lincoln (near Bismarck, North Dakota) with twelve hundred cavalrymen for the purpose of "reconnoitering a route to the Black Hills and of exploring their hither unattained mountainous interior" (Custer 1875). Among his many observations, Custer wrote: "Every step of our march that day was amid flowers of the most exquisite colors and perfume. So luxuriant in growth were they that men plucked them without dismounting from the saddle. . . . It was a strange sight . . . the men with beautiful bouquets in their hands, while the head-gear of the horses was decorated with wreaths of flowers fit to crown a queen of May. Deeming it a most fitting appelation, I named this Floral Valley." Custer also had his picture taken with a dead grizzly bear (Progulske 1974). He died two years later in the Battle of the Little Bighorn, in the centennial year of the United States. Much of the Black Hills was set aside by Congress as a forest reserve in 1897, and today most of the Black Hills and Bear Lodge Mountains are in the Black Hills National Forest (Green 1978).

The Black Hills are still an important source of gold, with North America's largest underground gold-producing mine located at Lead, South Dakota. The mountains are also an important source of wood, water, and livestock forage. Many tourists come to the area each summer, attracted by abundant wildlife, such interesting geologic fea-

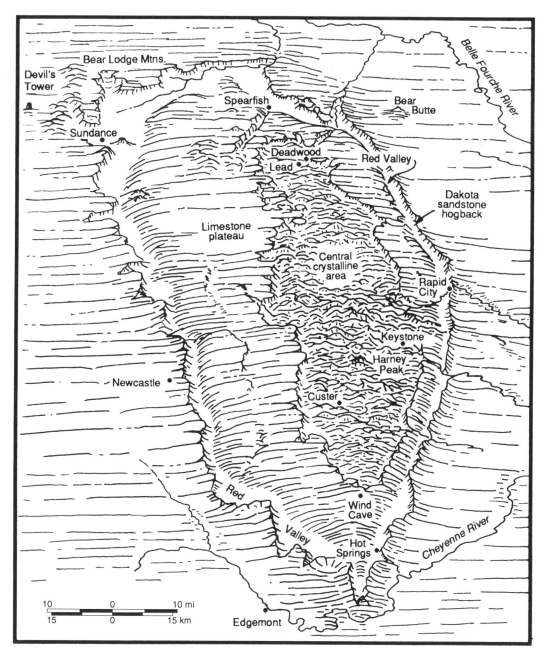

F I G. 16.1   Geomorphic features of the Black Hills. The elevation ranges from 955 m (3,140 ft) at Rapid City, 1,444 m (4,750 ft) at Sundance, and 1,318 m (4,334 ft) at Newcastle to 2,202 m (7,242 ft) on top of Harney Peak, the highest point in South Dakota. Adapted from A. N. Strahler 1969.

tures as extensive caverns, a summer climate that is cooler than the surrounding plains, beautiful forests and meadows, spectacular canyons, fossil beds, Devil's Tower, Wind Cave National Park, and of course, Mount Rushmore.

Botanically, one of the most interesting aspects of the Black Hills is the assemblage of plants found there (table 16.1; Hayward 1928; McIntosh 1931). As might be expected, many Rocky Mountain species occur in the Black Hills, such as dwarf huck-

F I G. 16.2   Devil's Tower flanked by
ponderosa pine savanna and mixed-
grass prairie. The historic fire-return
interval in this area was fourteen years
(Fisher et al. 1987). Blue grama, bluebunch wheatgrass, junegrass, green
needlegrass, little bluestem, and sideoats grama are common in this area
at 1,338 m (4,400 ft). Elevation at
the top of Devil's Tower is 1,556 m
(5,117 ft).

leberry, green gentian, heartleaf arnica, narrowleaf cottonwood, oregongrape, and ponderosa
pine. Their presence is best explained by the generally accepted conclusions that Rocky Mountain
forests were more widespread about ten thousand
years ago (Sears 1961; Marquis and Voss 1981)
and that the Black Hills provide an environment
where many of these species can still survive. Annual precipitation is considerably higher than on
the surrounding plains, ranging from 46 to 74 cm
(fig. 16.3).

More puzzling is the presence of species that
are typically found in the deciduous forests of
eastern North America (such as American elm,
bloodroot, boxelder, bur oak, hackberry, and
ironwood), or species from the boreal forests that
stretch across Canada (such as bunchberry dogwood, Canada scurvyberry, paper birch, and
white spruce). These species probably occurred

in the area even as the mountains were being
formed. With the uplifting of the Rocky Mountains, the climate became drier, and many forest
plants could not survive except in the Black Hills.

Alternatively, the eastern species could have
migrated westward along the moist tributaries of
the Cheyenne River (Buttrick 1914; Wright 1970),
where they still occur in favorable habitats. The
northern species may have migrated southward
during the cooler periods associated with the advance of the continental glaciers, though the Black
Hills were not glaciated. As the glaciers melted, the
plants persisted in the south wherever the climate
was suitable.

Geomorphic Regions

The anticlinal origin of the Black Hills and Bear
Lodge Mountains has led to the formation of four

TABLE 16.1   Some plants in the Black Hills that are representative of other floristic regions (based on Buttrick 1914, Hayward 1928, McIntosh 1931, Wright 1970, and Marriott 1985).

ROCKY MOUNTAIN SPECIES

Heartleaf arnica	*Arnica cordifolia*
Dwarf huckleberry	*Vaccinium scoparium*
Narrowleaf cottonwood	*Populus angustifolia*
Oregongrape	*Mahonia repens*
Ponderosa pine	*Pinus ponderosa*
Richardson geranium	*Geranium richardsonii*
Showy elkweed	*Frasera speciosa*

GREAT BASIN SPECIES

Big sagebrush	*Artemisia tridentata*
Bottlebrush squirreltail	*Sitanion hystrix*
Fendler threeawn	*Aristida fendleriana*
Gaillardia	*Gaillardia aristata*
Junegrass	*Koeleria macrantha*
Skunkbush sumac	*Rhus trilobata*
Threadleaf sedge	*Carex filifolia*
True mountain-mahogany	*Cercocarpus montanus*

EASTERN DECIDUOUS FOREST SPECIES

Bur oak	*Quercus macrocarpa*
American elm	*Ulmus americana*
Hackberry	*Celtis occidentalis*
Boxelder	*Acer negundo*
Ironwood	*Ostrya virginiana*
Green ash	*Fraxinus pennsylvanica*
Bloodroot	*Sanguinaria canadensis*
American columbine	*Aquilegia canadensis*
Yellow violet	*Viola pubescens*
Virginia creeper	*Parthenocissus quinquefolia*

BOREAL FOREST SPECIES

Bunchberry dogwood	*Cornus canadensis*
Canada scurvyberry	*Maianthemum canadense*
Paper birch	*Betula papyrifera*
Twinflower	*Linnaea americana*
White spruce	*Picea glauca*
Woodnymph	*Moneses uniflora*

SOUTHERN GREAT PLAINS

Bush morningglory	*Ipomoea leptophylla*
Pricklypoppy	*Argemone intermedia*
Red Threeawn	*Aristida longiseta*
Sand sagebrush	*Artemisia filifolia*
Smallflower gaura	*Gaura paryiflora*

geomorphic regions that strongly influence landscape patterns (fig. 16.1). On the perimeter is the Hogback Rim, composed of Lakota sandstone, Minnewasta limestone, and other erosion-resistant sedimentary strata that were tilted upward in the early Eocene. Although sharply defined to the east and south, the rim in Wyoming is more like a plateau. Ponderosa pine savannas predominate on rockier soils, grasslands on fine-textured soils. As is typical for the Rocky Mountain region, rivers

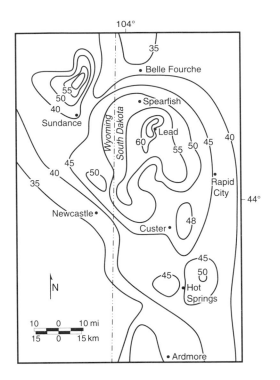

F I G. 16.3   Mean annual precipitation (cm) patterns in the Black Hills and Bear Lodge Mountains. Adapted from Boldt and Van Deusen 1974.

have eroded through the rim in several places.[1]

At the interior of the Hogback Rim is the Red Valley, also known as the race track because of the well-defined, oval-shaped valley that extends around much of the Black Hills. The valley was formed by the erosion of softer shales, siltstones, and sandstones underlying the more resistant strata that form the Hogback Rim. The name stems from the red shales of the Spearfish formation (known as the Chugwater farther west) that give a characteristic red color to the fine-textured soils. Grasslands predominated in the Red Valley during presettlement times, but now some of the area is cultivated. The elevation for the valley ranges from 1,000 to 1,100 m.

In the western part of the Black Hills and to the interior of the Red Valley, the sedimentary strata have not eroded significantly and form a comparatively flat area known as the Limestone Plateau. The limestone substrate is permeable to water, a geologic feature that has led to the formation of numerous caverns. Some creeks disappear into

the bedrock before emerging again (such as Spearfish and Boulder creeks in the northern Black Hills). The vegetation of the plateau consists primarily of ponderosa pine forests and savannas, with bur oak woodlands on drier sites and groves of white spruce in canyons or on north-facing slopes. Grasslands and foothill shrublands also occur on the plateau. Elevation ranges from 1,900 to 2,200 m.

The fourth geomorphic region is where the igneous core of the Hills has been exposed through the erosion of the sedimentary strata. Known as the Central Area, this extremely rugged region is entirely in South Dakota and has many pinnacles and canyons (fig. 16.4). The elevation is mostly 1,525–1,830 m, but five peaks composed of igneous rock rise above 2,130 m: Harney, 2,207 m; Bear Mountain, 2,184 m, Crook's Tower, 2,176 m; Terry Peak, 2,159 m; and Crow's Nest, 2,148 m. Again, the vegetation is predominately ponderosa pine forest, with groves of white spruce in the more mesic habitats. Several lakes occur in the area (such as Sylvan Lake), but all have been created by the damming of rivers.

The Bear Lodge Mountains, though having the same origin as the Black Hills, have less well-defined geomorphic regions (Lisenbee et al. 1981). They occupy a much smaller area and have few exposed igneous rocks.[2] The highest elevation is 2,029 m (Warren Peak). Except for the absence of white spruce, the vegetation of the Bear Lodge Mountains is similar to that of the Black Hills.

The Vegetation Mosaic

As in the Rocky Mountains, topography, elevation, moisture availability, and soil texture have a significant effect on Black Hills vegetation. These ecological relationships have been discussed in chapters 9 through 13. The vegetation of the Black Hills and Bear Lodge Mountains, however, is distinct in several ways that merit further discussion.

First, nowhere else in the region are ponderosa pine forests so extensive (fig. 16.5).[3] They occur on nearly all soil types and exposures and can be classified in seven habitat types (Hoffmann and Alexander 1987). Rocky Mountain juniper occurs with pine on drier locations (such as the Hogback Rim), and two shrubs, snowberry and bearberry,

F I G. 16.4   The granitic central area of the Black Hills has widespread forests of ponderosa pine, with groves of aspen and white spruce along ravines and in other mesic habitats. Elevation 1,525–1,830 m (5,000–6,000 ft).

are widespread associates in the Central Area and parts of the Limestone Plateau. Similar forests occur on escarpments in the Chadron National Forest in northwestern Nebraska and in Custer National Forest in southeastern Montana. Bur oak and ponderosa pine form a distinctive conifer-hardwood association in the Bear Lodge Mountains and parts of the northern Black Hills where annual precipitation is higher (fig. 16.6). Elsewhere where there is higher precipitation, on the Limestone Plateau as well as in the Central Area, ponderosa pine is found with aspen, chokecherry, common juniper, mountain ninebark, paper birch, and white spruce.

In all localities there seems little doubt that fire suppression has led to an increase in the density of ponderosa pine (Gartner and Thompson 1972; Progulske 1974; Bock and Bock 1984). This pine reproduces readily, sometimes forming dense dog-hair stands (Alexander and Edminster 1981; see chapter 10). Most of the ponderosa pine forests are second-growth, having been subjected to logging for many years. Mountain pine beetles and various diseases are known to occur in the Black Hills, but dwarf mistletoe is rare or absent (Boldt and Van Deusen 1974).

White spruce dominates two habitat types, one that is identified by an abundance of twinflower and the other by an abundance of dwarf huckleberry (Hoffmann and Alexander 1987). Spruce groves are found at higher elevations, in cool canyons and ravines, and on north-facing slopes (fig. 16.7). They occur on both igneous and limestone substrates with ponderosa pine, paper birch, and aspen. Engelmann spruce, subalpine fir, and Douglas-fir are not found in the Black Hills.

F I G. 16.5   Ponderosa pine forests are common in the Black Hills, the Laramie Mountains, and on east slopes of the Bighorn Mountains. Common juniper is the conspicuous shrub in this Black Hills forest. Elevation 1,946 m (6,400 ft).

Lodgepole pine and limber pine are found in only one or two localities (Rogers 1969; Thilenius 1970).

The Black Hills region has three kinds of deciduous woodland. The first, found in ravines and along the floodplains of rivers at lower elevations, is dominated by bur oak, boxelder, American elm, green ash, hackberry, and plains cottonwood. Shrub thickets with chokecherry, American plum, currant, wild rose, hawthorn, western snowberry, red-osier dogwood, and various willows are also found near these woodlands (see fig. 9.12). Historical photographs show that such streamside habitats were more abundant and more dense before livestock grazing and farming became so prevalent (Hoffmann and Alexander 1987).

The second type of deciduous woodland occurs on the upland and is dominated by bur oak (see fig. 9.14). It occurs in the northern and east-

ern parts of the Black Hills, mostly north of French Creek. Common associates include ironwood, ponderosa pine, and western snowberry, though bur oak may form a savanna with little understory on Mowry shale. Sometimes the oak occurs as a shrub in the understory of ponderosa pine forests (fig. 16.6), and because of its ability to sprout, the oak can become the dominant tree if the pines are killed by fire or harvesting. In fact, the reestablishment of pine may be difficult in some areas because of intense competition from the oak and other species (McIntosh 1931).

Aspen is the characteristic tree of the third type of deciduous woodland (Severson and Thilenius 1976). It too occurs in the northern and central Black Hills, where annual precipitation is higher, and occasionally it grows with paper birch. The aspen woodland often forms a narrow band between the pine or spruce forest above (on coarse-

F I G. 16.6   Bur oak sometimes grows in association with ponderosa pine in the northern Black Hills and Bear Lodge Mountains. Oak shrublands can be created if the pine are cut. Surface fires sometimes kill the oak above ground, but the root crowns survive and produce vigorous sprouts. Elevation 1,520 m (5,000 ft).

F I G. 16.7   White spruce in the Black Hills usually occurs at higher elevations or along valley bottoms characterized by cool-air drainage. This stand is at 2,098 m (6,900 ft).

textured soils) and the grasslands or meadows below (fig. 16.8; Hoffmann and Alexander 1987). Associated species include hazelnut, bracken fern, and wild sarsaparilla. Such woodlands can be seral to white spruce or ponderosa pine, and they commonly invade grasslands if browsing or grazing is not heavy. As with oak, the aspen and birch usually sprout from roots following disturbances, enabling them to regain dominance more quickly than the pine (see chapter 10).

The grasslands on the fringes of the Black Hills and Bear Lodge Mountains are typical mixed-grass prairie and are dominated by blue grama, green needlegrass, little bluestem, needle-and-thread grass, sideoats grama, western wheatgrass, and a variety of sedges and forbs (MacCracken et al., *Plant community variability,* 1983; see chapter 5). Big bluestem, Indian ricegrass, and prairie sandreed can be found where the soils are sandier. Streamside meadows in the same area have prairie

cordgrass, tufted hairgrass, wild iris, and a variety of sedges. Inland saltgrass is common where the soils are saline. The interesting interaction of prairie dogs and bison on the mixed-grass prairie of Wind Cave National Park has been discussed in chapter 9.

Mountain meadows are common at higher elevations in the Central Area and on the Limestone Plateau (Pase and Thilenius 1968). Four of the largest are Reynolds Prairie, Gillette Prairie, Slate Prairie, and Danby Park (fig. 16.9). In addition to mixed-grass prairie species, there is an abundance of the introduced Kentucky bluegrass and a variety of showy forbs, including fleabane, blackeyed coneflower, geranium, horsemint, larkspur, penstemon, sego lily, and wood lily. The absence of trees from these meadows is difficult to explain, but it is probably related to soil characteristics. Hayward (1928) suggested that fine-textured soils

F I G. 16.8   Aspen groves in the Black Hills and Bear Lodge Mountains frequently form a border separating meadows on fine-textured soils from ponderosa pine forests on coarse soils. Elevation 1,885 m (6,200 ft).

F I G. 16.9   Reynolds Prairie (west of Deerfield), one of several mixed-grass prairies that occur on fine-textured soils in the Black Hills. Elevation 1,824 m (6,000 ft).

favored meadows, whereas coarser soils favored forests, a conclusion that also seems appropriate for the Rocky Mountains (see chapter 12).

Foothill shrublands are common in the Black Hills region (MacCracken et al., ibid., 1983; Hoffmann and Alexander 1987), typically on the Hogback Rim and on lower parts of the Limestone Plateau. Mountain-mahogany, skunkbush sumac, sideoats grama, and an occasional Rocky Mountain juniper occur to the southwest. Various species of sagebrush (including big sagebrush, sand sagebrush, and silver sagebrush) are common on deeper soils, and snowbrush and russet buffaloberry are common in the northern foothills. Bur oak and skunkbush sumac occur together in the eastern foothills.

Land Management Issues

The forests of the Black Hills and Bear Lodge Mountains are the most productive in the region because of the longer, warmer growing season and comparatively high annual precipitation (including more abundant summer rainfall). Consequently, the forests have been heavily managed for timber production (Boldt and Van Deusen 1974). With an increased demand for outdoor recreational opportunities in the Black Hills and to help preserve biological diversity (as mandated by Congress), there is now considerable pressure to close some roads and preserve whatever remnants of old-growth forest remain.

The Black Hills also provide an abundance of livestock forage. Excessive grazing has occurred in some areas, but range managers are becoming more proficient in managing livestock. Still, problems in riparian zones frequently develop where cattle concentrate. Eroded stream banks and the heavy browsing of willows and other shrubs are obvious in many areas. Where shrublands were once abundant, meadows often predominate (Froiland 1990), a situation similar to that observed in Yellowstone National Park. Stream valleys are also a preferred location for summer homes because that is where private land is available. Except near rivers and creeks, the Black Hills are largely under federal control.

Perhaps the most subtle but far-reaching human effect on the Black Hills has been fire suppression. Suppression has been a guiding principle for land management there and remains so today, except in Wind Cave National Park, Devil's Tower, and a few other areas where prescribed burning has been used to effect presettlement environmental conditions. As discussed in chapter 10, ponderosa pine forests are characterized by surface fires every five to twenty-five years, depending on topographic position, elevation, and slope exposure. Burning kills most young trees but usually not the older trees, because of their thick bark, and it also maintains a more open forest with low amounts of fuel (Progulske 1974).

The importance of fires is well documented by fire-scarred trees (Gartner and Thompson 1972; Fisher et al. 1987).

Although surface fires are thought to have been characteristic of presettlement ponderosa pine forests, early explorers gave the impression that crown fires also occurred. In 1880, H. Newton and W. P. Jenney wrote:

The Black Hills have been subjected in the past to extensive forest fires, which have destroyed the timber over considerable area. Around Custer Peak and along the limestone divide, in the central portions of the Hills, on the headwaters of the Box Elder and Rapid Creeks, scarcely a living tree is to be seen for miles. . . . Some portions of the parks and valleys, now destitute of trees, show by the presence of

F I G. 16.10   An 1874 photograph of George Armstrong Custer's Seventh U.S. Cavalry camp near the confluence of Castle and Silver creeks in the Black Hills. Note the horses grazing beyond the tents, and the open ponderosa pine forests in the background. The photograph was taken in late July; elevation 1,829 m (6,000 ft). Photo by William H. Illingworth, provided by the South Dakota State University Cooperative Extension Service.

FIG. 16.11   This photograph is of the same area shown in fig. 16.10, but ninety-nine years later. The ponderosa pine are now more dense, probably because a longer time elapsed since the last fire. Aspen groves are conspicuous. Photo by Richard H. Sowell, provided by the South Dakota State University Cooperative Extension Service (originally published in Progulske 1974).

charred and decaying stumps that they were once covered by forest, but generally the pine springs up again as soon as it is burnt off, though sometimes it is succeeded for a time by thickets of small aspens.

Similarly, R. I. Dodge observed in 1876: "Throughout the Hills the number of trees which bear the marks of the thunderbolt is very remarkable. . . . The woods are frequently set on fire and vast damage is done. There are many broad belts of country covered with tall straight trunks of what was only a short time before a splendid forest of trees." Early photographs also suggest a more open forest with many dead standing trees (figs. 16.10, 16.11), and lightning strikes today are still a major cause of grassland and forest fires in the region (Lovaas 1976). Parts of the Black Hills

might have experienced a comparatively long fire-free period in the 1800s, which would have enabled fuel accumulation sufficient for the extensive crown fires suggested by the early explorers. Some areas might have been characterized by surface fires that maintained open savannas, while others at higher elevations had less frequent but more dramatic crown fires, as in the Rocky Mountains to the west. We should not expect all of the Black Hills and Bear Lodge Mountains to have a single fire regime, even if ponderosa pine is the most conspicuous tree throughout the area. Fisher et al. (1987) describe the effect that native Americans might have had on fire frequency in the Devil's Tower area (see chapter 10).

The effects of fire suppression are various and wide ranging: tree density has clearly increased, with trees often invading adjacent grasslands and

FIG. 16.12   With fire suppression, young ponderosa pine invade mixed-grass prairie. The trees in the background are bur oak. Elevation 1,672 m (5,500 ft).

causing a concomitant reduction in the amount of forage for livestock, deer, and elk (fig. 16.12; Thompson and Gartner 1971; Alexander and Edminster 1981).[4] Competition between domestic and wild herbivores thereby increases in certain seasons as their food base becomes less abundant. Thinning by timber harvesting can increase forage production, especially if the forest has not been in an excessively dense condition for so long that most of the understory plants have died (Thompson and Gartner 1971). If the understory has few native plants, then thinned forests are more susceptible to invasion by less desirable introduced species. Fire suppression on grasslands can lead to weed invasion as well (Schripsema 1977), primarily because the native species often tolerate burning better than introduced species (such as Japanese brome, Kentucky bluegrass, and sweetclover).

Increased tree density and forest expansion is also thought to have caused a reduction in streamflow (Orr 1968, 1975). Evapotranspiration is higher from forests and woodlands than it is from grasslands, and as discussed in chapter 11, watershed managers know that streamflow can be increased by tree harvesting. Beaver populations have declined in recent years (Froiland 1990), perhaps because of reduced streamflow caused by increased tree density on the upland.

The National Park Service as well as others now recognize the value of periodic fires for maintaining grasslands and savannas, reducing fuel amounts and the probability of hard-to-control crown fires, maintaining streamflow and forage availability, and preserving a more diverse landscape capable of sustaining a higher level of biological diversity (Lovaas 1976). Negative public opinion toward fire was fostered for many years by Smokey Bear, and by fires spreading onto private land from national parks and national forests. Areas where livestock grazing and timber harvesting were not allowed became known as the spawning grounds of destructive fires. Yet the negative effects of fire suppression have moderated those views, and prescribed fires are now becoming more common on both private and federal land.

Fire management plans, however, are not easily implemented. Some forests have become so dense that uncontrollable crown fires may easily develop. Tree thinning and the manual removal of fuel will often be necessary before fire can be used as a management tool. Such work can be expensive if the dense, small trees do not have a high market value. Prescribed burning in grasslands also poses significant challenges. For example, fires are usually kept small so that they can be extinguished easily, but the new regrowth of burned grasslands typically attracts bison and cattle. Excessive grazing can result if the burns are too small.

The challenges of resource management in a comparatively warm, humid landscape modified by a century of fire suppression are formidable, especially when uncertainty is involved. When and where will lightning-caused fires start? Is the trend toward increased tree density solely the effect of fire suppression? Could climatic change be a contributing factor? What are the effects of timber harvesting and other landscape manipulations on birds and small mammals as well as the game species? It remains to be seen whether timber harvesting and prescribed fires can be used effectively to reduce the flammability of the forests and preserve a landscape mosaic in a way acceptable to a diverse, knowledgeable public concerned about the area. The Black Hills, Bear Lodge Mountains, and Devil's Tower are beautiful landscapes that will be appreciated by a growing number of visitors for their natural value.

Sustainable
Land Management

CHAPTER 17

Using Wyoming Landscapes

The first Caucasian settlers in Wyoming Territory began building their homes in the 1820s. The human population at that time numbered fewer than ten thousand, 99 percent of whom were native Americans (Larson 1977). Bison, elk, deer, and antelope probably outnumbered people by 150 to 1. Just as they had been for ten thousand years or more, the primary human uses of the land were hunting and gathering. Even by 1860 the population of domesticated animals numbered only a few thousand—mostly horses and dogs—and tree cutting was restricted to small trees that were used for lodges, sleds, and weapons. The greatest human influence was probably the starting of fires that burned for weeks or months at a time. Tribal council meetings to discuss air and water quality were not necessary.

Great changes have occurred in the twentieth century. By 1985, the human population had increased dramatically to nearly five hundred thousand, mostly Caucasians.[1] Bison, elk, and antelope still outnumber people, but only by a ratio of 2 to 1. Livestock, mostly cattle and sheep, outnumber people by 5 to 1. With the industrial revolution, new chemical and mechanical tools have greatly enhanced the potential for environmental modification. Fossil fuels have replaced wood as the major source of energy. Technological advances and concentrations of people have benefited many, but they have also led to a need for environmental quality councils at federal, state, and local levels.

Three kinds of landscape now exist in the region: urban-industrial, agricultural, and natural or seminatural. In the strict sense, all are natural. Human needs, abilities, motivations, and innovations are surely a natural product of evolution. The term *natural,* however, is commonly used for landscapes that experience relatively little human influence. Today these landscapes are similar to those of thousands of years ago. Natural landscapes include grasslands, shrublands, woodlands, and forests dominated by native species, all of which would persist without human intervention. Wyoming rangelands, national forests, and national parks fall into this category (91 percent of the state). Significant human impacts on these seminatural landscapes include accelerated soil erosion in some areas, the introduction of undesirable exotic plants, fire suppression in low-elevation forests, and landscape fragmentation through timber harvesting and the planting of food crops.

Agricultural crops are grown on about 3–5 percent of Wyoming land. They are planted at the lower elevations with warmer and longer growing seasons and include wheat, alfalfa, sugar beets, barley, and pastures or hayland dominated by introduced species. Because of the frequent need for supplemental water, most of the agricultural land occurs in or near riparian zones. Crop and livestock production contributes about 3 percent of the state's gross state product, and as is true throughout North America, it depends heavily on fossil fuels, fertilizers, and pesticides. Croplands would soon disappear without human maintenance. Even with continued cultivation, current levels of production may not be sustainable for more than another century, because of high rates of soil erosion and a heavy dependence on fossil fuels.

The urban-industrial landscapes of Wyoming are towns and cities, large campgrounds, feedlots, power plants, mines, highways, oil and gas developments, reservoirs and surrounding mudflats, pipeline and power line corridors, and factories. Such landscapes constitute about 4–6 percent of the state. Properly designed, some of these fabricated environments can be pleasing and even unobtrusive, but considerable expenditures of time and energy are required for their maintenance. Without proper precautions, urban-industrial landscapes become the primary sources of water and air pollutants.

Despite significant changes in the past century, Wyoming landscapes are still predominantly natural or seminatural because of the state's cool, dry, short growing season. The climate alone explains why wildlife is so abundant, the air so clear, public land so common, and human population density so low. If the climate were warmer and more humid, Wyoming would have more people and less wildlife habitat.

People typically move to Wyoming in waves. One of the first occurred with the movement of thousands along the Oregon Trail, though most of these people were transients en route to the West Coast. Another wave of immigration followed the completion of the Union Pacific Railroad in 1869, and still another in the 1950s and 1970s, when Wyoming's oil, gas, uranium, coal, and soda ash were in great demand. Climatic conditions do not constrain this kind of industrial activity. Seasonally, tourists, campers, hunters, and anglers move to Wyoming during the summer and fall, when for them the climate is a great part of the attraction.

Attempts to hasten broad-based economic development and human population growth in Wyoming have been marginally successful at best. Water for irrigation is now more readily available because of reservoir construction, but except at the lowest elevations, summers are still too cool for most crops. Crop varieties might be developed that are better adapted to cool, short growing seasons, but the necessary increase in irrigation

would be accomplished only at high cost. Moreover, irrigation leads to the salinization of soil, groundwater, and rivers. Water development projects also inundate valuable riparian landscapes, and some reservoirs are short-lived because current rates of siltation are high. Invariably, there are significant costs associated with attempts to remove the constraints of natural environments. Modern technology often enables the lessening of ecological limitations, but sometimes undesirable ripple effects are difficult to avoid. As Garrett Hardin (1985) has observed, "We can never do merely one thing."

Significantly, the environmental constraints limiting some land uses also create new opportunities. As population pressures continue to increase along the front ranges of Colorado and Utah (fig. 17.1), more individuals, families, and industrial firms will look to sparsely populated mountains, plains, and basins for relief from urban-industrial living. Economic developers commonly offer economic incentives (such as tax breaks) to attract new businesses, and residents are well advised to capitalize on existing features. For example, what might be done to enhance the enjoyment of wide-open spaces? Highways could be made more pleasing by instituting strict controls over power lines, asphalt plants, abandoned automobiles, junkyards, and dilapidated billboards. State subsidies could promote tourism by constructing natural-history stops along major highways and new museums and field camps that emphasize cultural history and western ways of life. Some ranchers already supplement their income in this way.

Native American communities could provide more opportunities for learning about their heritage. Millions of people pay for the privilege of seeing static dioramas in city museums; many from around the world would also pay to experience more directly the dynamic landscapes and life of the sparsely populated North American West. Existing museums, dude ranches, rodeos, and fairs achieve these objectives to some extent, as do educational centers such as the Teton Science School, the Yellowstone Institute, the National Outdoor Leadership School, and national park visitor centers. But more can be done.

Some residents hesitate to encourage tourism, fearing that favorite places will become crowded

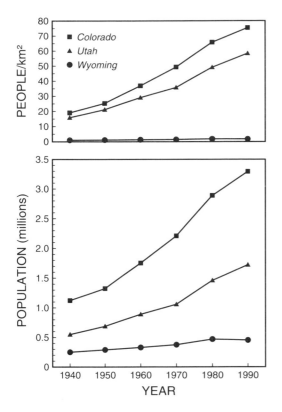

F I G. 17.1   The number of people per square kilometer is very low in Wyoming compared to Colorado and Utah. The top graph shows changes in the density of people (number/km²); the bottom graph shows changes in state population. Source: U.S. Census Bureau.

new problems. Nonetheless, making the region more attractive to visitors could also make western states more attractive to those industries that westerners would welcome in order to diversify their economy. Minimizing negative environmental impacts now will hasten desired future developments. Attractive towns and abundant opportunities to enjoy outdoor recreation will draw environmentally compatible industries as well as such talented people from other professions as teachers, nurses, and doctors.

Fossil fuels and uranium have put Wyoming in the position of subsidizing new developments through the Wyoming Permanent Mineral Trust Fund acquired from royalties and severance taxes. Nonrenewable resources, however, can be used only as the basis of a transitional economy. It is clear that Wyoming's economy should not be overly dependent on fossil fuels, agriculture, or industries that require more water than is currently available. Recognizing environmental constraints is fundamental to achieving sustainability.

Current Land Use Debates

Various environmental problems have developed in the western states during a century of agriculture, mining, water development, tourism, livestock grazing, timber harvesting, and urban-industrial growth. Some concerns have been addressed by state laws, illustrating the value placed by residents on, for example, land restoration after mining, the protection of air and water quality (including groundwater), the proper location of industrial activities, and the wise management of still-abundant fish and wildlife. Moreover, support is growing for the improved management of sensitive areas, such as riparian zones, and for the curtailment of the spread of exotic plants. Such problems are not controversial and simply require the application of improved technology and management systems. Other land management issues are more controversial, such as water development, the multiple use of federal lands, and ecosystem management.

WATER DEVELOPMENT

Not surprisingly, water availability at certain times of the year is a factor limiting human activ-

and preferred ways of life will change. Some towns may want to avoid the lucrative tourist-based economies that exist in Jackson, Cody, Estes Park, West Yellowstone, Gardiner, and elsewhere. Such change, however, will often be inevitable, and those residents who desire seclusion will probably know where and when they can find it. Population pressures elsewhere in North America, where climatic conditions have already attracted large numbers of people, will essentially force more people to Wyoming, even if only for vacations. Western states should take advantage of their remarkable environments through developments that are easy to sustain and less subject to boom-and-bust cycles.

Regardless of where people live, rising human population pressures lead to changes that create

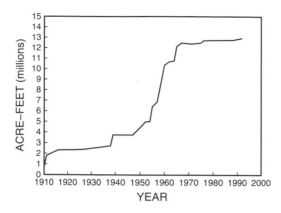

F I G. 17.2   The amount of water storage in major Wyoming reservoirs since the early 1900s. Little additional storage has been possible since the mid-1960s. Based on data from Ostresh et al. 1990.

ities throughout the Rocky Mountain region. The traditional solutions have been to construct reservoirs, irrigation systems, and transbasin diversions, thereby making water available wherever it is economically feasible to do so (fig. 17.2). Faced with interstate competition for water and the rights associated with prior use, a common attitude has been to "use it or lose it." Promoted by the Corps of Engineers, the Bureau of Reclamation, and local chambers of commerce, and heavily subsidized by state and federal grants, such projects have indeed stimulated the economy. Furthermore, flood control, hydroelectric power, and new recreational opportunities have been provided.

Still, rivers that once flowed freely long enough to erode the canyons through mountains are being replaced by regulated rivers and large reservoirs that will probably be silt-laden within a century or two. Silt can be removed, but only at great cost. Where should the silt be taken? New reservoirs and transbasin diversions are on the planning boards, and existing diversions are being upgraded. Constraints have been added in the form of state laws pertaining to instream flow and federal laws enabling the establishment of wild and scenic rivers. Such laws reflect the view of many people that unregulated rivers are a valuable resource. Instream flows are now considered another beneficial use of water.

Groundwater is also important in the region.

On the western Great Plains, extending from Nebraska and Wyoming into New Mexico and Texas, the Ogallala (High Plains) aquifer has supplied the needs of agriculture and municipalities for many years, but demands have led to significant drawdowns of the water table in some areas, even to the point of forcing farmers to revert to dryland farming (Kastner et al. 1989; Zwingle 1993). Furthermore, groundwater in some areas has been degraded through the deep percolation of polluted surface water, and some methods of fossil fuel extraction threaten to contaminate groundwater supplies that will surely become more important in the future. Clearly, human activities have the capacity to affect water supplies that lie a thousand feet or more below the surface— potentially one of the most important impacts of fossil fuel development and agriculture in the region.

The demand for water has led to the consideration of cloud seeding as a means of increasing precipitation. Experimentation in the region has shown that this method can increase rain and snowfall by 15 percent or more. Debates will continue, however, over whether increasing precipitation in one area causes a concomitant decline over downwind counties and states and whether the availability of more water in western environments creates as many problems as it solves (such as farming erodible land, salinization, and an adverse impact on other resources).

MULTIPLE USE OF PUBLIC LANDS

Until abuses were detected, little attention was given to land management activities on federal lands administered by the Forest Service and Bureau of Land Management. By the late 1950s, concerns were being expressed about excessive timber harvesting and excessive livestock grazing. Foresters and range managers were becoming increasingly sophisticated at the time, but traditions were difficult to change. A more balanced perspective was formally encouraged when Congress passed the Multiple Use–Sustained Yield Act in 1960, which emphasized the importance of recreation, rangeland, watershed, fish, and wildlife values as well as timber. A similar philosophy was gradually adopted by BLM, though it was not mandated by Congress until passage of the Federal Land Policy and Management Act in 1976.

Many observers are happy with the multiple-use concept. Wood harvesting is still possible, but national forests are now valued for more than their timber. Passage of the Wilderness Act in 1964 was controversial because some felt that the multiple-use concept would be abandoned. The Multiple Use–Sustained Yield Act, however, clearly stated that multiple use is "not necessarily the combination of uses that will give the greatest dollar return or the greatest unit output." Though often interpreted as meaning timber harvesting plus some other activity, the multiple-use concept does not require logging on every land parcel.

Over the years, lumbering has remained an important activity in the Rocky Mountains (fig. 17.3). Much of the wood is still harvested from virgin forests. Trees in some forests were cut in the era of mining and tie-hacking in the mid- and late 1800s (Veblen and Lorenz 1991), but uncut forests remained over large areas. Today, roadless, unharvested watersheds are rare, and they are valued by a growing segment of the public that increasingly looks to wildlands for outdoor recreation. Also, the old-growth forests often found in wildlands are now recognized as important habitats for certain species that contribute to the biological diversity of the region. The importance of preserving the diversity of habitats and organisms everywhere is emphasized by the Endangered Species Act and the National Forest Management Act. Both acts reflect the conviction of many people that all plants, animals, and microorganisms should be allowed to coexist in modern civilizations.

Much of the current debate over forest management in the Rocky Mountain region centers on the preservation of wildlands and on timber sales that cost the Forest Service more to arrange than they yield in revenues. Some managers argue that tree cutting is acceptable because it can be used to improve wildlife habitat while contributing to the local economy. They also maintain that forests left unmanaged by harvesting will eventually lead to large-scale fires or insect epidemics where such disturbances would be inappropriate, and that the roads associated with timber harvesting, though costly, serve many people besides the lumber industry.

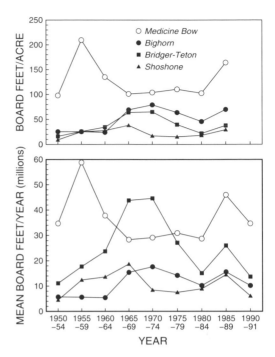

FIG. 17.3 Timber harvesting for the period 1950–91 on four national forests in Wyoming. Harvesting has been greater in the Medicine Bow National Forest, probably because of less rugged topography, less land area in designated wilderness, and more land area in forest. Harvesting in all national forests increased during the late 1980s, primarily because wood sold in previous periods was finally cut when the market improved during this time. Based on data provided by the U.S. Forest Service.

Opponents of the continued expansion of timber harvesting into virgin forests, however, argue that the remaining wildlands are now more valuable for wildlife habitat and dispersed recreation than they are as a source of wood, that there is already an adequate road system, and that the remaining wildlands should be large enough so that the inevitable fires and insect epidemics will not affect the entire wildland area. They have also expressed concern that national forest budgets will be strongly influenced by the amount of wood harvested,[2] and they wonder how many square kilometers of increasingly valuable roadless areas will remain when the lumber industry is able to obtain the wood that consumers need from second growth.

The debate on multiple use has generated credible statements from both sides, but some arguments are less convincing. For example, timber harvesting is not necessary to preserve the "health" of a forest. Forests existed for thousands of years before the initiation of management; Rocky Mountain forests are not plantations. Similarly, preserving small roadless areas is often difficult or impossible because fires are sure to occur, creating a landscape that may be less desirable for some uses than if the area had been subjected to careful timber harvesting. Moreover, wildlife populations will fluctuate, with or without tree cutting, and clearcuts will usually become forests again. Historical records suggest that Rocky Mountain forests can recover from significant disturbances (see figs. 2.8 and 2.9 and Amundson 1991).

Arguments over national forest management plans often arise when one or both sides adopt a narrow viewpoint. "Ecosystem management" shows the promise of a solution, but this concept too has been the subject of considerable debate.

ECOSYSTEM MANAGEMENT

The difficulties associated with modern land management in the West are attributable to the rising demands being placed on a limited resource base by a growing, more informed public with strong and diverse views about how landscapes should be managed. Ecosystem management is a concept that is advocated by many managers because it provides the potential to integrate, better than ever before, existing ecological information with the very real and difficult challenges faced by managers.

The concept of ecosystem management reinforces such commonly accepted practices as environmental impact assessment, public participation in the development of management plans, and the monitoring of resource values to ensure that goals are being achieved (Agee and Johnson 1988). Goals include managing for long-term sustainability. Ecosystems clearly have been sustainable over thousands of years without the aid of human management. Now, with more timber harvesting and extensive habitat fragmentation (fig 17.4), the challenge is to maintain them in a condition that will preserve as many options for future generations as exist today. That means

F I G. 17.4   Some national forests have been fragmented by timber harvesting, creating ecological effects that are still poorly understood.

maintaining the productive capacity of the soil (as mandated in the Multiple Use–Sustained Yield Act), preventing species extinction wherever possible (as mandated by the Endangered Species Act), and preserving the full range of current sustainable land uses. Some uses rely upon the preservation of noncommodity resources, such as scenic vistas, natural areas, and opportunities for dispersed outdoor recreation. Ecosystem management may also mean (1) habitat restoration so that as far as possible past uses will again become feasible, and (2) cutting back on the rate of resource extraction if present rates are judged to be higher than the ecosystem can sustain.

Achieving sustainability often requires planning and management over larger areas than before. Thus, agencies responsible for federal land near Yellowstone National Park have promoted the Greater Yellowstone Ecosystem concept to help coordinate their activities. Decisions made outside the park affect what happens inside and vice versa. Similarly, some species found most often in a park or national forest may depend on adjacent private land. Broadening the scale of management to include larger areas presents the manager with more complicated scenarios, but that seems necessary as pressures on natural resources increase.

How much land is necessary for ecosystem management? Like political boundaries, ecosystem boundaries are always somewhat arbitrary because animals, plants, water, nutrients, sediments, energy, and pollutants move or disperse across them regardless of where they are located. An entire state or region can be viewed as an ecosystem, and that view will probably become more common in the future. Now, however, while managers and scientists are still learning how to think from an ecosystem perspective, the area involved is the minimum area required to address the management issues that are clearly important to the public. For example, the area included in the Greater Yellowstone Ecosystem was determined to a large extent by considering the land involved in maintaining the widely dispersed elk and grizzly bear.

Ecosystem management also implies that the landscape mosaic will be considered in making decisions. Management decisions on national forests usually lead to changes in the juxtaposition of young and old forests in relation to, for example, water availability and feeding areas. Managers and scientists alike are asking about the effects of changing the mosaic of plant communities. Research has shown the effect of the mosaic on water yield (Leaf 1975), and biologists have long recognized the importance of edges between forests and meadows for wildlife. Still, many questions remain. What is the effect on biological diversity of retaining a large patch of old-growth forest instead of several smaller patches? In agricultural landscapes, what is the effect of maintaining parcels of native rangeland interspersed with wheat fields? The landscape mosaic has important influences on the number and abundance of species in a management unit as well as on water yield, the spread of disturbances, and other ecological phenomena (Franklin and Forman 1987; Knight and Wallace 1989; Chen et al. 1992).

The preservation of biological diversity is an important part of ecosystem management and stems from the realization that organisms have evolved over millennia of trial and error through natural selection. The result has been an amazing diversity of species with distinct chemical, physical, and biological characteristics. No one argues about the importance of preserving plants, animals, and microbes that contribute directly to health and prosperity, but those organisms not commonly seen or recognized are typically discounted because of the prevailing attitude that they are dispensable. Such attitudes imply that scientific research, in a mere century or less, has discovered the economic importance of every species.

In fact, we have much to learn about natural ecosystems and the species they harbor. For this reason, scientists have traditionally supported the establishment of natural areas where still unknown and poorly understood species and ecological processes can be maintained for further study and evaluation. In Wyoming, such natural areas are found in wilderness and research natural areas on lands administered by the National Park Service, the Forest Service, and the Bureau of Land Management. Some private organizations, such as the Nature Conservancy, purchase lands that their scientists conclude are essentially natural. Notably, some land uses, such as livestock grazing, are often allowed on Conservancy lands.

An emerging principle of ecosystem management is that preventing the extinction of some species is possible only when a network of natural areas exists in a larger area where the land is used in a manner sensitive to the needs of threatened species (Noss 1983). Single natural areas are often insufficient because they too will change—slowly through succession and abruptly following disturbances such as fires and windstorms. Fortunately, the establishment of a natural area network may be possible in Wyoming without adversely affecting economic development. A combined goal of ecosystem management is to prevent species from becoming endangered while shortening the list of those that are.

Additional features of ecosystem management are the adoption of a long-term perspective and the recognition that management activities are being superimposed on landscapes that will change with or without management. Climatic variation, disturbances, and changes due to natural succession occur whether planned for or not (see chapter 2). They should be understood and anticipated. Predicting the cumulative effects of timber harvesting, fire, insect epidemics, outdoor recreation, and air and water pollution is a central theme of ecosystem management, one that presents challenges for both managers and scientists.

Ecosystem management also recognizes that management activities are experiments in themselves. Every land use decision is based on less information than managers would like to have, but through the careful monitoring of key variables, managers can detect mistakes and avoid them in the future.

The principles of ecosystem management have been advocated by a series of congressional decisions that include the National Environmental Policy Act, the Clean Air and Clean Water acts, the Endangered Species Act, the National Forest Management Act, and the Farm Bills of 1985 and 1990 (Keiter 1988, 1989; Keiter and Boyce 1991). Arguments against ecosystem management come primarily from critics who feel that traditional ways of life are being eroded by congressional restric-

tions and from those who are not convinced that all restrictions are necessary to achieve sustainability. For example, the preservation of current modes of living is often viewed as more important than the preservation of biological diversity.

Clearly, human values differ depending on livelihood and culture. Managers should strive to achieve sustainability while avoiding wherever possible adverse impacts on those who depend on the landscapes of the region and who, indeed, can be important stewards of the land. The alternative to current land management practices may in some areas be subdivisions or other activities that are controversial for other reasons. Changes may be necessary, however, as greater demands are placed on the resources of the region. Making a living in a region with a semiarid, cool climate is problematic and must be viewed as experimental. The most appropriate livelihoods in western environments may not have been achieved in a mere century of trial and error. The best land management practices may not yet have been discovered. Though difficult to contemplate, some changes could lead to an economy that is more stable than the region has experienced since the late 1800s.

Epilogue

Traveling across western landscapes requires a major time commitment, but the exhilaration of wide-open spaces is impossible without long distances. The winters can be harsh and the summers cool, but such conditions maintain the vast beauty so many enjoy. Pleasing landscape patterns are created by drifting snow and the rugged topographic relief that extends from river valleys to distant mountaintops. There is satisfaction in understanding these patterns, the way they change during a lifetime, and how plants, animals, and microorganisms have evolved over millions of years to survive in rigorous environments.

Modern technology has greatly augmented resource availability, sometimes improving our enjoyment of life in ways that seem sustainable for many generations. Apparent successes have generated confidence. To illustrate, one building on the University of Wyoming campus is engraved with the words, "Strive on—the control of nature is won, not given." Another, completed twenty-four years later in 1950, greets students with, "The foundation of agriculture is not rooted in soil but rather in the vision and attainment of men."

Until recently, we have tended to downplay the limits to growth and development. The prevailing disposition has been to use resources. But resource depletion and adverse environmental impacts have led to the concept of living in harmony with the ecosystems on which we depend. Aldo Leopold wrote convincingly about the need for a land ethic: "We abuse land because we regard it as a commodity belonging to us. When we see land as a community to which we belong, we may begin to use it with love and respect." The research of scientists and the experience of land managers are providing insights on how long-term sustainability might be achieved, and Congress has passed laws to help ensure the nationwide application of certain restrictions. These laws complicate the lives of resource managers and landowners, but that should be expected as human pressures increase and many resources become less common. Resource management and economic development will never be as simple as they were fifty or a hundred years ago. Classroom discussions now probe this complexity, providing a more realistic perspective than what is conveyed by a few words etched in stone on buildings constructed when the world's human population was less than half what it is today.

Providing a high quality of life for so many, in a sustainable manner, may not be possible. Yet it is human nature to try. The Rocky Mountain region must contribute to the fossil fuels, minerals, wood fiber, and food required for the great human endeavor in which we are involved, but our land should be used with care and vision so that other valuable resources are not lost. We are still learning how to make a living from rugged western landscapes. Some changes will be necessary and should be discussed openly. Also, quoting from Wyoming's historian, T. A. Larson (1977), I think that we should ponder "whether what the world wants from Wyoming is worth more than what Wyoming already offers the world."

Latin Names for Plants Referred to in the Text by Common Name

Common Name	Latin Name	Common Name	Latin Name
Alder	*Alnus incana*	Basin big sagebrush	*Artemisia tridentata* ssp.
Alkali cordgrass	*Spartina gracilis*		*tridentata*
Alkali sacaton	*Sporobolus airoides*	Basin wildrye	*Elymus cinereus*
Alkali sagebrush	*Artemisia longifolia*	Bastardtoadflax	*Comandra umbellata*
Alkali wildrye	*Elymus simplex (Leymus*	Beaked sedge	*Carex rostrata*
	simplex)	Bearberry	*Arctostaphylus uva-ursi*
Alkaligrass	*Puccinellia nuttalliana*	Bearded wheatgrass	*Elymus (trachycaulus)*
Alpine avens	*Geum rossii*	Bebb willow	*Salix bebbiana*
Alpine bistort	*Polygonum bistortoides*	Bedstraw	*Galium* spp.
Alpine bluegrass	*Poa alpina*	Big bluestem	*Andropogon gerardii*
Alpine forget-me-not	*Eritrichium nanum*	Big sagebrush	*Artemisia tridentata*
Alpine kalmia	*Kalmia micropylla*	Bigtooth maple	*Acer grandidentatum*
Alpine mountainsorrel	*Oxyria digyna*	Bigflowered hymenoxys	*Hymenoxys grandiflora*
Alpine pennycress	*Thlaspi montanum*	Birch	*Betula* spp.
Alpine pussytoes	*Antennaria alpina*	Birchleaf spiraea	*Spiraea betulifolia*
Alpine sagewort	*Artemisia scopulorum*	Birdfoot sagewort	*Artemisia pedatifida*
Alpine timothy	*Phleum pratense*	Bitterbrush	*Purshia tridentata*
Alpinesedge	*Kobresia bellardii*	Black sagebrush	*Artemisia nova*
American elm	*Ulmus americana*	Blackeyed coneflower	*Rudbeckia hirta*
American licorice	*Glycyrrhiza lepidota*	Blackroot sedge	*Carex elynoides*
American plum	*Prunus americana*	Bloodroot	*Sanquinaria canadensis*
Antelope bitterbrush	*Purshia tridentata*	Blowout grass	*Redfieldia flexuosa*
Aralia	*Aralia nudicaulis*	Blue grama	*Bouteloua gracilis*
Arctic willow	*Salix arctica*	Blue spruce	*Picea pungens*
Arrowgrass	*Triglochin maritimum*	Blueberry willow	*Salix myrtillifolia*
Arrowleaf balsamroot	*Balsamorrhiza sagittata*	Bluebunch wheatgrass	*Agropyron spicatum (Elymus*
Ash	*Fraxinus pennsylvanica*		*spicatum)*
Aspen	*Populus tremuloides*	Bluegrass	*Poa* spp.
Aspen peavine	*Lathyrus lanszwertii*	Bluejoint reedgrass	*Calamagrostis canadensis*
		Blueleaf cinquefoil	*Potentilla diversifolia*
Ballhead sandwort	*Arenaria congesta*	Bog birch	*Betula glandulosa*
Balsam poplar	*Populus balsamifera*	Booth willow	*Salix boothii*
Baltic rush	*Juncus balticus*	Bottlebrush squirreltail	*Sitanion hystrix (Elymus*
Baneberry	*Actaea rubra*		*elymoides)*
Barrenground willow	*Salix brachycarpa*	Boxelder	*Acer negundo*

Notes: Numerous references are available for identifying the plants of Wyoming and adjacent states (Dorn 1992; Nelson and Williams 1992; and others); nomenclature here follows Dorn 1992.

APPENDIX A *(continued)*

Common Name	Latin Name	Common Name	Latin Name
Bracken fern	*Pteridium aquilinum*	Eastern juniper	*Juniperus virginiana*
Bristlecone pine	*Pinus aristata*	Ebony sedge	*Carex ebenea*
Broom snakeweed	*Guttierezia sarothrae*	Elephanthead lousewort	*Pedicularis groenlandica*
Buckwheat	*Eriogonum* spp.	Elk sedge	*Carex geyeri*
Bud sagewort	*Artemisia spinescens*	Elkweed	*Frasera speciosa*
Buffaloberry	*Shepherdia canadensis*	Engelmann spruce	*Picea engelmannii*
Buffalograss	*Buchloe dactyloides*		
Bunchberry dogwood	*Cornus canadensis*	False sagebrush	*Sphaeromeria capitatum*
Bur oak	*Quercus macrocarpa*	Farr willow	*Salix farriae*
Bush morningglory	*Ipomoea leptophylla*	Fendler threeawn	*Aristida purpurea* ssp. *fendleriana*
Buxbaum sedge	*Carex buxbaumii*	Fewflowered spikerush	*Eleocharis pauciflora (quinqueflora)*
California brome	*Bromus carinatus*	Fireweed	*Epilobium angustifolium*
Canada scurvyberry	*Maianthemum canadense*	Fireweed summer-cypress	*Kochia scoparia*
Canada wildrye	*Elymus canadensis*		
Canby bluegrass	*Poa canbyi*	Fleabane	*Erigeron* spp.
Canyon maple	*Acer grandidentatum*	Fourwing saltbush	*Atriplex canescens*
Cascades willow	*Salix cascadensis*	Foxtail barley	*Hordeum jubatum*
Cattail	*Typha.*	Fringed sagewort	*Artemisia frigida*
Ceanothus	*Ceanothus velutinus*		
Cheatgrass	*Bromus tectorum*	Gaillardia	*Gaillardia aristata*
Chokecherry	*Prunus virginiana*	Galleta hilaria	*Hilaria jamesii*
Cinquefoil	*Potentilla* spp.	Gambel oak	*Quercus gambelii*
Comandra	*Comandra umbellata*	Gardner saltbrush	*Atriplex gardneri*
Common chokecherry	*Prunus virginiana*	Geranium	*Geranium* spp.
Common juniper	*Juniperus communis*	Geyer willow	*Salix geyeriana*
Common snowberry	*Symphoricarpos albus*	Goldenweed	*Haplopappus* spp.
Cordgrass	*Spartina* spp.	Gooseberry (Mountain)	*Ribes montigenum*
Cottonwood	*Populus* spp.		
Coyote willow	*Salix exigua*	Goosefoot	*Chenopodium* spp.
Curlleaf mountain-mahogany	*Cercocarpus ledifolius*	Grayleaf willow	*Salix glauca*
		Greasewood	*Sarcobatus vermiculatus*
Currant	*Ribes* spp.	Great Basin wildrye	*Elymus cinereus*
Cusick bluegrass	*Poa cusickii*	Green ash	*Fraxinus pennsylvanica*
		Green gentian	*Frasera speciosa*
		Green needlegrass	*Stipa viridula*
Dandelion	*Taraxacum officinale*	Greenmolly summer-cypress	*Kochia americana*
Danthonia	*Danthonia* spp.		
Dogwood	*Cornus* spp.	Ground juniper	*Juniperus communis*
Douglas-fir	*Pseudotsuga menziesii*	Grouse whortleberry	*Vaccinium scoparium*
Douglas hawthorn	*Crateagus douglasii*	Gumweed	*Grindelia* sp.
Douglas rabbitbrush	*Chrysothamnus viscidiflorus*		
Drummond sedge	*Carex drummondiana*		
Drummond willow	*Salix drummondiana*	Hackberry	*Celtis occidentalis*
Dunhead sedge	*Carex phaeocephala*	Hairy goldenaster	*Heterotheca villosa*
Dwarf clover	*Trifolium nanum*	Halogeton	*Halogeton glomeratus*
Dwarf huckleberry	*Vaccinium scoparium*	Hawthorn	*Crateagus* spp.
Dwarf mistletoe	*Arceuthobium americanum*	Hazelnut	*Corylus cornuta*

APPENDIX A *(continued)*

Common Name	Latin Name	Common Name	Latin Name
Heartleaf arnica	*Arnica cordifolia*	Monolepis	*Monolepis nuttalliana*
Hoary balsamroot	*Balsamorhiza incana*	Monticola willow	*Salix monticola*
Hood phlox	*Phlox hoodii*	Monument plant	*Frasera speciosa*
Hood sedge	*Carex hoodii*	Moss silene	*Silene acaulis*
Hooker sandwort	*Arenaria hookeri*	Mountain big sage-	*Artemisia tridentata* var.
Horsebrush	*Tetradymia canescens*	brush	*vaseyana*
Horsetail	*Equisetum* spp.	Mountain bluebells	*Mertensia ciliata*
Huckleberry	*Vaccinium* spp.	Mountain gooseberry	*Ribes montigenum*
		Mountain-mahogany	*Cercocarpus* spp.
Idaho fescue	*Festuca idahoensis*	Mountain ninebark	*Physocarpus monogynus*
Indian ricegrass	*Oryzopsis hymenoides*	Mountain silver sage-	*Artemisia cana* var. *viscidula*
Indiangrass	*Sorgastrum nutans*	brush	
Inland saltgrass	*Distichlis stricta*	Mountain snowberry	*Symphoricarpos oreophilus*
Iris	*Iris missouriensis*	Mouseear chickweed	*Cerastium arvense*
Ironwood	*Ostrya virginiana*	Mud sedge	*Carex limosa*
		Mulesear wyethia	*Wyethia amplexicaulis*
Japanese brome	*Bromus japonicus*	Mutton bluegrass	*Poa fendleriana*
Junegrass	*Koeleria macrantha*		
Juniper	*Juniperus* spp.	Narrowleaf	*Populus angustifolia*
		cottonwood	
Kentucky bluegrass	*Poa pratensis*	Nebraska sedge	*Carex nebrascensis*
King spikefescue	*Leucopoa kingii*	Needle-and-thread	*Stipa comata*
Kobresia	*Kobresia bellardii*	grass	
Kochia	*Kochia scoparia*	Needleleaf sedge	*Carex stenophylla*
		Ninebark	*Physocarpus* spp.
Lambert locoweed	*Oxytropis lambertii*	Nodding bluegrass	*Poa reflexa*
Lanceleaf cottonwood	*Populus acuminata*	Nuttall goldenweed	*Machaeranthera grindelioides*
Larkspur	*Delphinium* spp.	Nuttall saltbush	*Atriplex gardneri*
Leafybract aster	*Aster foliaceus*		
Lightpetal dryad	*Dryas octopetala*	Obtuse sedge	*Carex obtusata*
Lily (Wood)	*Lilium philadelphicum*	Onion	*Allium textile*
Limber pine	*Pinus flexilis*	Orchardgrass	*Dactylis glomerata*
Little bluestem	*Andropogon scoparius*	Oregongrape	*mahonia repens*
Locoweed	*Oxytropus* spp.		
Lodgepole pine	*Pinus contorta* var. *latifolia*	Pachistima	*Paxistima myrsinites*
Longstock clover	*Trifolium longipes*		*(Pachistima myrsinites)*
Louisiana sagewort	*Artemisia ludoviciana*	Pacific willow	*Salix lasiandra*
Lousewort	*Pedicularis racemosa*	Pale agoseris	*Agoseris glauca*
Low sagebrush	*Artemisia arbuscula*	Paper birch	*Betula papyrifera*
Lupine	*Lupinus* spp.	Parry clover	*Trifolium parryi*
		Parry lousewort	*Pedicularis parryi*
Maple	*Acer* spp.	Pasque flower	*Anemone patens*
Marsh marigold	*Caltha leptosepala*	Patterson bluegrass	*Poa pattersonii*
Meadow barley	*Hordeum brachyantherum*	Peachleaf willow	*Salix amygdaloides*
Meadowrue	*Thalictrum* spp.	Penstemon	*Penstemon* spp.
Mentzelia	*Mentzelia* spp.	Pepperweed	*Lepidium densiflorum*
Milkvetch	*Astragalus* spp.	Phlox	*Phlox* spp.
Mistletoe	*Arceuthobium americana*	Pine reedgrass	*Calamagrostis rubescens*
		Pinyon pine	*Pinus edulis*

A P P E N D I X   A   *(continued)*

Common Name	Latin Name	Common Name	Latin Name
Plains cottonwood	*Populus deltoides*	Saltwort	*Salicornia rubra*
Plains pricklypear	*Opuntia polyacantha*	Sand dropseed	*Sporobolus cryptandrus*
Plains silver sagebrush	*Artemisia cana var. cana*	Sand lily	*Leucocrinum montanum*
Planeleaf willow	*Salix planifolia*	Sand lovegrass	*Eragrostis trichodes*
Ponderosa pine	*Pinus ponderosa*	Sand muhly	*Muhlenbergia arenicola*
Prairie cordgrass	*Spartina pectinata*	Sand sagebrush	*Artemisia filifolia*
Prairie dropseed	*Sporobolus heterolepis*	Sandbar willow	*Salix exigua*
Prairie junegrass	*Koeleria macrantha*	Sandberg bluegrass	*Poa sandbergii*
Prairie sandreed	*Calamovilfa longifolia*	Sandhill muhly	*Muhlenbergia pungens*
Prairie smoke	*Geum triflorum*	Sarsaparilla aralia	*Aralia nudicaulis*
Pricklypear cactus	*Opuntia polyacantha*	Saskatoon serviceberry	*Amelanchier alnifolia*
Pricklypoppy	*Argemone* spp.	Scarlet gilia	*Ipomopsis aggregata*
Princesplume	*Stanleya pinnata*	Scarlet globemallow	*Sphaeralcea coccinea*
Purple reedgrass	*Calamagrostis purpurascens*	Scribner wheatgrass	*Elymus scribneri (Agropyron scribneri)*
Pussytoes	*Antennaria* spp.		
Pyrola	*Pyrola* spp. *(Orthilia secunda)*	Scurfpea	*Psoralea tenuiflora*
		Seablight	*Suaeda nigra*
Rabbitbrush	*Chrysothamnus* spp.	Seaside arrowgrass	*Triglochin maritimum*
Red-osier dogwood	*Cornus sericea*	Sedge	*Carex* spp.
Red threeawn	*Aristida purpurea (var. longiseta)*	Seepweed	*Suaeda nigra*
		Segolily	*Calochortus nuttallii*
Redtop bentgrass	*Agrostis stolonifera*	Selaginella	*Selaginella densa*
Reedgrass	*Calamagrostis* spp.	Serviceberry	*Amelanchier alnifolia*
Richardson geranium	*Geranium richardsonii*	Shadscale saltbush	*Atriplex confertifolia*
Rock sedge	*Carex rupestris*	Sheep fescue	*Festuca saximontana*
Rocky Mountain glasswort	*Salicornia rubra*	Sheep sedge	*Carex illota*
		Showy elkweed	*Frasera speciosa*
Rocky Mountain juniper	*Juniperus scopulorum*	Shrubby cinquefoil	*Pentaphylloides floribunda (Potentilla fruticosa)*
Rocky Mountain maple	*Acer glabrum*	Sibbaldia	*Sibbaldia procumbens*
		Sideoats grama	*Bouteloua curtipendula*
Rocky Mountain nailwort	*Paronychia pulvinata*	Silky locoweed	*Oxytropis sericea*
		Silky lupine	*Lupinus sericeus*
Rose	*Rosa* spp.	Silver buffaloberry	*Shepherdia argentea*
Rosecrown stonecrop	*Sedum rhodanthum*	Silver sagebrush	*Artemisia cana*
Ross bentgrass	*Agrostis rossiae*	Silverberry	*Elaeagnus commutata*
Ross sedge	*Carex rossi*	Silvery lupine	*Lupinus argenteus*
Rubber rabbitbrush	*Chrysothamnus nauseosus*	Skeletonplant	*Lygodesmia juncea*
Russet buffaloberry	*Shepherdia canadensis*	Skunkbush sumac	*Rhus trilobata*
Russian olive	*Elaeagnus angustifolia*	Slender wheatgrass	*Elymus trachycaulus*
Russian thistle	*Salsola australis (S. kali)*	Slimflower scurfpea	*Psoralea tenuiflora*
Rusty lupine	*Lupinus pusillus*	Smallflower gaura	*Gaura parviflora*
		Smooth brome	*Bromus inermis*
Sagebrush	*Artemisia* spp.	Snakeweed	*Gutierrezia sarothrae*
Salicornia	*Salicornia rubra*	Snow willow	*Salix reticulata*
Salina wildrye	*Elymus salinus*	Snowberry	*Symphoricarpos* spp.
Saltbush	*Atriplex* spp.	Snowbrush ceanothus	*Ceanothus velutinus*
Saltcedar	*Tamarix chinensis*	Soapweed	*Yucca glauca*
Saltgrass	*Distichlis stricta*	Spike trisetum	*Trisetum spicatum*

APPENDIX A *(continued)*

Common Name	Latin Name	Common Name	Latin Name
Spiny aster	*Machaeranthera tanacetifolia*	Utah juniper	*Juniperus osteosperma*
Spiny hopsage	*Grayia spinosa*	Utah serviceberry	*Amelanchier utahensis*
Spiraea	*Spiraea betulifolia*		
Spruce	*Picea* spp.	Vetch	*Astragalus* spp.
Squarestem phlox	*Phlox muscoides*	Virginia creeper	*Parthenocissus vitacea*
Starry cerastium	*Cerastium arvense*		
Sticky geranium	*Geranium viscosissimum*	Water birch	*Betula occidentalis*
Sticky polemonium	*Polemonium viscosum*	Water sedge	*Carex aquatilis*
Stonecrop	*Sedum lanceolatum*	Wax currant	*Ribes cereum*
Subalpine fir	*Abies lasiocarpa*	Weedy milkvetch	*Astragalus miser*
Subalpine needle-grass	*Stipa nelsonii*	Western coneflower	*Rudbeckia occidentalis*
		Western snowberry	*Symphoricarpos occidentalis*
Sulfurflower buck-wheat	*Eriogonum umbellatum*	Western wheatgrass	*Elymus smithii (Agropyron smithii)*
Sumac (skunkbush)	*Rhus trilobata*	Wheeler bluegrass	*Poa nervosa*
Summercypress	*Kochia* spp.	White fir	*Abies concolor*
Sweetclover	*Melilotus officinalis*	White spruce	*Picea glauca*
Switchgrass	*Panicum virgatum*	Whitebark pine	*Pinus albicaulis*
		Wild geranium	*Geranium* spp.
Thickspike wheatgrass	*Elymus lanceolatus (Agropyron dasystachyum)*	Wild iris	*Iris missouriensis*
Thinleaf alder	*Alnus incana*	Wild onion	*Allium textile*
Threadleaf sedge	*Carex filifolia*	Wild rose	*Rosa* spp.
Threeawn	*Aristida* spp.	Wild sarsaparilla	*Aralia nudicaulis*
Threetip sagebrush	*Artemisia tripartita*	Winterfat	*Krascheninnikovia lanata (Ceratoides lanata)*
Thurber fescue	*Festuca thurberi*		
Timberline bluegrass	*Poa ripicola*	Wolf willow	*Salix wolfii*
Timothy	*Phleum pratense*	Wood lily	*Lilium philadelphicum*
Tineleaved milkvetch	*Astragalus pectinatus*	Wood nymph	*Moneses uniflora*
Trisetum	*Trisetum spicatum*	Wood rose	*Rosa woodsii*
True mountain-mahogany	*Cercocarpus montanus*	Woody aster	*Xylorhiza glabriuscula*
Tufted evening prim-rose	*Oenothera caespitosa*	Wyoming big sagebrush	*Artemisia tridentata* var. *wyomingensis*
Tufted fleabane	*Erigeron caespitosus*	Yarrow	*Achillea millefolium* var. *lanulosa*
Tufted hairgrass	*Deschampsia caespitosa*	Yellow beeplant	*Cleome lutea*
Tufted phlox	*Phlox pulvinata*	Yellow salsify	*Tragopogon dubius*
Twinflower	*Linnaea borealis*	Yellow wildrye	*Elymus flavescens*
Two-grooved milk-vetch	*Astragalus bisulcatus*	Yellow willow	*Salix lutea*
		Yucca	*Yucca glauca*

Latin Names for Birds, Mammals, Reptiles, Amphibians, and Invertebrates Referred to in the Text by Common Name

Common Name	Latin Name	Common Name	Latin Name

Birds[a]

Common Name	Latin Name	Common Name	Latin Name
Bald eagle	*Haliaeetus leucocephalus*	Meadowlark, Western	*Sturnella neglecta*
Belted kingfisher	*Ceryle alcyon*	Mountain bluebird	*Sialia currucoides*
Blackbird, Brewer's	*Euphagus cyanocephalus*	Nighthawk, Common	*Chordeiles minor*
Blackbird, Red-winged	*Agelaius phoeniceus*	Northern flicker	*Colaptes auratus*
Blackbird, Yellow-headed	*Xanthocephalus xanthocephalus*	Northern harrier	*Circus cyaneus*
Chickadee, Black-capped	*Parus atricapillus*	Nuthatch, Red-breasted	*Sitta canadensis*
Chickadee, Mountain	*Parus gambeli*	Nuthatch, White-breasted	*Sitta carolinensis*
Clark's nutcracker	*Nucifraga columbiana*	Owl, Boreal	*Aegolius funereus*
Crossbill, Red	*Loxia curvirostra*	Owl, Burrowing	*Athene cunicularia*
Dipper, American	*Cinclus mexicanus*	Owl, Great-horned	*Bubo virginianus*
Finch, Rosy	*Leucosticte arctoa*	Pipit, American (water)	*Anthus rubescens*
Golden eagle	*Aquila chrysaetos*		
Goldfinch, American	*Carduelis tristis*	Prairie falcon	*Falco mexicanus*
Goshawk, Northern	*Accipiter gentilis*	Ptarmigan, White-tailed	*Lagopus leucurus*
Grosbeak, Pine	*Pinicola enucleator*		
Grouse, Blue	*Dendragapus obscurus*	Raven, Common	*Corvus corax*
Grouse, Ruffed	*Bonasa umbellus*	Robin, American	*Turdus migratorius*
Grouse, Sage	*Centrocercus urophasianus*	Sage thrasher	*Oreoscoptes montanus*
Hawk, Red-tailed	*Buteo jamaicensis*	Snipe, Common	*Gallinago gallinago*
Hawk, Swainson's	*Buteo swainsoni*	Solitaire, Townsend's	*Myadestes townsendi*
Horned lark	*Eremophila alpestris*	Sparrow, Brewer's	*Spizella breweri*
Jay, Gray	*Perisoreus canadensis*	Sparrow, Lark	*Chondestes grammacus*
Jay, Stellar's	*Cyanocitta stelleri*	Sparrow, Lincoln's	*Melospiza lincolnii*
Jay, Pinyon	*Gymnorhinus cyanocephalus*	Sparrow, Sage	*Amphispiza belli*
Junco, Dark-eyed	*Junco hyemalis*	Sparrow, Savanna	*Passerculus sandwichensis*
Kestrel, American	*Falco sparverius*	Sparrow, Song	*Melospiza melodia*
Killdeer	*Charadrius vociferus*	Sparrow, Vesper	*Pooecetes gramineus*
Kinglet, Ruby-crowned	*Regulus calendula*	Stellar jay	*Cyanocitta stelleri*
Lark bunting	*Calamospiza melanocorys*	Swallow, Bank	*Riparia riparia*
Longspur, McCown's	*Calcarius mccownii*	Swallow, Northern rough-winged	*Stelgidopteryx serripennis*
Magpie, Black-billed	*Pica pica*		

[a]The Wyoming Game and Fish Department has a list of all bird species known to occur in Wyoming.

273

APPENDIX B *(continued)*

Common Name	Latin Name	Common Name	Latin Name
Swallow, Tree	*Tachycineta bicolor*	Turkey vulture	*Cathartes aura*
Swallow, Violet-green	*Tachycineta thalassina*	Woodpecker, Downy	*Picoides pubescens*
Tanager, Western	*Piranga ludoviciana*	Woodpecker, Hairy	*Picoides villosus*
Towhee, Green-tailed	*Pipilo chlorurus*	Wren, House	*Troglodytes aedon*

Mammals[b]

Common Name	Latin Name	Common Name	Latin Name
Badger	*Taxidea taxus*	Lemming, Collared	*Dicrostonyx groenlandicus*
Barrenground caribou	*Rangifer arcticus*	Mink	*Mustela vison*
Bear, Black	*Ursus americanus*	Moose	*Alces alces shirasi*
Bear, Grizzly	*Ursus arctos horribilis*	Mountain lion	*Felis concolor*
Beaver	*Castor canadensis*	Mouse, Deer	*Peromyscus maniculatus*
Bighorn sheep	*Ovis canadensis canadensis*	Mouse, Northern grasshopper	*Onychomys leucogaster*
Bison	*Bison bison*		
Black bear	*Ursus americanus cinnamomum*	Muskox	*Ovibos moschatus*
Black-footed ferret	*Mustela nigripes*	Muskrat	*Ondatra zibethicus*
Bobcat	*Lynx rufus*	Myotis, Little brown	*Myotis lucifugus*
Buffalo	*Bison bison*	Packrat	*Neotoma cinerea*
Bushy-tailed woodrat	*Neotoma cinerea*	Peccary, Collared	*Pecari angulatus*
Chipmunk, Least	*Tamias minimus*	Pika	*Ochotona princeps*
Collared lemming	*Dicrostonyx groenlandicus*	Pocket gopher, Northern	*Thomomys talpoides*
Cottontail, Desert	*Sylvilagus audubonii*	Porcupine	*Erethizon dorsatum*
Cottontail, Nuttall's	*Sylvilagus nuttalliana*	Prairie dog, Black-tailed	*Cynomys ludovicianus ludovicianus*
Cougar	*Felis concolor*		
Coyote	*Canis latrans*	Prairie dog, White-tailed	*Cynomys leucurus*
Deer, Mule	*Odocoileus hemionus*		
Deer, White-tailed	*Odocoileus virginianus*	Pronghorn antelope	*Antilocapra americana americana*
Elk	*Cervus elaphus nelsoni*		
Fox, Red	*Vulpes vulpes*	Raccoon	*Procyon lotor*
Fox, Swift	*Vulpes velox*	Red squirrel	*Tamiasciurus hudsonicus*
Ground squirrel, Thirteen-lined	*Spermophilus tridecemlineatus*	Snowshoe hare	*Lepus americanus*
		Skunk, Striped	*Mephitis mephitis*
Ground squirrel, Wyoming	*Spermophilus elegans*	Vole, Southern red-backed	*Clethrionomys gapperi*
Jackrabbit, Blacktailed	*Lepus californicus melanotis*	Wapiti	*Cervus elaphus nelsoni*
Jackrabbit, Whitetailed	*Lepus townsendii*	Wolf, Gray	*Canis lupus*
Kangaroo rat (Ord's)	*Dipodomys ordii*		

Reptiles

Common Name	Latin Name	Common Name	Latin Name
Bullsnake	*Pituophis melanoleucus sayi*	Rattlesnake, Prairie	*Crotalus viridis*
Lizard, Northern sagebrush	*Sceloporus graciosus*	Snake, Red-sided garter	*Thamnophis sirtalis parietalis*
		Snake, Valley garter	*Thamnophis sirtalis fitchi*
Lizard, Spiny	*Sceloporus undulatus*	Snake, Wandering garter	*Thamnophis elegans vagrans*

Amphibians

Common Name	Latin Name		
Boreal chorus frog	*Pseudacris trisceriata*		

[b]Clark and Stromberg (1987) provide descriptions of all mammals known to occur in Wyoming.

APPENDIX B *(continued)*

Common Name	Latin Name	Common Name	Latin Name
Invertebrates			
Army worms	Order: Lepidoptera; Family: Noctuidae	Mealy bugs	Family: Pseudococcidae
		Mites	Order: Acarina
Aroga moth	*Aroga websteri*	Mormon cricket	*Anabrus simplex*
Douglas-fir beetle	*Dendroctonus pseudotsuga*	Mountain pine beetle	*Dendroctonus ponderosae*
Gall midges	Family: Cecidomyiidae	Nematodes	Phylum: Nematoda
Grasshoppers	Order: Orthoptera (>100 species)	Western spruce beetle	*Dendroctonus rufipennis*
		Western spruce bud-worm	*Choristoneura occidentali*
Harvester ant	*Pogonomyrmex occidentalis*		

Notes

Chapter 1: Introduction

1. A small herd of elk persists in the desert northeast of Rock Springs, where adequate habitat is available near Steamboat Mountain and the Killpecker Sand Dunes.
2. The effects of landscape patterns are not well known, and with a few exceptions, are not addressed in this book.

Chapter 2: Landscape History

1. The ancestors of pronghorn antelope appeared in the fossil record of the Miocene, about 20 million years ago, while bison did not appear until the Pleistocene, about 1 million years ago.
2. The word *laramide* is used because the development of the Rocky Mountains was first described based on research done in the mountains near Laramie.
3. Two periods of glaciation widely recognized in Wyoming are the Pinedale (about 18,000 years ago) and the Bull Lake (about 30,000 years ago) (Mears 1974).
4. The Holocene Epoch is defined as the last 10,000 years and is characterized by the proliferation of human activity. In Wyoming, the earliest known evidence of humans dates back about 12,000 years, based on the presence of Clovis points and other artifacts (Frison 1991; Frison and Todd 1986).
5. Most of the bison are in Yellowstone National Park.
6. The effects of grazing along the Oregon Trail might have been minimal if the livestock had passed through the region during midsummer or fall and if the animals had grazed the rangeland for only a short time.
7. In January 1992 there were 1,320,000 cattle and 850,000 sheep in Wyoming; horses numbered about 48,000 in 1987 (Wyoming State Department of Agriculture). Although experts agree that the number of cattle was much higher in the 1880s, accurate data are not available. Sheep numbers peaked in 1908 at 6 million.
8. Veblen and Lorenz (1986) describe forest recovery following mining in the Colorado Front Range.
9. Other repeat-photo studies have been done by Phillips (1963) for the northern Great Plains; McGinnies et al. (1991) for the eastern plains of Colorado; Progulske (1974) for the Black Hills; Veblen and Lorenz (1991) for the Front Range of Colorado; Gruell (1979, 1980a, 1980b) for the Bridger-Teton National Forest near Jackson; Houston (1982) for northern Yellowstone National Park; and the Wyoming State Historical Society (1976) for the Bighorn Mountains.

Chapter 3: Modern Environments

1. Martner (1986) showed how droughts occurred in the mid-1930s, mid-1950s, and the early 1960s.
2. Common salts in the region include sodium and magnesium chloride, sodium sulfate, calcium sulfate, and magnesium sulfate, and bicarbonates of calcium and magnesium.
3. The grazing of large herbivores such as bison, elk, cattle, and sheep becomes a disturbance in some grasslands only if too many animals are concentrated for too long in an area that is too small.

Chapter 4: Riparian Landscapes

1. Youngblood et al. (1985) noted that riparian soils could include Entisols, Inceptisols, Mollisols, Histosols, and rarely, Alfisols (see table 3.1).
2. American elm occurs primarily in the Black Hills and has been killed in many areas on the Great Plains by the Dutch elm disease.
3. Baker (1990), for example, concluded that cool, moist climatic conditions during the last half of the nineteenth century enabled the expansion of narrowleaf cottonwood along the Animas River in southwestern Colorado.
4. Some research suggests that riparian woodlands can be damaged by floods (Currier 1982; Knopf and Scott 1990). Thus, while the distribution of cottonwood probably depends on flooding, the floods must occur in a manner that does not cause damage to the trees.
5. Russian olive is an introduced tree that is becoming very common along some rivers (Olson and Knopf 1986). This species may become the dominant tree as plains cottonwood dies along some low-elevation rivers (Currier 1982).
6. Saltcedar is an introduced riparian shrub found only at low elevations in Wyoming, particularly along the Bighorn River (Merkel and Hopkins 1957; Robinson 1965).
7. Along the South Platte River, Sedgwick and Knopf (1989) concluded that cottonwood seedling mortality was caused by depth to the water table and competition from other herbaceous plants. They suggested that cottonwood will probably be replaced by ash, Russian olive, or eastern juniper. In contrast, along the North Platte River in Nebraska, cottonwood regeneration is now abundant and apparently does not depend on the formation of point bars. Woodlands are expanding in area (W. Carter Johnson, pers. comm.), reducing the amount of habitat available for the thousands of sandhill cranes migrating through this area in the spring.
8. The suppression of disturbances along some rivers allows the development of a self-perpetuating woodland, for example, along the Snake River south of Jackson, where blue spruce and red-osier dogwood are common. In eastern Wyoming, ash and boxelder are becoming more common (Miller 1979). Unlike pioneer species such as cottonwood and some willows, these species, where they occur, can reproduce successfully away from the channel.
9. Beaver are able to obtain construction material for their dams and lodges over distances of 50–100 m (Bruce H. Smith, pers. comm.; Call 1970).
10. As on the adjacent uplands, grazing can be a minor factor in riparian landscapes if carefully managed and if it occurs late in the growing season (Sedgwick and Knopf 1991).
11. Sedgwick and Knopf (1991) found that livestock preferentially grazed on recently fallen cottonwood leaves during late autumn along the South Platte River in Colorado.
12. Fifty-one percent flows to the Missouri River, 14% to the Colorado River, 32% to the Columbia River, and 3% to the Great Basin (Ostresh et al. 1990).
13. The Jackson Lake dam on the Snake River was finished in 1906, enlarging Jackson Lake and creating the oldest large reservoir in Wyoming (Ostresh et al. 1990).
14. Defining riparian zones has become especially important because they are considered wetlands, and federal policies have advocated a "no net loss" of land area in wetlands. Most scientists are skeptical that new "wetlands" can be created through irrigation to replace the losses due to the drainage of natural wetlands, but wetland restoration—even wetland creation where that is possible—is being attempted more often than ever before. The high cost of wetland restoration suggests that those wetlands that remain should be protected.

Chapter 5: Grasslands

1. In Wyoming, buffalo grass is most common in the shortgrass prairie east and south of Cheyenne, though it is found sporadically northward into Montana.
2. Detailed studies on the plant species composition of mixed-grass prairie include Hanson (1955), Wright and Wright (1948), Weaver (1954), and Singh et al. (1983).
3. In contrast, most plants of tropical rainforests have their buds above ground (trees and shrubs). Based on this observation, Raunkiaer proposed that the position of perennial tissue during the unfavorable season is an adaptation to the plant's environment.
4. Wright (1971) found that squirreltail is more tolerant to burning than needle-and-thread grass because less fuel accumulates at the base of the plant.
5. Richards and Caldwell (1985) found that most of the energy required for regrowth in two species of wheatgrass was derived from current photosynthate by remaining leaves rather than from soluble carbohydrates stored in roots.
6. Nematodes are a highly diverse group of organisms. Some grassland species feed on plant roots, others

feed on fungi, bacteria, and protozoans. Notably, some fungi have evolved adaptations for trapping and digesting nematodes.

7. The weight of all microbial organisms in the soil has been estimated as about 55% the weight of the aboveground plant biomass in a Canadian mixed-grass prairie (Paul et al. 1979).

8. Aboveground NPP is surely between 100–300 g/m²/yr, but belowground estimates are unreliable because of the problems associated with root growth measurement.

9. There is some evidence to suggest that N-fixation occurs in or adjacent to the roots of some grasses (Wullstein 1979).

10. Clipping plants uniformly for experimental studies on the effects of herbivory may not give the same results as actual grazing—which typically leaves some parts of a plant uneaten, or at least cut at different heights (Stroud et al. 1985). The novel idea that plant growth-promoting substances might occur in the saliva of herbivores has not been supported by most research (Detling and Dyer 1981).

11. Rauzi and Smith (1973) and Gifford and Hawkins (1978) have reviewed the effects of livestock grazing on water infiltration into the soil.

12. Succession following the abandonment of agricultural land is described by Lang (1973) and Reichhardt (1982).

13. About 336 kg/ha (300 lb/acre) of aboveground biomass (dry weight) are required to burn a grassland dominated by blue grama and needle-and-thread grass, while more fuel is required in grasslands dominated by bunchgrasses (J. Dodd, pers. comm.).

14. Woody plants on the fringe of burning grasslands are sometimes ignited and could smolder long enough to sustain a fire through a rainstorm.

15. The distribution and characteristics of different species of grasshoppers in Wyoming are summarized by Lockwood et al. (1993).

16. Van Vuren (1987) and Urness (1989) did not agree with Mack and Thompson's explanation, suggesting instead that bison numbers were sometimes low in the Great Basin because of intensive hunting by native Americans.

17. The beneficial effects of burrowing mammals on soil structure have been discussed by Ellison and Aldous 1952, McGinnies 1960, Mielke 1977, Inouye et al. 1987, and Whicker and Detling 1988.

18. Oscillations in prairie dog populations are typical and can be caused by plague—the same bacterial disease that has caused severe epidemics in human populations and that was first reported in North America in the early 1900s. Transmittal to humans

is now rare because of more sanitary living conditions.

Chapter 6: Sagebrush Steppe

1. Some evidence suggests that drier years may favor sagebrush seedling establishment in the Great Basin (R. F. Miller et al. 1993).

2. Other species of *Artemisia* will be discussed in subsequent chapters. Fisser (1987) has prepared a detailed, cross-referenced bibliography on Wyoming shrublands.

3. A second subspecies, mountain silver sagebrush (ssp. *viscidula*), occurs in moist mountain meadows.

4. The focus of this section is on big sagebrush. Although other species of *Artemisia* have many of the same adaptations, generalizing from big sagebrush must be done with caution.

5. Some big sagebrush plants are browsed more heavily than others (Welch et al. 1991), suggesting that the production of antiherbivory chemicals is genetically controlled and varies from plant to plant. These aromatic compounds (terpenes, including camphor, pinene, and cincole) could also inhibit the growth of neighboring plants or the germination of their seeds. Some laboratory experiments have demonstrated allelopathic effects of big sagebrush water extracts and volatile vapors on other plants (Hoffman and Hazlett 1977; Weaver and Klarich 1977; Groves and Anderson 1981), but until field experiments are conducted, it seems prudent to assume that the volatile oils reduce herbivory rather than the growth rates of neighboring plants. There is some evidence that the sagebrush compounds reduce the rate of nitrification (West 1991).

6. The abundance of bison in the intermountain basins may have varied from place to place depending on the amount of grassland available (Van Vuren 1987; Urness 1989; R. F. Miller et al. 1993).

7. Mehringer and Wigand (1990) found that, over the last 5,500 years, the abundance of charcoal has been inversely related to the abundance of sagebrush pollen and directly related to grass pollen in the Great Basin. This observation suggests that grasslands burned more frequently than sagebrush steppe.

8. The soil may have sufficient cheatgrass seed to produce more than 2,000 seedlings per square meter (Young and Evans 1975; Mack and Pyke 1983; Hassan and West 1986).

9. Hillsides in parts of the Great Basin of Utah and

Idaho, now largely shrub-dominated, were once dominated by bluebunch wheatgrass and sandberg bluegrass, with only scattered big sagebrush and Utah juniper (R. F. Miller et al. 1993).

10. Great year-to-year variation in shadscale-dominated desert shrublands has been documented in the Great Basin through annual photographs taken of the same area by Sharp et al. (1990).

11. Van Vuren (1987) and Urness (1989) emphasize the importance of intensive hunting by native Americans.

Chapter 7: Desert Shrublands and Playas

1. The most common salts in Wyoming desert shrublands and playas are magnesium and sodium sulfates, sodium chloride, and bicarbonates of calcium and magnesium. All soils with high concentrations of these salts can be referred to as saline, with the term *alkaline* being used for soils where hydroxides cause a pH of 7.4 or above. The term *sodic* is used for soils where 15% or more of the exchange capacity is sodium. Alkaline soils commonly have low rates of water infiltration and air permeability, and an electrical conductivity of 4 dS/m or above.

2. The morphological and physiological adaptations of halophytes and other desert shrubland species are described in detail by Jefferies (1981) and Smith and Nowak (1990).

3. Mixed desert shrubland is also referred to as sagebrush desert shrubland or sagebrush semidesert (West 1983).

4. Also referred to as salt desert shrubland.

Chapter 8: Sand Dunes, Badlands, Mud Volcanoes, and Mima Mounds

1. Prairie sandreed and blowout grass are restricted to dunes in the eastern half of the state (Hallsten et al. 1987), perhaps because of their intolerance of the drier dunes to the west.

2. Wolfe (1973) studied the effects of fire on stabilized sand dunes.

3. High infiltration and storage in the Nebraska sandhills provides much of the groundwater contained within the Ogallala aquifer, on which many Great Plains municipalities and farmers depend (Zwingle 1993).

4. Most Wyoming coal comes from the Fort Union formation.

5. Vegetation banding caused by geologic bedrock is found elsewhere as well, such as near Lander and Kemmerer.

Chapter 9: Escarpments and the Foothill Transition

1. The two species of mountain-mahogany occur together in at least one locality on the west side of Flaming Gorge Reservoir.

2. True mountain-mahogany is thought to sprout more readily than curlleaf mountain-mahogany (Young 1983).

3. Soils were shallow in all of Wight and Fisser's 45 study areas (10–91 cm; mean, 40 cm), with 30%–87% sand and a pH ranging from 6.8 to 8.1. Soil conductivity was low (0.45–1.5 dS/m).

4. Fire return intervals of 30–40 years may be adequate to prevent juniper from invading sagebrush steppe in the Great Basin (Burkhardt and Tisdale 1976).

5. Ponderosa pine sometimes is found with Rocky Mountain juniper on outcrops of shale that are rich in bentonite, such as near Upton and Moorcroft. Elsewhere it is found on outcrops of sandstone (e.g., southwest of Casper) or granite (e.g., west of Cheyenne and Wheatland).

6. Skunkbush sumac is commonly associated with scoria buttes or the rugged topography of badlands and river valleys (Sanford 1970; Brown 1971). It is also found on the lee sides of some slopes—possibly where seeps develop—and commonly in foothill grasslands on the eastern slopes of the Bighorn and Laramie mountains.

7. Major and Rejmanek (1992) view the two as one species, citing several taxonomic studies.

8. A recent calculation argues that an aspen grove resulting from cloning may be the heaviest organism on earth.

9. Elk also can damage mature trees through antler rubbing or the eating of bark, which facilitates the spread of certain fungal diseases such as aspen canker (Hinds 1976 in Veblen and Lorenz; also Parker and Parker 1983).

Chapter 10: Mountain Forests

1. Veblen and Lorenz (1986) describe human-caused disturbances in relation to gold and silver mining in the Front Range of Colorado.

2. White pine blister rust *(Cronartium ribicola)*, which kills whitebark pine, is an introduced disease.

3. Using Daubenmire's habitat-type approach, each forest is first classified into a *series* based on the tree species that all available evidence suggests would dominate the climax forest in the absence of large-scale disturbances. The series is further subdivided into *habitat types* based on the abundance of certain

shrubs, forbs, and grasses in the understory. If information is insufficient for estimating the nature of the climax forest, then the forest is labeled a *community type* instead of habitat type. Each habitat type or community type may be subdivided into *phases* based on understory species composition. Keys for identifying habitat and community types, along with a brief description of each, are available for Wyoming forests specifically (Alexander 1986a) and for the Rocky Mountains in general (Alexander 1985b).

4. In Colorado, Goldblum and Veblen (1992) estimated a mean fire return interval of 69 years.

5. Unlike bark beetles, the western spruce budworm is a moth. In the Rocky Mountains, the larvae feed primarily on the buds and young leaves of Douglas-fir. Outbreaks appear to occur when many of the trees are in a state of low vigor.

6. Various investigators have estimated the amount of canopy and bole scorching that different species can tolerate (Peterson and Arbaugh 1986; Wyant et al. 1986; Ryan et al. 1988).

7. White spruce is common in the Black Hills; white fir is known from only two localities in southwestern Wyoming (in the foothills of the Uinta Mountains near Lonetree [Uinta County] and on Little Mountain in Sweetwater County).

Chapter 11: The Forest Ecosystem

1. The understory vegetation beneath snowbanks that persist until late June or early July is often much different than on adjacent areas, probably because of a shorter growing season (Holloway and Ward 1963; Kyte 1975; Knight et al. 1977).

2. In contrast, Severson and Kranz (1976) found that understory production was not correlated with aspen basal area or density.

3. As with other trees bearing nutritious seeds, lodgepole pine is characterized by having mast years, i.e., years when an above-average amount of seed is produced. Mast years occur at 3–5 year intervals and help ensure that some seed will escape consumption by small mammals and birds (predation).

4. The characteristics of high- and low-vigor lodgepole pine trees are described by Kaufmann and Watkins (1990).

5. Wood quality is apparently not affected by the fungus. In fact, some people prefer the more varied grain pattern of blue-stained wood.

6. One common root disease is in the genus *Armillaria,* a fungal pathogen that can cause large gaps

in the forest canopy (Shaw and Kile 1991). Root rots have similar effects in the western hemlock forests of Washington and Oregon (Matson and Boone 1984; Dickman and Cook 1989).

7. Plants capable of growing tall are more competitive because their chlorophyll-bearing leaves are lifted above those of other plants, but they must also allocate more energy to the maintenance of woody stems, branches, and roots. The net effect is more respiration and a lower NPP than might be expected with the same amount of energy (Tilman 1988).

8. Edminster (1987) reviews literature on wood production in the central Rocky Mountains.

9. High populations of porcupines lower forest NPP by feeding on the bark of conifers (Harder 1979), which interferes with the flow of carbohydrates to the roots.

10. Ten cm of water, for example, is the amount of water that will cover any area 10 cm deep; thus, 10 cm of water over 1 hectare (10,000 m²), whether as rain or the water equivalent of snow, is an input of 1,000,000 liters.

11. DeByle et al. (1989) found that forage quality (nutrient content) increased after burning an aspen forest.

12. Veblen et al. (1991) found increased tree growth after spruce beetle outbreaks in Colorado.

13. Weaver and Forcella (1977) provide estimates on the losses that could occur with harvested wood.

Chapter 12: Mountain Meadows and Snowglades

1. Subtle differences in topography can affect the establishment of many plants, especially in rigorous environments. Concave surfaces, or depressions, often have a different species composition than adjacent convex surfaces such as ridges or hilltops.

2. Timber harvesting has increased the amount of edge habitat in the Rocky Mountains, to the point where it probably is not a limiting factor for wildlife abundance in most areas.

Chapter 13: Upper Treeline and Alpine Tundra

1. Krummholz "tree islands" may move slowly downwind (Marr 1977; Benedict 1984), the result of abrasion on the upwind side and layering on the downwind side.

2. Bristlecone pine, a common treeline species in Colorado, northern New Mexico, and westward (Baker 1992), also has flexible branches.

3. Billings (1974, 1978) estimated that about 50% of the alpine flora of the Beartooth Mountains consists of species also found in the Arctic.

4. Several exceptions are the tiny annuals studied by Reynolds (1984) in the Beartooth Mountains: *Koenigia islandica, Polygonum confertiflorum,* and *P. douglasii.*

5. Briggs and MacMahon (1983) studied wetland vegetation at high elevations in the Uinta Mountains.

6. Though freeze-thaw cycles still affect alpine tundra, some of the patterned ground may be remnants from the Pleistocene. Solifluction terraces have been observed in mountain foothills, and polygonal ground is apparent on the floor of some high basins (see chap. 3).

7. The shrubs of avalanche tracks are often the preferred summer food of moose.

Chapter 14: The Yellowstone Plateau

1. Christiansen (1984) identified three main epidsodes of "caldera-forming rhyolitic volcanism" in the Yellowstone region over the last 2.2 million years. The most recent is the present-day caldera, which developed about 630,000 years ago.

2. Glaciers persisted in some high mountain valleys until about 10,00–8,500 years ago (Mears 1974).

3. Eversman and Carr (1992) provide a roadside guide to YNP ecology.

4. Aspen seedlings were abundant after the 1988 fires, but it remains to be seen whether the area occupied by aspen will increase significantly.

5. Schullery (1992) reviews the ecology and management of bears in YNP.

6. If climatic warming occurs, Romme and Turner (1991) predict that the alpine zone will be reduced in area and that the whitebark pine woodlands will become more fragmented. In contrast, foothill grasslands, shrublands, and Douglas-fir woodlands will become more common. Ryan (1991) predicts that fire frequency will increase.

Chapter 15: Jackson Hole and the Tetons

1. In 1879 Thomas Moran wrote in his journal, "The fires in the surrounding mountains had become so dense as almost to obscure the peaks of the Tetons and the sun went down in fiery redness."

2. The Waterfalls Canyon fire was the first major fire

allowed to burn in a national park system under the new prescribed natural burn policy (Steve Cain, pers. comm.).

3. While there are now about 15,000 elk in the Jackson Elk herd, only about 8,500 are typically found in the valley (Mark Boyce, pers. comm.; Boyce 1989). Moose are most abundant during the winter.

4. The brucellosis bacteria must be consumed by cattle when they are exposed to aborted elk or bison fetuses. Regional newspapers covered this issue in early 1992. The same issue is significant in YNP, and for this reason bison leaving the park for Montana are routinely shot (see chap. 14).

Chapter 16: The Black Hills, Bear Lodge Mountains, and Devil's Tower

1. The name "Buffalo Gap" stems from the observation that buffalo commonly used this break in the Hogback Ridge for travel between the plains and mountains.

2. Devil's Tower, Sundance Mountain, and Inyan Kara Mountain are igneous or volcanic intrusions that lie just outside the Bear Lodge Mountains proper.

3. The dark color of the Black Hills is apparent from a distance and is caused by the dense cover of ponderosa pine. The Indian name for the Black Hills is Paha Sapa, which translates to "hills that are black" (Froiland 1990).

4. Uresk and Painter (1985) found that shrubs and trees (bur oak and ponderosa pine) in the northern Black Hills made up 37% of cattle diets in September. During the grazing season, the diet was 54% grasses, 17% forbs, and 28% shrubs and trees. Similarly, cattle consume recently fallen cottonwood leaves in September and October along the South Platte River in Colorado (Sedgwick and Knopf 1991).

Chapter 17: Using Wyoming Landscapes

1. By 1990 the state's population had dropped to about 470,000 because of a reduced demand for Wyoming oil and other mineral resources. The native American population is divided into two tribes, the Shoshone and the Arapaho, and totals approximately 15,000.

2. Because of the Knutsen-Vandenberg Act of 1930 (U.S. News and World Report, 1 July 1991).

Glossary

A-horizon | The surface layer of soil where decomposing detritus is mixed with mineral soil.

Acre-foot | The amount of water that will cover 1 acre of land 1 foot deep; 1 acre-foot is equal to 43,560 ft³ or 1,232.75 m³.

Actinorhizae | A symbiotic association between plant roots and a group of bacteria known as actinomycetes.

Adventitious root | A root that grows from a stem rather than from another root.

Aeolian | *See* Eolian.

Alfisols | Soils with a leached surface mineral layer and a subsurface accumulation of clay; formerly known as gray-brown podzols.

Allelopathic | The production of chemicals by an organism which are toxic to other organisms.

Alluvial soils (alluvium) | Soils that have developed on material transported by water.

Altithermal | A comparatively warm, dry, postglacial period that occurred 7,500–4,000 years ago.

Anaerobic | An environment without oxygen.

Andesite | A volcanic rock composed primarily of plagioclase feldspar.

Anion | A negatively charged ion, e.g., nitrate.

ANPP | Aboveground net primary productivity. *See* Primary productivity.

Anthropogenic | Caused by humans.

Anticline (anticlinal) | An area where the earth's crust has been folded upward, forming a convex surface.

Apical dominance | The inhibition of growth in lower plant buds by hormones produced in terminal or apical buds; lower buds may begin growing if the apical buds are removed.

Apomixis | The production of viable seeds without fertilization.

Aridisol | A soil type with distinct horizons that occurs in desert basins and that has accumulations of clay, calcium carbonate, gypsum, and/or soluble salts.

Arroyo | A small, steep-sided water course or gulch with a nearly flat floor; usually dry except after heavy rains.

AUM | Animal unit month, equal to one cow and a calf, or five sheep, feeding on a unit of land for one month. Typical stocking rates on Wyoming mixed-grass prairie are 0.1–0.3 AUM's per acre.

Autotroph | An organism capable of using light as a source of energy (e.g., green plants capable of photosynthesis) or using the energy in the chemical bonds of certain inorganic molecules (e.g., certain bacteria).

Bentonite | A soft, pliable, porous, light-colored rock composed essentially of clay minerals in the montmorillonite group and having the ability to absorb large quantities of water, causing expansion to eight times its dehydrated volume.

Biomass   Plant, animal, and microbial organic matter (living, dead, or both combined), expressed as dry weight per unit area.

Blowout   An area of renewed wind erosion in a sand dune that had been stabilized by vegetation.

Bole   The trunk of a tree.

Braided stream   A stream with several channels that sometimes shift during periods of high water (see fig. 4.6).

C_3 plants   All plants where the first stable product of photosynthesis is a 3-carbon compound; usually more abundant in cool, moist environments.

C_4 plants   All plants where the first stable product of photosynthesis is a 4-carbon compound, usually more common in warm, dry environments because they use water more efficiently.

Caldera   A large, basinlike depression resulting from the explosion and collapse of the center of a volcano.

Caliche   A layer of calcium carbonate accumulation in arid-land soils at the average depth of water percolation; a cemented layer (hardpan) that restricts root growth.

Cambium   A meristematic tissue, where new xylem and phloem are produced by cell division and differentiation, that causes the growth in diameter of woody stems and roots.

Canopy gap   A small opening in the forest canopy created by the death of one or more neighboring trees.

Catena   The gradual change in soil characteristics along sloping land, such as from a hilltop to a swale.

Cation   A positively charged ion, e.g., potassium, calcium, and ammonium.

Cellulose   A chemically inert carbohydrate that is the primary constituent of plant cell walls and therefore of most plant biomass.

cfs   Cubic feet per second.

Chronosequence   A series of stands found in different places that are very similar except for the time since the last major disturbance.

Cirque   A steep-walled, mountain landform carved by a glacier at its origin (see fig. 2.5).

Climax   An ecosystem that is changing more slowly than the seral ecosystems that preceded it; dominant plant species appear to be capable of reproducing or persisting in the community.

Clone (clonal)   A plant or group of plants that has developed from root or stem sprouts rather than from seeds and that has the same genetic composition as the parent plant.

Colluvial   Referring to soil material or rock that is transported downhill by gravity; not alluvial or eolian.

Community   The assemblage of living organisms in an ecosystem.

Conductance   A measure of salt concentration expressed as dS/m (see below).

Congeliturbation   The churning, heaving, and thrusting of soil material resulting from pressures created by frost.

Convectional   Vertical air movement caused by increasing temperatures at the earth's surface.

Coppice dune   A dune that develops because of the deposition of eolian material around the base of shrubs and that grows larger as the shrubs become larger or more numerous.

Corm   A modified stem that facilitates asexual reproduction (cloning).

Cover (canopy cover)   The proportion of land area covered by a particular kind of plant or by all plants combined.

Crown fire   A hot fire that burns through the tops (crowns) of trees.

Cryopedogenesis   Soil development where frequent freezing and thawing (congeliturbation) is an important phenomenon, such as in alpine eco-systems.

Cryostatic pressure Pressure that develops in wet soils when they freeze.

Cryoturbation The mixing of soil material caused by alternate freezing and thawing; congeliturbation.

Cryptogamic crust A layer of lichens, algae, and bacteria that sometimes develops on the surface of northern desert soils.

Crystalline rock Igneous or metamorphic rock formed from magma deep in the earth's crust (e.g., granite), where cooling was slow, enabling the formation of large crystals.

Cyanobacteria Photosynthetic, oxygen-producing bacteria (formerly known as bluegreen algae).

Debris jam A small dam across a creek created by fallen trees and other floating detritus.

Decomposer Any organism that breaks detritus into smaller pieces of organic matter (reduction) or converts organic matter into inorganic compounds (mineralization, typically bacteria and fungi).

Deflation hollow A basin that has developed because of wind erosion.

Denitrification The production of nitrogen gas from molecules containing nitrogen (e.g., nitrate or ammonium).

Desert pavement The stony surface that develops following the wind erosion of fine soil particles in arid environments.

Detritivore A heterotrophic animal that uses detritus as a source of food.

Detritus Dead organic matter; also referred to as litter, forest floor, or mulch (in grasslands).

dS/m Decisiemens per meter; a measure of soil salinity; 1 dS/m = 1 mmhos/cm.

Dune An accumulation of sand or silt formed by wind.

Ecosystem Any area where plants, animals, and other organisms interact with each other and their physical environment; boundaries are defined based on research or management objectives.

Ecotype A population of a species that is genetically different owing to the environmental conditions where it occurs.

Edaphic Related to or caused by soil conditions.

Entisols Soils that are young and have little or no profile development, such as those that occur on eroding slopes and along ephemeral streams; found at all elevations.

Eolian Transported by wind, e.g., the dune sand and loess.

Eutrophication The process of nutrient enrichment, whether caused by humans or other factors.

Evapotranspiration (ET) The sum of water lost from an ecosystem owing to evaporation and transpiration.

Even-aged A description of forests where most of the trees became established in a comparatively short time (perhaps fifty years) following a stand-replacing disturbance, such as a crown fire.

Exclosure An area surrounded by a fence that prevents grazing.

Fellfield Alpine tundra where soil does not yet cover all the rocks; intermediate between a boulder field with no soil and an alpine meadow or turf with deep soils.

Forb An herbaceous plant that is not a grass or sedge.

Forest floor The layer of decaying leaves, twigs, and branches that covers the mineral soil of forest ecosystems; an accumulation of detritus.

Frost hummock A mound formed by the sorting of fine and coarse soil material by frequent freezing and thawing, such as occurs in alpine tundra; formed by congeliturbation.

Gap *See* Canopy gap.

Genotype The genetic composition of a specific organism that determines its morphological and physiological characteristics.

Genus (genera) A category for classifying organisms, designated by a Latin name, that usually includes more than one species; the first name of an organism's scientific name, e.g., *Artemisia* (the genus for sagebrush).

Geomorphic Pertaining to physical phenomena that cause changes in the earth's surface.

Geyserite A rock that is rich in silica that develops around geysers and hot springs.

Glacial till Sand, gravel, and rock deposited by melting glaciers.

Graminaceous (graminoid) Describes grasses or sedges, or plants that look like them.

Gross state product (GSP) A measure of the total sales by businesses and industry during a specific period.

Gypsiferous High in gypsum (hydrated calcium sulfate).

Halophyte (halophytic) A plant that can tolerate environments with high concentrations of salts, such as plants that grow on saline soils.

Heterotroph All organisms that require organic matter as the source of nutrition, in contrast to autotrophs.

Holocene The last 10,000 years of geologic history.

Hydrothermal Environments that are influenced by the warm water originating deep in the earth's crust, such as the geyser basins of Yellowstone National Park.

Hyphae Multicellular filaments that constitute the biomass (mycellium) of fungi.

Hypsodont dentition Teeth that continue growing, thereby replacing enamel lost to abrasion during the chewing of plant tissues.

Igneous Describes rocks solidified from molten or partly molten material, i.e., from magma.

Inceptisols Soils typically found in the Rocky Mountains that have weakly developed horizons; subsoil horizons are brown, and in the alpine, have a surface mineral layer with a low pH and an accumulation of humus.

Incised stream A stream channel with little if any potential for lateral movement because of topographic constraints, such as in a narrow V-shaped valley.

Inflorescence The cluster of flowers that occurs on some plants.

Insectivorous A description of plants capable of trapping insects, from which some nutrients are obtained as the insects decompose.

Intercalary meristem A tissue capable of cell division that is found at the base of some grass leaves and stems.

Interception The hydrologic process whereby some rain is held on plant or detrital surfaces and subsequently evaporated, without infiltrating into the soil.

Internode That part of a stem separating nodes.

kJ/m² Kilojoules or 1,000 joules per square meter; a measure of energy equal to 239 calories per square meter.

Layering A form of asexual, vegetative reproduction (cloning) that can occur in some woody plants when lower branches sprout new roots after being pressed against the soil (typically owing to the weight of the snow).

Leaching The process whereby nutrients, salts, and other soil materials are transported by water to greater depths or into streamwater and groundwater.

Levee A natural or man-made deposit of sand and mud created by floodwaters along some streams and rivers.

Lignin An organic compound that, along with cellulose, is an important constituent of wood and some other plant tissues.

Lignite A form of coal with a relatively low energy content.

Ligule A bractlike appendage found at the top of the sheath of some grasses.

Litter — Plant material that falls or accumulates on the soil surface; detritus that contributes to the mulch of grasslands or the forest floor.

Loess — Fine soil material that has been dispersed by wind; calcareous and usually buff to light yellow.

Megagram (Mg) — 1 metric ton, 1,000 kg, or 1,000,000 g.

Meristem — A plant tissue capable of cell division; primary meristems (such as in seeds and buds) lead to plant elongation, and secondary meristems (the cambium) lead to increases in the diameter of stems and roots.

Mesic — Intermediate between wet and dry; the term *more mesic* implies that relatively more moisture is available.

Metamorphic rocks — Rocks that have been changed owing to intense heat and pressure (e.g., slate from shale, quartzite from sandstone, schist from rhyolite, marble from limestone, and gneiss from granite).

Metric ton — 1,000 kilograms; 1 megagram (Mg)

Microphytic crust — A layer of lichens, algae, and bacteria that sometimes forms on the surface of desert soils; also known as cryptogamic crust.

Mima mound — Low mounds of puzzling origin (see chapter 8).

Mollisol — Soil with a thick, dark accumulation of humus in the surface mineral layer; typical of moist grasslands, meadows, and some aspen forests; formerly known as chernozem soils.

Monoculture — A plant community where most of the plants belong to one species, such as cropland and some tree plantations.

Moraine — A deposit of soil and rock that forms as glaciers melt.

MPa — Megapascals; a measure of pressure equal to 0.1 bars or 0.101 atmospheres.

Mutualism (mutualistic) — A symbiotic association between two species that is beneficial to both.

Mycorrhizae — A symbiotic, mutualistic association between plant roots and certain species of fungi that facilitates the acquisition of water and nutrients by the plant.

Mycorrhizal fungi — Those fungi capable of forming mycorrhizae.

Needle ice — Small needle-shaped ice crystals that often form in the alpine tundra and contribute to cryopedogenesis.

Nematode — Microscopic, unsegmented worms in the phylum Nematoda; roundworms.

Net primary productivity (NPP) — The amount of new plant material (biomass) that is formed per unit area per unit time; often expressed as $g/m^2/yr$.

Niche — The role that an organism plays in an ecosystem (e.g., herbivores fill a different niche than do carnivores).

Nitrification — The conversion of ammonium to nitrate by certain species of bacteria.

Nitrogen fixation — The conversion of nitrogen gas to ammonium and nitrate by certain bacteria.

Nivation hollow — An area where the vegetation and soil are different because of deep snow accumulation year after year due to drifting; typically on the lee side of some ridges.

Node — An area on a stem where buds or leaves originate, separated by internodes.

NPP — *See* Net primary productivity.

Open system — A system with flows of energy and nutrients across its boundaries or membranes; all ecosystems and organisms.

Orogeny — The process of mountain building.

Oxbow — A U-shaped landform created when the channel of a river cuts across a meander.

Pangaea — A huge landmass believed to

have existed before the continents began to drift apart 300–200 million years ago (see fig. 2.2).

Parasitic — Obtaining energy and nutrients from living organisms.

Park — A term commonly used in western North America for mountain grasslands or meadows.

Perched water table — An area where soil conditions prevent rapid drainage, causing an accumulation of water that is above the typical groundwater level.

Perennating buds — Plant parts such as rhizomes, bulbs, and buds that enable many plants to live more than one year.

Periglacial — Occurring near glaciers.

Perturbation — A disturbance that causes a sudden change in some aspect of an ecosystem.

Phenotype — The morphologic and physiologic expression of a genotype; the form and processes that characterize a specific organism.

Pheromone — An aromatic chemical that facilitates communication between insects.

Phloem — The plant tissue through which carbohydrates and other organic compounds are transported; the inner bark of trees and shrubs.

Phreatophyte — A plant that obtains much of its annual water requirement from groundwater generated from precipitation in the nearby mountains or foothills, such as along creeks and rivers, rather than from soil water derived from rain or snow in the immediate area; usually woody plants, e.g., alder, cottonwood, saltcedar, water birch, and many species of willow.

Plasticity — The production of a wide range of phenotypes from one genotype.

Playa — A flat area with slow drainage, fine-textured soils, and high salt concentrations on the surface; usually in a depression.

Pleistocene — An epoch of geologic history dating from 2 million to 10,000 years ago (see fig. 2.1).

Point bar — A deposit of sand or silt that rises above the water level, such as a sand bar; common along meandering rivers on the "point" across from a cutbank (see figs. 4.6, 4.7).

ppm — A measure of concentration, expressed as parts per million; 1 ppm = 1 mg/l.

Primary productivity — Gross primary productivity, i.e., the total dry weight of carbohydrates produced by photosynthesis in an ecosystem in a specific period; commonly expressed as g/m^2/yr. *See* Net primary productivity.

Pubescence — The short, fine hairs that are found on some leaf and stem surfaces.

Quaternary — The period in geologic history from about 2 million years ago to the present, comprised of the Pleistocene and Holocene epochs (see fig. 2.1).

Rainshadow — That area on the leeward side of a mountain that is drier because of precipitation over the mountain and the fact that the air is warming as it descends after crossing the mountain.

Ramet — A branch produced by sprouting or some other form of vegetative or asexual reproduction.

Reabsorbtion — *See* Retranslocation.

Refugium — An area where some species survive a period of unfavorable environmental conditions, such as unglaciated areas surrounded by glaciers.

Retranslocation — The process of transferring some nutrients from leaves to twigs before the leaves fall; a mechanism for conserving nutrients within perennial plants.

Rhizomatous — Referring to a plant capable of producing rhizomes for vegetative reproduction.

Rhizome — A belowground, horizontal stem that facilitates the asexual reproduction and spread of some plants.

Rhyolite (rhyolitic)
A fine-grained igneous rock rich in silica; the volcanic equivalent of granite.

Ribbon forest
A narrow band of trees found at high elevations, caused by the interaction of wind and snow accumulation.

Riparian
Describes, generally, land that is adjacent to creeks and rivers where plants benefit from the streamflow by having more water than is available on the adjacent upland; typically defined by the presence of phreatophytes such as cottonwoods and certain willows or by other plants that require substantially more water than upland species.

Root crown
The perennial tissue of grasses, sedges, and forbs located near the soil surface from which all roots of a plant emanate and from which new herbaceous stems and leaves grow.

Root-shoot ratio
The ratio between the dry weight of roots and the dry weight of shoots.

Saltate (saltation)
The intermittent movement of sand and gravel particles across a surface, driven by wind or water.

Saprophytic
Referring to organisms that obtain energy from decomposing organic matter (detritus), such as do many fungi.

Sapwood
The outer wood of trees and shrubs through which water and nutrients are transported; the outer xylem.

Scoria
A hard rock that develops following the intense heat created by burning coal seams; often forms the crests of buttes and pinnacles in the Powder River Basin.

Seleniferous
Describes soils or rocks having high concentrations of selenium.

Senescent (senescing)
Describes plants or plant parts that are about to be shed or die from old age.

Seral
Referring to an ecosystem that is still developing after a large-scale disturbance such as fire or plowing; not a climax community.

Serotinous cone
A type of cone that remains closed, preventing seed dispersal, until the resin bonds of the cone scales are broken, such as often happens during the heat of a forest fire; found on some individuals of lodgepole pine.

Sessile
Lacking a petiole or stem.

Sinter
Hard deposits around springs or geysers that are rich in silica or calcareous matter.

Sinuosity
A measure of the amount of meandering in creeks and rivers.

Snag
A dead, standing tree; often used by hawks, eagles, and cavity-nesting birds.

Snowglade
A meadow that is devoid of trees because of snowdrifts that persist until late summer.

Solifluction
The slow, viscous downslope flow of waterlogged soil underlain by frozen ground (not necessarily permafrost) that produces solifluction lobes, such as occurs in alpine tundra (fig. 13.7); flow rates are normally 0.5–5.0 cm/yr.

Sphagnum
A group of mosses that are typically found in bogs.

Spotting
The spread of fire by wind-blown embers (known as fire brands).

Sprout(ing)
New plant parts originating from older plant parts, not from a new seed, e.g., new aspen shoots that develop from aspen roots; vegetative reproduction.

Stand
A specific area or ecosystem from which ecological data are gathered.

Steppe
An extensive treeless plain; applied originally to the grasslands of Russia and Asia but now commonly used when referring to the extensive sagebrush shrublands of western North America. The terms *grassland* and *prairie* are used where grasses predominate rather than shrubs.

Stochastic
Applied to events that occur ir-

Stolon

Stratification

Sublimation

Succession

Succulent

Surface fire

Swale

Symbiotic

regularly or randomly but with a certain probability.

An aboveground, horizontal stem that facilitates the asexual reproduction and spread of some plants (e.g., strawberry and buffalograss).

Exposing seeds to a cold period that stimulates seed germination by breaking seed dormancy.

The conversion of a solid, such as snow or ice, to a vapor without melting.

The process of ecosystem change or development over time after large-scale disturbances such as fires; changes occur rapidly at first but then slow considerably as the climax or mature ecosystem is reached.

Describes plant organs that appear to be swollen, such as the pads of a cactus or the leaves of stonecrop and greasewood.

A fire that burns the detritus and understory vegetation of a forest but that often does not kill trees with thick bark (e.g., ponderosa pine and Douglas-fir).

A depression where soils are deeper and finer textured than on the surrounding upland; typically water is more readily available.

Describes an association between two species that live together in direct contact; when beneficial to both organisms, the symbiosis is referred to as

Syncline

Terpene

Till

Translocation

Transpiration

Travertine

Trona

Turgid

Ungulate

Vernal transpiration

Xerophyte (xerophytic)

Xylem

mutualism, such as in the case of mycorrhizae or lichens.

A geomorphic feature where the earth's crust is folded downward, forming a basin or concave surface.

A hydrocarbon produced by some plants.

See Glacial till.

The movement of carbohydrates and other organic compounds through the phloem.

The loss of water from plants in the form of vapor, primarily through stomata on leaves.

A dense, finely crystalline form of limestone, typically deposited with evaporation near springs, especially hot springs; usually white, tan, or cream colored.

A grayish or yellowish mineral; hydrous sodium carbonate and bicarbonate.

Describes whole plants, plant organs, or plant cells that are comparatively firm because of water pressure on the interior of cell walls; not wilted.

A mammal with hooves.

Transpiration that occurs in Rocky Mountain forests while there is still snow on the ground.

A plant that is well adapted for dry (xeric) environments.

The plant tissue through which water and nutrients are transported; wood is composed largely of xylem produced by the vascular cambium.

References

Abbott, E. C., and H. H. Smith. 1955. We pointed them north. Univ. Oklahoma Press, Norman. (New ed.)

Agee, J. K., and D. R. Johnson, eds. 1988. Ecosystem management for parks and wilderness. Univ. Washington Press, Seattle.

Agnew, W., D. W. Uresk, and R. M. Hansen. 1986. Flora and fauna associated with prairie dog colonies and adjacent ungrazed mixed-grass prairie in western South Dakota. J. Range Mgmt. 39:135–39.

Ahlbrandt, T. S. 1973. Sand dunes, geomorphology and geology, Killpecker Creek area, northern Sweetwater County, Wyoming. M.S. thesis, Univ. Wyoming, Laramie.

Akashi, Y. 1988. Riparian vegetation dynamics along the Bighorn River, Wyoming. M.S. thesis, Univ. Wyoming, Laramie.

Albertson, F. W., and G. W. Tomanek. 1965. Vegetation changes during a thirty-year period on grassland communities near Hays, Kansas. Ecol. 46:714–20.

Albertson, F. W., and J. E. Weaver. 1944. Nature and degree of recovery of grassland from the great drought of 1933 to 1940. Ecol. Monogr. 14:393–479.

Alexander, R. R. 1985a. Diameter and basal area distributions in old-growth spruce-fir stands in Colorado. U.S. For. Ser. Res. Note RM-451.

———. 1985b. Major habitat types, community types, and plant communities in the Rocky Mountains. U.S. For. Ser. Gen. Tech. Rep. RM-123.

———. 1986a. Classification of the forest vegetation of Wyoming. U.S. For. Ser. Res. Note RM-466.

———. 1986b. Silvicultural systems and cutting methods for old-growth lodgepole pine forests in the central Rocky Mountains. U.S. For. Ser. Gen. Tech. Rep. RM-127.

———. 1987a. Ecology, silviculture, and management of the Engelmann spruce-subalpine fir type in the Central and Southern Rocky Mountains. U.S. For. Ser. Agric. Handbook 659.

———. 1987b. Silvicultural systems, cutting methods, and cultural practices for Black Hills ponderosa Pine. U.S. For. Ser. Gen. Tech. Rep. RM-139.

Alexander, R. R., and C. B. Edminster. 1981. Management of ponderosa pine in even-aged stands in the Black Hills. U.S. For. Ser. Res. Paper RM-228.

Allen, E. B. 1979. The competitive effects of introduced annual weeds on some native and reclamation species in the Powder River Basin, Wyoming. PhD. diss., Univ. Wyoming, Laramie.

Allen, E. B., and D. H. Knight. 1984. The effects of introduced annuals on secondary succession in sagebrush-grassland, Wyoming. Southwest. Natural. 29:407–21.

Allen, M. F. 1991. The ecology of mycorrhizae. Cambridge Univ. Press, Cambridge.

Allen, M. F., E. B. Allen, and N. E. West. 1987. Influence of parasitic and mutualistic fungi on *Artemisia tridentata* during high precipitation years. Bull. Torrey Bot. Club 114:272–79.

Allen, M. F., and J. A. MacMahon. 1985. Impact of disturbance on cold desert fungi: Comparable microscale dispersion patterns. Pedobiologia 28: 215–24.

Alley, H. P., and D. W. Bohmont. 1958. Big sagebrush control. Wyo. Agric. Exp. Sta. Bull. 354.

Allred, B. W. 1941. Grasshoppers and their effect on sagebrush on the Little Powder River in Wyoming and Montana. Ecol. 22:387–92.

Amman, G. D. 1977. The role of mountain pine beetle in lodgepole pine ecosystems: Impact on succession. In W. J. Mattson, ed., The role of arthropods in forest ecosystems, pp. 3–18. Springer-Verlag, New York.

———. 1978. The biology, ecology, and causes of outbreaks of mountain pine beetle in lodgepole pine forests. In M. A. Berryman, G. D. Amman, and R. W. Stark, eds., Theory and practice of mountain pine beetle management in lodgepole pine forests, 39–53. Univ. Idaho Experiment Station, Moscow.

———, compiler. 1989. Symposium on the management of lodgepole pine to minimize losses to the mountain pine beetle. U.S. For. Ser. Gen. Tech. Rep. INT-262.

Amman, G. D., and W. E. Cole. 1983. Mountain pine beetle dynamics in lodgepole pine forests. Part II. Population dynamics. U.S. For. Ser. Gen. Tech. Rep. INT-145.

Amman, G. D., and K. C. Ryan. 1991. Insect infestation of fire-injured trees in the Greater Yellowstone Area. U.S. For. Ser. Res. Note INT-398.

Amundson, M. A. 1991. Wyoming: Time and again. Pruett Publishing, Boulder.

Anderson, D. C. 1987. Below-ground herbivory in natural communities: A review emphasizing fossorial animals. Quart. Rev. Biol. 62:261–86.

Anderson, D. C., K. T. Harper, and R. C. Holmgren. 1982. Factors influencing development of cryptogamic soil crusts in Utah deserts. J. Range Mgmt. 35:180–85.

Anderson, D. C., K. T. Harper, and S. R. Rushforth. 1982. Recovery of cryptogamic soil crusts from grazing on Utah winter ranges. J. Range Mgmt. 35:355–59.

Anderson, E. 1974. A survey of the Late Pleistocene and Holocene mammal fauna of Wyoming. In M. Wilson, ed., Applied Geology and Archaeology: The Holocene history of Wyoming, 79–90. Geol. Surv. Wyo. Rep. Invest. 10.

Anderson, J. E., and K. E. Holte. 1981. Vegetation development over twenty-five years without grazing on sagebrush-dominated rangeland in southeastern Idaho. J. Range Mgmt. 34:25–29.

Anderson, J. E., and M. L. Shumar. 1986. Impacts of black-tailed jackrabbits at peak population densities on sagebrush steppe vegetation. J. Range Mgmt. 39:152–56.

Anderson, N. L. 1961. Seasonal losses in rangeland vegetation due to grasshoppers. J. Economic Entomology 54:369–78.

———. 1964. Some relationships between grasshoppers and vegetation. Ann. Entomol. Soc. Am. 57:736–742.

Anderson, R. V., C. R. Tracy, and Z. Abramsky. 1979. Habitat selection in two species of short-horned grasshoppers: The role of thermal and hydric stresses. Oecologia 38:359–74.

Ansley, R. J., and R. H. Abernethy. 1984. Seed pretreatments and their effects on field establishment of spring-seeded gardner saltbush. J. Range Mgmt. 37:509–13.

Antos, J. A., B. McCune, and C. Bara. 1983. The effect of fire on an ungrazed western Montana grassland. Am. Midl. Nat. 110:354–64.

Aplet, G. H., R. D. Laven, and F. W. Smith. 1988. Patterns of community dynamics in Colorado Engelmann spruce-subalpine fir forests. Ecol. 69:312–19.

Archer, S., M. G. Garrett, and J. K. Detling. 1987. Rates of vegetation change associated with prairie dog (Cynonomys ludoviciana) grazing in North American mixed-grass prairie. Vegetatio 72:159–66.

Arkley, R. J., and H. C. Brown. 1954. The origin of mima mound (hogwallow) microrelief in the far western states. Soil Sci. Soc. Am. Proc. 18:195–99.

Arno, S. F. 1979. Forest regions of Montana. U.S. For. Ser. Res. Paper INT-218.

———. 1980. Forest fire history of the northern Rockies. J. For. 78:460–65.

———. 1985. Ecological effects and management implications of Indian fires. In J. E. Lotan, B. M. Kilgore, W. C. Fischer, and R. W. Mutch, tech. coords., Symposium and workshop on wilderness fire, pp. 81–86. U.S. For. Ser. Gen. Tech. Rep. INT-182.

Arno, S. F., and G. E. Gruell. 1983. Fire history at the forest-grassland ecotone in southwestern Montana. J. Range Mgmt. 36:332–36.

———. 1986. Douglas fir encroachment into mountain grasslands in southwestern Montana. J. Range Mgmt. 39:272–76.

Arno, S. F., and R. P. Hammerly. 1984. Timberline: Mountain and arctic forest frontiers. The Mountaineers, Seattle.

Arno, S. F., and R. J. Hoff. 1989. Silvics of whitebark pine (Pinus albicaulis). U.S. For. Ser. Gen. Tech. Rep. INT-253.

Arno, S. F., and A. E. Wilson. 1986. Dating past fires in currleaf mountain-mahogany communities. J. Range Mgmt. 39:241–43.

Aro, R. S. 1971. Evaluation of pinyon-juniper conversion to grassland. J. Range Mgmt. 24:188–97.

Austin, D. D., and P. J. Urness. 1980. Response of curlleaf mountain mahogany to pruning treatments in northern Utah. J. Range Mgmt. 33:275–77.

Axelrod, D. I. 1948. Climate and evolution in western North America during middle Pleistocene time. Evol. 2:127–44.

———. 1968. Tertiary floras and topographic history

of the Snake River Basin, Idaho. Geol. Soc. Am. Bull. 79:713–34.

———. 1985. Rise of the grassland biome, central North America. Bot. Rev. 51:163–201.

Baker, R. G. 1970. Pollen sequence from late Quaternary sediments in Yellowstone Park. Science 168:1449–50.

———. 1976. Late Quaternary vegetation history of the Yellowstone Lake basin, Wyoming. U.S.G.S. Prof. Paper 729-E.

———. 1983. Holocene vegetational history of the western United States. In H. E. Wright, Jr., ed., Late-Quaternary environments of the United States, vol. 2, pp. 109–27. Univ. Minnesota Press, Minneapolis.

Baker, W. L. 1984. A preliminary classification of the natural vegetation of Colorado. Great Basin Nat. 44:647–76.

———. 1988. Size-class structure of contiguous riparian woodlands along a Rocky Mountain river. Phys. Geog. 9:1–14.

———. 1989. Macro- and micro-scale influences on riparian vegetation in western Colorado. A. Assoc. Am. Geogr. 79:65–78.

———. 1990. Climatic and hydrologic effects on the regeneration of *Populus angustifolia* James along the Animas River, Colorado. J. Biogeogr. 17:59–73.

———. 1992a. Structure, disturbance, and change in the bristlecone pine forests of Colorado, USA. Arct. Alp. Res. 24:17–26.

———. 1992b. Effects of settlement and fire suppression on landscape structure. Ecol. 73:1879–87.

Baker, W. L., and S. C. Kennedy. 1985. Presettlement vegetation of part of northwestern Moffat Country, Colorado, described from remnants. Great Basin Nat. 45:747–77.

Baker, W. L., and T. T. Veblen. 1990. Spruce beetles and fires in the nineteenth-century subalpine forests of western Colorado, USA. Arct. Alp. Res. 22:65–80.

Bamberg, S. A. 1971. Plant ecology of alpine tundra areas in Montana and adjacent Wyoming. M.A. thesis, Univ. Colorado, Boulder.

Bamberg, S., and J. Major. 1968. Ecology of the vegetation and soils associated with calcareous parent materials in three alpine regions of Montana. Ecol. Monogr. 38:127–67.

Barker, J. R., and C. M. McKell. 1986. Differences in big sagebrush (*Artemisia tridentata*) plant stature along soil-water gradients: Genetic components. J. Range Mgmt. 39:147–51.

Barnes, B. V. 1966. The clonal growth habit of American aspens. Ecol. 47:439–47.

Barnes, P. W., and A. T. Harrison. 1982. Species distri-

bution and community organization in a Nebraska Sandhills mixed prairie as influenced by plant/soil-water relationships. Oecologia 52:192–201.

Barrett, S. W., and S. F. Arno. 1982. Indian fires as an ecological influence in the northern Rockies. J. For. 80:647–51.

Bartos, D. L., and G. D. Amman. 1989. Microclimate: An alternative to tree vigor as a basis for mountain pine beetle infestations. U.S. For. Ser. Res Paper INT-400.

Bartos, D. L., and W. F. Mueggler. 1979. Influence of fire on vegetation production in the aspen ecosystems in western Wyoming. In M. S. Boyce and L. D. Hayden-Wing, eds., North American elk: Ecology, behavior and management, pp. 75–78. Univ. Wyoming, Laramie.

Bartos, D. L., W. F. Mueggler, and R. B. Campbell, Jr. 1991. Regeneration of aspen by suckering on burned sites in western Wyoming. U.S. For. Ser. Res. Paper INT-448.

Baumgartner, D. M., R. G. Krebill, J. T. Arnott, and G. F. Weetman, eds. 1985. Lodgepole pine: The species and its management. Coop. Ext. Ser. Wash. St. Univ.

Beard, K. C., L. Krishtalka, and R. K. Stucky. 1991. First skulls of the Early Eocene primate *Shoshonius cooperi* and the anthropoid-tarsier dichotomy. Nature 349:64–67.

Beath, O. A. 1937. The occurrence of selenium and seleniferous vegetation in Wyoming. Wyo. Agric. Exp. Sta. Bull. 221:29–64.

Beath, O. A., and H. F. Eppson. 1947. The form of selenium in some vegetation. Univ. Wyo. Agric. Exp. Sta. Bull. 278:1–5.

Beath, O. A., C. S. Gilbert, and H. F. Eppson. 1941. The use of indicator plants in locating seleniferous areas in western United States. Am. J. Bot. 28:887–900.

Beauchamp, H., R. Lang, and M. May. 1975. Topsoil as a seed source for reseeding strip mine spoils. Wyo. Agric. Exp. Sta. Res. J. 90.

Beetle, A. A. 1950a. Buffalograss: Native of the short-grass prairie. Wyo. Agric. Exp. Sta. Bull. 293.

———. 1950b. Range condition classes on the Laramie Plains, Wyoming. Wyo. Agric. Exp. Sta. Circ. 37.

———. 1956. Range survey in Wyoming's Big Horn Mountains. Wyo. Agric. Exp. Sta. Bull. 341.

———. 1960. A study of sagebrush: The section Tridentae of Artemisia. Wyo. Agric. Exp. Sta. Bull. 368.

———. 1968. Range survey in Teton County, Wyoming. Pt. 3, Trends in vegetation. Wyo. Agric. Exp. Sta. Res. J. 26:1–16.

———. 1974a. Holocene changes in Wyoming vegetation. In M. Wilson, ed., Applied geology and archaeology: The Holocene history of Wyoming, pp. 71–73. Geol. Surv. Wyo. Rep. Invest. 10.

———. 1974b. Range survey in Teton County Wyoming. Pt. 4, Quaking aspen. Wyo. Agric. Exp. Sta. Sci. Monogr. 27.

———. 1974c. The zootic disclimax concept. J. Range Mgmt. 27:30-32.

Beetle, A. A., and K. L. Johnson. 1982. Sagebrush in Wyoming. Univ. Wyo. Agric. Exp. Sta. Bull. 779.

Beetle, A. A., W. M. Johnson, R. L. Lang, M. May, and D. R. Smith. 1961. Effect of grazing intensity on cattle weights and vegetation on the Bighorn Experimental Pastures. Wyo. Agric. Exp. Sta. Bull. 373.

Behan, M. J. 1957. The vegetation and ecology of Dry Park in the Medicine Bow Mountains. M.S. thesis, Univ. Wyoming, Laramie.

Beiswenger, J. M. 1991. Late Quaternary vegetational history of Grays Lake, Idaho. Ecol. Monogr. 6:165–82.

Beiswenger, J. M., and M. Christensen. 1989. Fungi as indicators of past environments. Curr. Res. Pleistocene 6:54–56.

Bell, K. L. 1974. Autumn, winter, and spring phenology of some Colorado alpine plants. Am. Midl. Nat. 91:460–64.

Bell, K. L., and L. C. Bliss. 1979. Autecology of *Kobresia bellardii*: why winter snow accumulation limits local distribution. Ecol. Monogr. 49:377–402.

Bell, R. H. V. 1971. A grazing ecosystem in the Serengeti. Sci. Am. 225:86–93.

Belsky, A. J. 1986. Does herbivory benefit plants? A review of the evidence. Am. Nat. 127:870–92.

———. 1987. The effects of grazing: Confounding of ecosystem, community and organism scales. Am. Nat. 129:777–83.

Benedict, J. B. 1984. Rates of tree-island migration, Colorado Rocky Mountains, USA. Ecol. 65:820–23.

Benkman, C. W., R. P. Balda, and C. C. Smith. 1984. Adaptations for seed dispersal and the compromises due to seed predation in limber pine. Ecol. 65:632–42.

Berg, A. W. 1990. Formation of mima mounds: A seismic hypothesis. Geol. 18:281–84.

Betancourt, J. L., W. S. Schuster, J. B. Mitton, and R. S. Anderson. 1991. Fossil and genetic history of a pinyon pine (*Pinus edulius*) isolate. Ecology 72:1685–97.

Billings, W. D. 1949. The shadscale vegetation zone of Nevada and eastern California in relation to climate and soils. Am. Midl. Nat. 42:1225–28.

———. 1954. Temperature inversions in the pinyon-juniper zone of a Nevada mountain range. Butler Univ. Bot. Studies 12.

———. 1969. Vegetational pattern near alpine timberline as affected by fire-snowdrift interactions. Vegetatio 19:192–207.

———. 1973. Arctic and alpine vegetations: similarities, differences, and susceptibility to disturbance. BioScience 23:697–704.

———. 1974. Adaptations and origins of alpine plants. Arct. Alp. Res. 6:129–42.

———. 1990. *Bromus tectorum,* a biotic cause of ecosystem impoverishment in the Great Plains. In G. M. Woodwell, ed., The earth in transition: Patterns and processes of biotic impoverishment, pp. 301–22. Cambridge Univ. Press, New York.

Billings, W. D., and L. C. Bliss. 1959. An alpine snowbank environment and its effect on vegetation, plant development, and productivity. Ecol. 40:388–97.

Billings, W. D., and H. A. Mooney. 1959. An apparent frost hummock-sorted polygon cycle in the alpine tundra of Wyoming. Ecol. 40:16–20.

———. 1968. The ecology of arctic and alpine plants. Biol. Rev. 43:481–529.

Birkby, J. L. 1983. Interaction of western harvester ants with southeastern Montana soils and vegetation. M.S. thesis, Montana State Univ., Bozeman.

Bissell, J. K. 1973. Geomorphological influence on vegetation in northwest Wyoming. M.S. thesis, Univ. Wyoming, Laramie.

Black, A. L., and J. R. Wight. 1979. Range fertilization: Nitrogen and phosphorus uptake and recovery over time. J. Range Mgmt. 32:349–53.

Blackstone, D. L., Jr. 1988. Traveler's guide to the geology of Wyoming (2d edition). Bulletin 67. Geol. Surv. Wyo., Laramie.

Blackwelder, E. 1912. The Gros Ventre slide: An active earth-flow. Geol. Soc. Am. Bull. 23:487–92.

Blair, N. 1987. The history of wildlife management in Wyoming. Wyo. Game and Fish Dept., Cheyenne.

Blaisdell, J. P. 1949. Competition between sagebrush seedlings and reseeded grasses. Ecol. 30:512–19.

———. 1950. Effects of controlled burning on bitterbrush on the upper Snake River Plains. U.S. For. Ser. Res. Paper 20.

———. 1953. Ecological effects of planned burning of sagebrush-grass range on the upper Snake River Plains. U.S. Dep. Agric. Tech. Bull. 1075.

Blaisdell, J. P., and R. C. Holmgren. 1984. Managing intermountain rangelands: Salt desert shrub ranges. U.S. For. Ser. Gen. Tech. Rep. INT-163.

Blaisdell, J. P., and W. F. Mueggler. 1956. Sprouting of bitterbrush (*Purshia tridentata*) following burning or top removal. Ecol. 37:365–69.

Blaisdell, J. P., R. B. Murray, and E. D. McArthur. 1982. Managing intermountain rangelands: Sagebrush-grass ranges. U.S. For. Ser. Gen. Tech. Rep. INT-134.

Blake, I. H. 1945. An ecological reconnaissance in the

Medicine Bow Mountains. Ecol. Monogr. 15:207–47.

Blauer, A. C., A. P. Plummer, E. D. McArthur, R. Stevens, and B. C. Giunta. 1976. Characteristics and hybridization of important intermountain shrubs. II. Chenopod family. U.S. For. Ser. Res. Paper INT-177.

⨯ Bleed, A., and C. Flowerday, eds. 1990. The atlas of the sand hills. Conservation and Survey Division, Institute of Agriculture and Natural Resources, Univ. Nebraska, Lincoln.

Bliss, L. C. 1962. Adaptations of arctic and alpine plants to environmental conditions. Arctic 15:117–144.

Bock, J. H., and C. E. Bock. 1984. Effect of fires on woody vegetation in the pine-grassland ecotone of the southern Black Hills, USA. Am. Midl. Nat. 112:35–42.

Boggs, K. 1984. Riparian communities of the lower Yellowstone River, Montana. M.S. thesis, Montana State Univ., Bozeman.

Boldt, C. E., and J. L. Van Deusen. 1974. Silviculture of ponderosa pine in the Black Hills: The status of our knowledge. U.S. For. Ser. Res. Paper RM-124.

Bonham, C. D. 1972. Vegetation analysis of grazed and ungrazed alpine hairgrass meadows. J. Range Mgmt. 25:276–79.

Bonham, C. D., and A. C. Lerwick. 1976. Vegetation changes induced by prairie dogs on shortgrass range. J. Range Mgmt. 29:221–25.

Bonnicksen, T. 1989. Fire gods and federal policy. Am. Forests 95:14–16, 66–68.

Booth, D. T. 1985. The role of fourwing saltbush in mined land reclamation: A viewpoint. J. Range Mgmt. 38:562–65.

Boutton, T. W., A. T. Harrison, and B. N. Smith. 1980. Distribution of biomass of species differing in photosynthetic pathway along an altitudinal gradient in southeastern Wyoming. Oecologia 45:287–98.

Bowman, R. A., D. M. Mueller, and W. J. McGinnies. 1985. Soil and vegetation relationships in a central plains saltgrass meadow. J. Range Mgmt. 38:325–28.

Boyce, M. S. 1989. The Jackson elk herd: Intensive wildlife management in North America. Cambridge Univ. Press, New York.

Boyce, M. S., and L. D. Hayden-Wing. 1979. North American elk: Ecology, behavior and management. Univ. Wyoming, Laramie.

Bradley, C. E., and D. G. Smith. 1986. Plains cottonwood recruitment and survival on a prairie meandering river floodplain, Milk River, southern Alberta and northern Montana. Can. J. Bot. 64:1433–42.

Bradley, F. J. 1873. Report of Frank H. Bradley, geologist, U. S. Geological Survey of the Territories. U. S. Gov. Printing Office, Washington, D.C.

Bragg, T. B. 1982. Seasonal variations in fuel and fuel consumption by fires in a bluestem prairie. Ecol. 63:7–11.

Bramble-Brodahl, M. K. 1978. Classification of Artemesia vegetation in the Gros Ventre area, Wyoming. M.S. thesis, Univ. Idaho, Moscow.

Brandegee, T. S. 1899. Teton Forest Reserve. 55th Congress, 3d sess. Serial 3763. House Doc. 5.

Branson, F. A. 1985. Vegetation changes on western rangelands. Range Monogr. 2. Soc. Range Mgmt., Denver.

Branson, F. A., and R. F. Miller. 1981. Effects of increased precipitation and grazing management on northeastern Montana rangelands. J. Range Mgmt. 34:3–10.

Branson, F. A., R. F. Miller, and I. S. McQueen. 1965. Plant communities and soil moisture relationships near Denver, Colorado. Ecol. 46:311–19.

———. 1967. Geographic distribution and factors affecting the distribution of salt desert shrubs in the United States. J. Range Mgmt. 20:287–96.

———. 1970. Plant communities and associated soil and water factors on shale-derived soils in northeastern Montana. Ecol. 51:391–407.

———. 1976. Moisture relationships in twelve northern desert shrub communities near Grand Junction, Colorado. Ecol. 57:1104–24.

Breckle, S. W. 1975. Zur Ökologie und zu den Mineralstoffverhältnissen absalzender und nichtabsalzender Xerohalophyten. Postdoctoral diss., Univ. Bonn.

Briggs, G. M., and J. A. MacMahon. 1983. Alpine and subalpine wetland plant communities of the Uinta Mountains, Utah. Great Basin Nat. 43:523–30.

Brock, T. D. 1978. Thermophilic microorganisms and life at high temperatures. Springer-Verlag, New York.

Brookes, M. H., R. W. Campbell, J. J. Colbert, R. G. Mitchell, and R. W. Stark, tech. coords. 1987. Western spruce budworm. U.S. For. Ser. Tech. Bull. 1694. (See also Tech Bull. 1695 and 1696.)

Brooks, A. C. 1962. An ecological study of Cercocarpus montanus and adjacent communities in part of the Laramie Basin. M.S. thesis, Univ. Wyoming, Laramie.

Brosz, D. J. 1986. Increasing irrigation water use efficiencies and resulting return flows. In D. J. Brosz and J. D. Rogers, coords, Proceedings: Wyoming Water 1986 and Streamside Zone Conference, Casper, Wyoming, 28–30 April 1986, pp. 136–41. Wyoming Water Research Center, Laramie.

Brotherson, J. D., D. L. Anderson, and L. A. Szyska. 1984. Habitat relations of Cercocarpus montanus (true mountain mahogany) in central Utah. J. Range Mgmt. 37:321–24.

Brown, J. K. 1975. Fire cycles and community dynamics in lodgepole pine forests. In D. M. Baumgartner, ed., Management of lodgepole pine ecosystems, pp. 429–56. Washington State Univ. Coop. Ext. Ser., Pullman, Washington.

———. 1991. Should management ignitions be used in Yellowstone National Park? In R. B. Keiter and M. S. Boyce, eds., The Greater Yellowstone Ecosystem: Redefining America's wilderness heritage, pp. 137–48. Yale Univ. Press, New Haven.

Brown, J. K., and N. V. DeByle. 1989. Effects of prescribed fire on biomass and plant succession in western aspen. U.S. For. Ser. Res. Paper INT-412.

Brown, R. W. 1962. Paleocene flora of the Rocky Mountains and Great Plains. U.S. Geol. Surv. Prof. Paper 375.

———. 1971. Distribution of plant communities in southeastern Montana badlands. Am. Midl. Nat. 85:458–77.

Brubaker, L. B. 1986. Responses of tree populations to climatic change. Vegetatio 67:119–30.

Brunsfeld, S. J., and F. D. Johnson. 1985. Field guide to the willows of east-central Idaho. Forest, Wildlife, and Range Exp. Sta. Bull. 39, Univ. Idaho, Moscow.

Buckner, D. L. 1977. Ribbon forest development and maintenance in the central Rocky Mountains of Colorado. Ph.D. diss., Univ. Colorado, Boulder.

Budd, K. J. 1991. Ecosystem management: Will national forests be "managed" into national parks. In R. B. Keiter and M. S. Boyce, eds., The Greater Yellowstone Ecosystem: Redefining America's wilderness heritage, pp. 65–76. Yale Univ. Press, New Haven.

Bunting, S. C., B. M. Kilgore, and C. L. Bushey. 1987. Guidelines for prescribed burning sagebrush-grass rangelands in the northern Great Plains. U.S. For. Ser. Gen. Tech. Rep. INT-231.

Burkart, M. R. 1976. Pollen biostratigraphy and late Quaternary vegetation history of the Bighorn Mountains, Wyoming. Ph.D. diss., Univ. Iowa, Iowa City.

Burke, I. C. 1987. Distribution and turnover of nitrogen in a sagebrush landscape. Ph.D. diss., Univ. Wyoming, Laramie.

———. 1989. Control of N mineralization in a sagebrush steppe. Ecol. 70:1115–26.

Burke, I. C., W. A. Reiners, and R. K. Olson. 1989. Topographic control of vegetation in a mountain big sagebrush steppe. Vegetatio 84:77–86.

Burke, I. C., W. A. Reiners, and D. S. Schimel. 1989. Organic matter turnover in a sagebrush steppe landscape. Biogeochemistry 7:11–31.

Burke, I. C., W. A. Reiners, D. L. Sturges, and P. A. Matson. 1987. Herbicide treatment effects on properties of mountain big sagebrush soils after fourteen years. Soil Sci. Soc. Am. J. 51:1337–43.

Burkhardt, J. W., and E. W. Tisdale. 1976. Causes of juniper invasion in southwestern Idaho. Ecol. 57:472–84.

Burkhardt, M. R. 1976. Pollen biostratigraphy and late Quaternary vegetation history of the Bighorn Mountains, Wyoming. Ph.D. diss., Univ. Iowa, Iowa City.

Burzlaff, D. F. 1962. A soil and vegetation inventory and analysis of three Nebraska sandhill range sites. Univ. Nebraska Exp. Sta. Res. Bull. 206.

Butler, D. R. 1979. Snow avalanche path terrain and vegetation, Glacier National Park, Montana. Arct. Alp. Res. 11:17–32.

———. 1985. Vegetation and geomorphic change on snow avalanche paths, Glacier National Park, Montana, USA. Great Basin Nat. 45:313–17.

Buttrick, P. L. 1914. The probable origin of the forests of the Black Hills of South Dakota. For. Quart. 12:223–27.

Cain, R. H. 1974. Pimple mounds: A new viewpoint. Ecol. 55:178–82.

Cain, S. A. 1943. Sample-plot technique applied to alpine vegetation in Wyoming. Am. J. Bot. 30:240–47.

Caine, N. 1974. The geomorphic processes of the alpine environment. In J. K. Ives and R. G. Barry, eds., Arctic and alpine environments, pp. 721–48. Methuen, London.

Caldwell, M. M. 1979. Physiology of sagebrush. In The sagebrush ecosystem: A symposium, pp. 74–85. Utah State Univ., Logan.

Caldwell, M. M., J. H. Richards, D. A. Johnson, R. S. Nowak, and R. S. Dzurec. 1981. Coping with herbivory: Photosynthetic capacity and resource allocation in two semiarid Agropyron bunchgrasses. Oecologia 50:14–24.

Caldwell, M. M., R. Robberecht, and W. D. Billings. 1980. A steep latitudinal gradient of solar ultraviolet-B radiation in the arctic-alpine life zone. Ecol. 61:600–611.

Call, M. W. 1966. Beaver pond ecology and beaver-trout relationships in southeastern Wyoming. Ph.D. diss., Univ. Wyoming, Laramie.

Callison, J., and K. T. Harper. 1982. Temperature in relation to elevation in the Uinta Mountains, Utah. Encyclia 59:21–27.

Campbell, C. J., and W. Green. 1968. Perpetual succession of stream-channel vegetation in a semiarid region. Ariz. Acad. Sci. J. 5:86–98.

Campbell, G. S., and G. A. Harris. 1977. Water relations and water use patterns for Artemisia tridentata Nutt. in wet and dry years. Ecol. 58:652–59.

Carlson, D. C. 1986. Effects of prairie dogs on mound soils. M.S. thesis, South Dakota State Univ., Brookings.

Carpenter, A. T., and N. E. West. 1987. Indifference to mountain big sagebrush growth to supplemental water and nitrogen. J. Range Mgmt. 40:448–51.

Carrera, P. E., D. A. Trimble, and M. Ruben. 1991. Holocene treeline fluctuations in the northern San Juan Mountains, Colorado, usa, as indicated by radiocarbon-dated conifer wood. Arct. Alp. Res. 23:233–46.

Cary, L. E. 1966. A study of forest margins. M.S. thesis, Univ. Wyoming, Laramie.

Cawker, K. B. 1980. Evidence of climatic control from population age structure of Artemisia tridentata Nutt. in southern British Columbia. J. Biogeogr. 7:237–48.

Chadde, S. W., and C. E. Kay. 1991. Tall willow communities on Yellowstone's northern range: A test of the "natural regulation" paradigm. In R. R. Keiter and M. S. Boyce, eds., The greater Yellowstone ecosystem: Redefining America's wilderness heritage, pp. 231–62. Yale Univ. Press, New Haven.

Chadde, S. W., P. L. Hansen, and R. D. Pfister. 1988. Wetland plant communities of the Northern Range, Yellowstone National Park. Final Report to Yellowstone National Park, Mammoth, Wyoming.

Chadwick, H. W., and P. D. Dalke. 1965. Plant succession on dune sands in Freemont County, Idaho. Ecol. 46:765–80.

Charley, J. L. 1977. Mineral cycling in rangeland ecosystems. In R. E. Sosebee, ed., Rangeland plant physiology, pp. 215–56. Range Science Series 4. Soc. Range Mgmt., Denver.

Charley, J. L., and N. West. 1975. Plant induced soil chemical patterns in some shrub dominated semidesert ecosystems of Utah. J. Ecol 63:

Chase, A. 1986. Playing God in Yellowstone: The destruction of America's first national park. Atlantic Monthly Press, New York.

Choate, C. M., and J. R. Habeck. 1967. Alpine plant communities at Logan Pass, Glacier National Park. Proceedings, Montana Acad. Sci. 27:36–54.

Choate, G. A. 1963. The forests of Wyoming. U.S. For. Ser. Res. Bull. int-2.

Choudhuri, G. N. 1968. Effect of soil salinity on germination and survival of some steppe plants in Washinton. Ecol. 49:465–71.

Christensen, E. M. 1964. Succession in a mountain brush community in central Utah. Proc. Utah Acad. Sci. Arts Letters 41:10–13.

Christensen, N. L., J. K. Agee, P. F. Brussard, J. Hughes, D. H. Knight, G. W. Minshall, J. M. Peek, S. J. Pyne, F. J. Swanson, J. W. Thomas, S. Wells, S. E. Williams, and H. A. Wright. 1989. Interpreting the Yellowstone fires of 1988. BioScience 39:678–83.

Christiansen, E., R. H. Waring, and A. A. Berryman. 1987. Resistance of conifers to bark beetle attack: Searching for general relationships. For. Ecol. Mgmt. 22:89–106.

Churchill, E. D., and H. C. Hanson. 1958. The concept of climax in arctic and alpine vegetation. Bot. Rev. 24:127–91.

Cid, M. S., J. K. Detling, A. D. Whicker, and M. A. Brizuela. 1991. Vegetational response of a mixed-grass prairie site following exclusion of prairie dogs and bison. J. Range Mgmt. 44:100–5.

Clark, F. E. 1977. Internal cycling of 15-Nitrogen in shortgrass prairie. Ecol. 58:1322–33.

Clark, F. E., C. V. Cole, and R. A. Bowman. 1980. Nutrient cycling. In A. I. Breymeyer and G. M. Van Dyne, eds., Grasslands, systems analysis and man, pp. 659–712. Cambridge Univ. Press, Cambridge.

Clark, R. G., C. M. Britton, and F. A. Sneva. 1982. Mortality of bitterbrush after burning and clipping in eastern Oregon. J. Range Mgmt. 35:711–14.

Clark, T. W., and M. R. Stromberg. 1987. Mammals of Wyoming. Univ. Kansas Press, Lawrence.

Clary, W. P., and B. F. Webster. 1990. Riparian grazing guidelines for the intermountain region. Rangelands 12:209–212.

Clary, W. P., E. D. McArthur, D. Bedunah, and C. L. Wambolt, compilers. 1992. Symposium on ecology and management of riparian shrub communities. U. S. For. Serv. Gen. Tech. Rep. int-289.

Cluff, G. J., R. A. Evans, and J. A. Young. 1983. Desert saltgrass seed germination and seedbed ecology. J. Range Mgmt. 36:419–23.

Cochran, P. H., and C. M. Berntsen. 1973. Tolerance of lodgepole and ponderosa pine seedlings to low night temperatures. For. Sci. 19:272–80.

Conway, T. M. 1982. Response of understory vegetation to varied lodgepole pine (Pinus contorta) spacing intervals in western Montana. M.S. thesis, Montana State Univ., Bozeman.

Cooke, W. B. 1955. Subalpine fungi and snowbanks. Ecol. 36:124–30.

Cooper, H. W. 1953. Amounts of big sagebrush in plant communities near Tensleep, Wyoming, as affected by grazing treatment. Ecol. 34:186–89.

Coppock, D. L., and J. K. Detling. 1986. Alteration of bison/prairie dog grazing interaction by prescribed burning. J. Wildl. Mgmt. 50:452–55.

Coppock, D. L., J. K. Detling, J. E. Ellis, and M. I. Dyer. 1983a. Plant-herbivore interactions in a North American mixed-grass prairie. I. Effects of black-tailed prairie dogs on intraseasonal aboveground plant biomass and nutrient dynamics and plant species diversity. Oecologia 56:1–9.

Coppock, D. L., J. E. Ellis, J. K. Detling, and M. I. Dyer. 1983b. Plant-herbivore interactions in a North American mixed-grass prairie. II. Responses of bison to modification of vegetation by prairie dogs. Oecologia (Berlin) 56:10–15.

Costello, D. F. 1944. Important species of the major forage types in Colorado and Wyoming. Ecol. Monogr. 14:107–34.

Coughenour, M. B. 1985. Graminoid responses to grazing by large herbivores: Adaptations, exaptations and interacting processes. Ann. Missouri Bot. Gard. 72:852–63.

———. 1991a. Spatial components of plant-herbivore interactions in pastoral, ranching, and native ungulate ecosystems. J. Range Mgmt. 44:530–42.

———. 1991b. Biomass and nitrogen responses to grazing of upland steppe on Yellowstone's northern winter range. J. Appl. Ecol. 28:71–82.

———. 1993. Interactions between herbaceous plants and large herbivores on the Northern Yellowstone elk winter range: integrating landscape, climate, plant growth, elk nutritional requirements, and population responses. In D. G. Despain and R. H. Hamre, eds. Plants and their environments: First Biennial Scientific Conference on the Greater Yellowstone Ecosystem. National Park Serv. Tech. Rep., U. S. Gov. Printing Office, Denver.

Coughenour, M. B., and F. J. Singer. 1991. The concept of overgrazing and its application to Yellowstone's Northern Range. In R. B. Keiter and M. S. Boyce, eds., The Greater Yellowstone Ecosystem: Redefining America's wilderness heritage, pp. 209–30. Yale Univ. Press, New Haven.

Coughenour, M. B., F. J. Singer, and J. Reardon. 1993. The Parker transects revisited: long-term herbaceous vegetation trends on the northern winter range, 1954–89. In D. G. Despain and R. H. Hamre, eds. Plants and their environments: First Biennial Scientific Conference on the Greater Yellowstone Ecosystem. National Park Serv. Tech. Rep., U. S. Gov. Printing Office, Denver.

Coupland, R. T., and G. M. Van Dyne. 1979. Systems synthesis. In R. T. Coupland, ed., Grassland ecosystems of the world: Analysis of grasslands and their uses, pp. 97–106. Cambridge Univ. Press, New York.

Covington, W. W. 1975. Altitudinal variation of chlorophyll concentration and reflectance of the bark of *Populus tremuloides*. Ecol. 56:715–20.

Cowan, F. T. 1929. Life history, habits, and control of the mormon cricket. U.S. Dept. Agric. Tech. Bull. 161.

Cox, G. W. 1984. Mounds of mystery. Natural History 93 (June): 36–45.

———. 1990a. Comment and reply to "Formation of mima mounds: A seismic hypothesis." Geol. 18: 1259–60.

———. 1990b. Form and dispersion of mima mounds in relation to slope steepness and aspect on the Columbia Plateau. Great Basin Nat. 50:21–31.

Cox, G. W., and D. W. Allen. 1987a. Sorted stone nets and circles of the Columbia Plateau: A hypothesis. Northw. Sci. 61:179–85.

———. 1987b. Soil translocation by pocket gophers in a mima moundfield. Oecologia 72:207–10.

Cox, G. W., and C. G. Gakahu. 1986. A latitudinal test of the fossorial rodent hypothesis of mima mound origin in western North America. Zeitschrift für Geomorphologie 30:485–501.

———. 1987. Biogeographical relationships of rhizomyid mole rats with Mima mound terrain in the Kenya highlands. Pedobiologia 30:263–75.

Cox, G. W., C. G. Gakahu, and D. W. Allen. 1987. The small stone content of mima mounds in the Columbia Plateau and Rocky Mountain regions: Implications for mound origin. Great Basin Nat. 47:609–19.

Cox, G. W., and V. G. Roig. 1986. Argentinian mima mounds occupied by ctenomyid rodents. J. Mammol. 67:428–32.

Crowe, D. M. 1990. The first century: A hundred years of wildlife conservation in Wyoming. Wyoming Game and Fish Department, Cheyenne.

Cui, Muyi. 1990. The ecophysiology of seedling establishment in subalpine conifers of the Central Rocky Mountains, USA. Ph.D. diss., Univ. Wyoming, Laramie.

Culver, D. C., and A. J. Beattie. 1983. Effects of ant mounds on soil chemistry and vegetation patterns in a Colorado montane meadow. Ecol. 64:485–92.

Cunningham, H. 1971. Soil-vegetation relationships of a bitterbrush-sagebrush association in northwestern Colorado. M.S. thesis, Colorado State Univ., Fort Collins.

Curl, H., Jr., J. T. Hardy, and R. Ellermeir. 1972. Spectral absorption of solar radiation in alpine snowfields. Ecol. 53:1189–1194.

Currie, P. O., and D. L. Goodwin. 1966. Consumption of forage by black-tailed jackrabbits on salt desert ranges of Utah. J. Wildl. Mgmt. 30:304–11.

Cushman, M. J. 1981. The influence of recurrent snow avalanches on vegetation patterns in the Washington Cascades. Ph.D. diss., Univ. Washington, Seattle.

Custer, G. A. 1875. Expedition to the Black Hills. 43d Congress, 2d sess. Senate Exec. Doc. 32.

Dalquest, W. W., and V. B. Scheffer. 1942. The origin of the mima mounds of western Washington. J. Geol 50:68–84.

D'Antonio, C. M., and P. M. Vitousek. 1992. Biological invasions by exotic grasses, the grass fire cycle, and global change. Ann. Rev. Ecol. Sys. 23:63–87.

Daubenmire, R. 1943a. Soil drought as a factor determining lower altitudinal limits of trees in the Rocky Mountains. Bot. Gaz. 105:1–13.

———. 1943b. Vegetational zonation in the Rocky Mountains. Bot. Rev. 9:325–93.

———. 1954. Alpine timberlines in the Americas and their interpretation. Butler Univ. Bot. Studies 11: 119–36.

———. 1968a. Ecology of fire in grasslands. Adv. Ecol. Res. 5:209–66.

———. 1968b. Soil moisture in relation to vegetation distribution in the mountains of northern Idaho. Ecol. 49:431–38.

———. 1975. Ecology of Artemisia tridentata ssp. tridentata in the state of Washington. Northw. Sci. 49:24–35.

Davis, G. V. 1959. A vegetative study of three relic areas located within Fort Laramie National Monument. M.S. thesis, Univ. Wyoming, Laramie.

Day, R. J. 1972. Stand structure, succession, and use of southern Alberta's Rocky Mountain forest. Ecol. 53:472–78.

Day, T. A., and J. K. Detling. 1990a. Grassland patch dynamics and herbivore grazing preference following urine deposition. Ecol. 71:180–88.

———. 1990b. Changes in grass leaf water relations following urine deposition. Am. Midl. Nat. 123: 171–78.

Dealy, J. E. 1975. Ecology of curlleaf mountain mahogany (Cercocarpus ledifolius Nutt.) in eastern Oregon and adjacent areas. Ph.D. diss., Oregon State Univ., Corvallis.

DeByle, N. V. 1979. Potential effects of stable versus fluctuating elk populations in the aspen ecosystem. In M. S. Boyce and L. D. Hayden-Wing, eds., North American elk: Ecology, behavior and management, pp. 13–19. Univ. Wyoming, Laramie.

DeByle, N. V., C. D. Bevins, and W. C. Fischer. 1987. Wildfire occurrence in aspen in the interior western United States. West. J. Appl. For. 2:73–76.

DeByle, N. V., P. J. Urness, and D. L. Blank. 1989. Forage quality in burned and unburned aspen communities. U.S. For. Ser. Res. Paper INT-404.

DeByle, N. V., and R. P. Winokur, eds. 1985. Aspen: Ecology and management in the western United States. U.S. For. Ser. Gen. Tech. Rep. RM-119.

Del Moral, R. 1984. The impact of the Olympic marmot

on subalpine vegetation structure. Am. J. Bot. 71: 1228–36.

Del Moral, R., and D. C. Deardorff. 1976. Vegetation of the mima mounds, Washington State. Ecol. 57:520–30.

DePuit, E. J., and M. M. Caldwell. 1973. Seasonal pattern of net photosynthesis of Artemisia tridentata. Am. J. Bot. 60:426–35.

Despain, D. G. 1973. Vegetation of the Big Horn Mountains, Wyoming in relation to substrate and climate. Ecol. Monogr. 43:329–55.

———. 1983. Nonpyrogenous climax lodgepole pine communities in Yellowstone National Park. Ecol. 64:231–34.

———. 1990. Yellowstone vegetation: Consequences of environment and history in a natural setting. Roberts Rinehart, Boulder.

Despain, D. G., D. Houston, M. Meagher, and P. Schullery. 1986. Wildlife in transition: Man and nature on Yellowstone's Northern Range. Roberts Rinehart, Boulder.

Despain, D., A. Rodman, P. Schullery, and H. Schovic. 1989. Burned area survey of Yellowstone National Park: The fires of 1988. Internal report, Yellowstone National Park, Mammoth, Wyoming.

Despain, D. G., and R. E. Sellers. 1977. Natural fire in Yellowstone National Park. West. Wildlands 4:20–24.

Detling, J. K. 1979. Processes controlling blue grama production on the shortgrass prairie. In N. R. French, ed., Perspectives in grassland ecology, pp. 25–42. Springer-Verlag, New York.

———. 1988. Grasslands and savannas: Regulation of energy flow and nutrient cycling by herbivores. In L. R. Pomeroy and J. J. Alberts, eds., Concepts of ecosystem ecology, pp. 131–48. Springer-Verlag, New York.

Detling, J. K., and M. I. Dyer. 1981. Evidence for potential plant growth regulators in grasshoppers. Ecol. 62:485–88.

Detling, J. K., M. I. Dyer, C. Procter-Gregg, and D. T. Winn. 1980. Plant-herbivore interactions: Examination of potential effects of bison saliva on regrowth of Bouteloua gracilis (H.B.K.) Lag. Oecologia 45:26–31.

Detling, J. K., M. I. Dyer, and D. T. Winn. 1979a. Effect of simulated grasshopper grazing on CO_2 exchange rates of western wheatgrass leaves. J. Econ. Entomol. 72:403–6.

Detling, J. K., and L. G. Klikoff. 1973. Physiological response to moisture stress as a factor in halophyte distribution. Am. Midl. Nat. 90:307–18.

Detling, J. K., and E. L. Painter. 1983. Defoliation responses of western wheatgrass populations with diverse histories of prairie dog grazing. Oecologia 57:65–71.

Detling, J. K., E. L. Painter, and D. L. Coppock. 1986. Ecotypic differentiation resulting from grazing pressure: Evidence for a likely phenomenon. In P. J. Joss, P. W. Lynch, and O. B. Williams, eds., Rangelands: A resource under siege, pp. 431–33. Second International Rangelands Congress, Australian Academy of Science, Canberra.

Detling, J. K., W. J. Parton, and H. W. Hunt. 1979. Simulation model of *Bouteloua gracilis* biomass dynamics on the North American shortgrass prairie. Oecologia 38:167–91.

Detling, J. K., C. W. Ross, M. H. Walmsley, D. W. Hilbert, C. A. Bonilla, and M. I. Dyer. 1981. Examination of North American bison saliva for potential plant growth regulators. J. Chem. Ecol. 7:239–46.

Dickman, A., and S. Cook. 1988. Fire and fungus in a mountain hemlock forest. Can. J. Bot. 67:2005–16.

Dirks, R. A., and B. E. Martner. 1982. The climate of Yellowstone and Grand Teton National Parks. U.S. National Park Service, Occasional Paper 6.

Dix, R. L. 1958. Some slope-plant relationships in the grasslands of the Little Missouri badlands of North Dakota. J. Range Mgmt. 11:88–91.

Dodd, J. L., and W. K. Lauenroth. 1975. Response of *Opuntia polyacantha* to water and nitrogen perturbations in the shortgrass prairie. In M.K. Wali, ed., Prairie: A multiple view, pp. 229–40. Univ. North Dakota Press, Grand Forks.

———. 1979. Analysis of the response of a grassland ecosystem to stress. In N. R. French, ed., Perspectives in grassland ecology, pp. 43–58. Springer-Verlag, New York.

Dodge, R. I. 1876. The Black Hills. James Miller Publ., New York. (Reprinted 1965 by Ross and Haines, Minneapolis, Minn.)

Doering, W. R., and R. G. Reider. 1992. Soils of Cinnabar Park, Medicine Bow Mountains, Wyoming, U.S.A.: Indicators of park origin and persistence. Arct. Alp. Res. 24:27–39.

Doescher, P. S., R. F. Miller, J. Wang, and J. Rose. 1990. Effects of nitrogen availability on growth and photosynthesis of *Artemisia tridentata* ssp. *wyomingensis*. Great Basin Nat. 50:9–19.

Dorf, E. 1942. Upper Cretaceous floras of the Rocky Mountain region ii: Flora of the Lance Creek formation at the type locality, Niobrara County, Wyoming. Carnegie Institution of Washington Publ. 508:79–159.

———. 1964. The petrified forests of Yellowstone Park. Sci. Am. 210:107–14.

Dorn, R. D. 1986. The Wyoming landscape, 1805–1878. Mountain West Publishing, Cheyenne, Wyo.

———. 1992. Vascular plants of Wyoming. Mountain West Publishing, Cheyenne, Wyo.

Dorn, R. D., and J. L. Dorn. 1977. Flora of the Black Hills. (Published by the authors.)

Duckham, A. N., and G. B. Masfield. 1971. Farming systems of the world. Chatto and Windus, London.

Duncan, E. 1975. The ecology of curlleaf mountain mahogany in southwestern Montana with special reference to use by mule deer. M.S. thesis, Montana State Univ., Bozeman.

Dunnewald, T. J. 1930. Grass and timber soils distribution in the Big Horn Mountains. J. Am. Soc. Agron. 22:577–86.

Dunwiddie, P. W. 1977. Recent tree invasion of subalpine meadows in the Wind River Mountains, Wyoming. Arct. Alp. Res. 9:393–99.

Dyer, M. I., and U. G. Bokhari. 1976. Plant-animal interactions: Studies of the effects of grasshopper grazing on blue grama grass. Ecol. 57:762–72.

Eaton, F. D. 1971. Soil moisture depletion, actual and potential evapotranspiration in an Engelmann spruce-subalpine fir forest. M.S. thesis, Utah State Univ., Logan.

Eckert, Jr., R. E., F. F. Peterson, M. S. Meurisse, and J. L. Stephens. 1986. Effects of soil-surface morphology on emergence and survival of seedlings in big sagebrush communities. J. Range Mgmt. 39:414–20.

Edminster, C. B. 1987. Growth and yield of subalpine conifer stands in the central Rocky Mountains. In C. A. Troendle, M. R. Kaufmann, R. H. Hamre, and R. P. Winokur, tech. coords., Management of subalpine forests: Building on fifty years of research, pp. 33–40. U.S. For. Ser. Gen. Tech. Rep. RM-149.

Elliott, E. T., and D. C. Coleman. 1988. Let the soil work for us. Ecol. Bull. (Sweden) 39:23–32.

Elliott, P. F. 1974. Evolutionary responses of plants to seedeaters: Pine squirrel predation on lodgepole pine. Evol. 28:221–31.

Elliott-Fisk, D. L., B. S. Adkins, and J. L. Spaulding. 1983. A re-evaluation of the postglacial vegetation of the Laramie Basin. Great Basin Nat. 43:377–84.

Ellison, L. 1954. Subalpine vegetation of the Wasatch Plateau, Utah. Ecol. Monogr. 24:89–184.

———. 1960. Influence of grazing on plant succession of rangelands. Bot. Rev. 26:1–78.

Ellison, L. and C. M. Aldous. 1952. Influence of pocket gophers on vegetation of subalpine grassland in central Utah. Ecol. 33:177–86.

Ellison, L., and E. J. Woolfolk. 1937. Effects of drought in vegetation near Miles City, Montana. Ecol. 18:329–36.

Engle, D. M., and P. M. Bultsma. 1984. Burning of northern mixed prairie during drought. J. Range Mgmt. 37:398–401.

Everett, R. L., compiler. 1987. Proceedings: Pinyon-juniper conference. U.S. For. Ser. Gen. Tech. Rep. INT-215.

Everitt, B. L. 1968. Use of the cottonwood in an investigation of the recent history of a floodplain. Am. J. Sci. 266:417–39.

Eversman, S. T. 1968. A comparison of plant communities and substrates of avalanche and non-avalanche areas in south-central Montana. M.S. thesis, Montana State Univ., Bozeman.

Eversman, S., and M. Carr. 1992. Yellowstone ecology: A road guide. Mountain Press Publishing, Missoula, Montana.

Fagerstone, K. A., H. P. Tietjen, and O. Williams. 1981. Seasonal variation in the diet of black-tailed prairie dogs. J. Mammal. 62:820–24.

Fahey, T. J. 1983. Nutrient dynamics of aboveground detritus in lodgepole pine *(Pinus contorta* ssp. *latifolia)* ecosystems. Ecol. Monogr. 53:51–72.

Fahey, T. J., and D. H. Knight. 1986. Lodgepole pine ecosystems. BioScience 36:610–17.

Fahey, T. J., J. B. Yavitt, and G. Joyce. 1988. Precipitation and throughfall chemistry in *Pinus contorta* ssp. *latifolia* ecosystems, southeastern Wyoming. Can. J. For. Res. 18:337–45.

Fahey, T. J., J. B. Yavitt, J. A. Pearson, and D. H. Knight. 1985. The nitrogen cycle in lodgepole pine forests, southeastern Wyoming. Biogeochem. 1:257–75.

Fallat, C. L. Schoene, B. Lundberg, P. Sandene, and F. Porter. 1987. Wyoming land inventory 1987. Geol. Surv. Wyo. Map Series 24. Laramie.

Farnsworth, R. B., E. M. Romney, and A. Wallace. 1978. Nitrogen fixations by microfloral-higher plant associations in arid to semiarid environments. In N. E. West and J. J. Skujins, eds. Nitrogen in desert ecosystems, pp. 17–19. Dowden, Hutchinson, and Ross, Inc., Stroudsberg, Pennsylvania.

Fenneman, N. M. 1931. Physiography of the western United States. McGraw-Hill, New York.

Fenner, P. W. W. Brady, and D. R. Patton. 1985. Effects of regulated water flows on regeneration of Fremont cottonwood. J. Range Mgmt. 38:135–38.

Ferguson, C. W. 1964. Annual rings in big sagebrush. Paper no. 1, Laboratory of tree-ring research. Univ. Arizona Press, Tucson.

Fetcher, N., and M. J. Trlica. 1980. Influence of climate on annual production of seven cold desert forage species. J. Range Mgmt. 33:35–37.

Fichtner, F. A. 1959. *Festuca idahoensis* on the Big Horn

Mountains of Wyoming. M.S. thesis, Univ. Wyoming, Laramie.

Fischer, W. C., and A. F. Bradley. 1987. Fire ecology of western Montana forest habitat types. U.S. For. Ser. Gen. Tech. Rep. INT-223.

Fischer, W. C., and B. Clayton. 1983. Fire ecology of Montana forest habitat types east of the continental divide. U.S. For. Ser. Gen. Tech. Rep. INT-141.

Fisher, F. M., and J. R. Gosz. 1986. Effects of trenching on soil processes and properties in a New Mexico mixed-conifer forest. Biol. Fertil. Soils 2:35–42.

Fisher, R. F., M. J. Jenkins, and W. F. Fisher. 1987. Fire and the prairie-forest mosaic of Devil's Tower National Monument. Am. Midl. Nat. 117:250–57.

Fisser, H. G. 1962. An ecological study of the *Artemisia tripartita* ssp. *rupicola* and related shrub communities in Wyoming. Ph.D. diss., Univ. Wyoming, Laramie.

———. 1964. Range survey in Wyoming's Bighorn Basin. Wyo. Agric. Exp. Sta. Bull. 424R.

———. 1987. Wyoming shrubland ecology: A chronologic supplemented multiple source annotated bibliography. Wyo. Agric. Exp. Sta., Laramie.

Fisser, H. G., K. L. Johnson, K. S. Moore, and G. E. Plumb. 1989. Fifty-one-year change in the shortgrass prairie of eastern Wyoming. Proc. Eleventh No. Am. Prairie Conf., pages 29–31.

Flowers, S. 1934. Vegetation of the Great Salt Lake region. Bot. Gaz. 95:353–418.

Flowers, S., and F. R. Evans. 1966. The flora and fauna of the Great Salt Lake region, Utah. In H. Boyko, ed., Salinity and aridity. Monographics biologicae 16, pp. 367–93. W. Junk, The Hague.

Fogel, R. and J. M. Trappe. 1978. Fungus consumption (mycophagy) by small animals. Northw. Sci. 52:1–31.

Forcella, F. 1977. Flora, chorology, biomass and productivity of the *Pinus albicaulis/Vaccinium scoparium* association. M.S. thesis, Montana State Univ., Bozeman.

Foster, M. A., and J. Stubbendieck. 1980. Effects of the plains pocket gopher *(Geomys bursarius)* on rangeland. J. Range Mgmt. 33:74–78.

Foster, R. H. 1968. Distribution of the major plant communities in Utah. Ph.D. diss., Brigham Young Univ., Provo, Utah.

Frank, D. A., and S. J. McNaughton. 1992. The ecology of plants, large mammalian herbivores, and drought in Yellowstone National Park. Ecology 73:2043–58.

———. 1993. Plant-ungulate ecology of grasslands on the northern range of Yellowstone National Park. In D. G. Despain and R. H. Hamre, eds. Plants and their environments: First Biennial Scientific Conference on the Greater Yellowstone Ecosystem. National Park Serv. Tech. Rep., U. S. Gov. Printing Office, Denver.

Franklin, J. F., and R. T. T. Forman. 1987. Creating landscape patterns by forest cutting: Ecological consequences and principles. Landscape Ecol. 1:5–18.

Freeland, R. O. 1944. Apparent photosynthesis in some conifers during winter. Plant Phys.19:179–85.

Fremont, J. C. 1845. Report of the exploring expedition to the Rocky Mountains in the year 1842, and to Oregon and North California in the years 1843–44. 28th Congress, 2d sess. Serial 461. Senate Exec. Doc. 174.

French, N. R., ed. 1979. Perspectives in grassland ecology. Springer-Verlag, New York.

Frischknecht, N. C., and M. F. Baker. 1972. Voles can improve sagebrush rangelands. J. Range Mgmt. 25: 466–68.

Frischknecht, N. C., and A. P. Plummer. 1955. A comparison of seeded grasses under grazing and protection on a mountain brush burn. J. Range Mgmt. 8:170–75.

Frison, G. C. 1975. Man's interaction with Holocene environments on the plains. Quaternary Res. 5:289–300.

———. 1984. The Carter/Kerr-McGee Paleoindian site: Cultural resource management and archaeological research. Am. Antiquity 49:288–314.

———. 1991. Prehistoric hunters of the high plains. 2d edition. Academic Press, New York.

Frison, G. C., and L. C. Todd. 1986. The Colby mammoth site: Taphonomy and archaeology of a Clovis kill in northern Wyoming. Univ. New Mexico Press, Albuquerque.

Frison, G. C., D. N. Walker, S. D. Webb, and G. M. Zeimens. 1978. Paleo-indian procurement of *Camelops* on the northwestern plains. Quaternary Res. 10:385–400.

Froiland, S. G. 1990. Natural history of the Black Hills and Badlands. The Center for Western Studies, Augustana College, Sioux Falls, S.D.

Frost, L. A., ed. 1979. With Custer in '74: James Calhoun's diary of the Black Hills expedition. Brigham Young Univ. Press, Provo.

Fryxell, F. 1932. Thomas Moran's journey to the Tetons in 1879. Augustana Historical Society Publication 2:3–12.

Fuller, D. P. 1958. The effect of chemical control of big sagebrush (*Artemisia tridentata* Nutt.) upon the composition and production of associated native species. M.S. thesis, Univ. Wyoming, Laramie.

Furniss, M. M., and R. G. Krebill. 1972. Insects and diseases of shrubs on western big game ranges. In D.

M. McKell, J. P. Blaisdell, and J. R. Goodin, eds., Wildland shrubs—their biology and utilization, pp. 218-26. U.S. For. Ser. Gen. Tech. Rep. INT-1

Galbraith, A. F. 1971. The soil water regime of a shortgrass prairie ecosystem. Ph.D. diss., Colorado State Univ., Fort Collins.

Ganskopp, D. C. 1986. Tolerances of sagebrush, rabbitbrush, and greasewood to elevated water tables. J. Range Mgmt. 39:334–37.

Gara, R. I., W. R. Littke, J. K. Agee, D. R. Geiszler, J. D. Stuart, and C. H. Driver. 1985. Influence of fires, fungi, and mountain pine beetles on development of a lodgepole pine forest in south-central Oregon. In D. M. Baumgartner, R. G. Krebill, J. T. Arnott, and G. F. Weetman, eds., Lodgepole pine: The species and its management, pp. 153–62. Washington State Univ. Coop. Ext. Ser., Pullman.

Garcia-Moya, E., and C. M. McKell. 1970. Contribution of shrubs to the nitrogen economy of a desert-wash plant community. Ecol. 51:81–88.

Garland, C. B., III. 1972. Ecology of arid land vegetation in western Wyoming. M.S. thesis, Univ. Wyoming, Laramie.

Gartner, F. R. 1967. Microclimate, vegetation and soils along a vertical gradient on Elk Mountain, Wyoming. M.S. thesis, Univ. Wyoming, Laramie.

Gartner, F. R., and W. W. Thompson. 1972. Fire in the Black Hills forest-grass ecotone. Tall Timbers Fire Ecol. Conf. Proc. 12:37–68.

Gary, H. L. 1972. Rime contributes to water balance in high-elevation aspen forests. J. For. 70:93–97.

———. 1976. Crown structure and distribution of biomass in a lodgepole pine stand. U.S. For. Ser. Res. Paper RM-165.

———. 1979. Duration of snow accumulation in-after harvesting in lodgepole pine in Wyoming and Colorado. U.S. For. Ser. Res. Note RM-366.

Gary, H. L., and C. A. Troendle. 1982. Snow accumulation and melt under various stand densities in lodgepole pine in Wyoming and Colorado. U.S. For. Ser. Res. Note RM-417.

Gates, D. H. 1964. Sagebrush infested by a leaf defoliation moth. J. Range Mgmt. 17:209–310.

Gates, D. H., L. A. Stoddart, and C. W. Cook. 1956. Soil as a factor in influencing plant distribution on salt deserts of Utah. Ecol. Monogr. 26:155–75.

Gaylord, D. R. 1982. Geologic history of the Ferris dune field, south-central Wyoming. Geol. Soc. Am. Spec. Paper 192, pp. 65–82.

———. 1983. Recent eolian activity and paleoclimate fluctuations in the Ferris Lost Soldier area, south-central Wyoming. Ph.D. diss., Univ. Wyoming, Laramie.

Geils, B. W., and W. R. Jacobi. 1984. Incidence and severity of *Comandra* blister rust on lodgepole pine in northwestern Wyoming. Plant Dis. 68:1049–51.

———. 1987. *Comandra* blister rust: A threat to lodgepole pine. In C. A. Troendle, M. R. Kaufmann, R. H. Hamre, and R. P. Winokur, tech. coords., Management of subalpine forests: Building on fifty years of research, pp. 216–17. U.S. For. Ser. Gen. Tech. Rep. RM-149.

———. 1990. Development of comandra blister rust on lodgepole pine. Can. J. For. Res. 20:159–65.

Geiszler, D. R., R. I. Gara, C. H. Driver, V. F. Gallucci, and R. E. Martin. 1980. Fire, fungi, and beetle influences on a lodgepole pine ecosystem of south-central Oregon. Oecologia (Berlin) 46:239–43.

Gerhart, W. A., and R. A. Olson. 1982. Handbook for evaluating the importance of Wyoming's riparian habitat to terrestrial wildlife. Wyoming Game and Fish Department, Cheyenne.

Gibbens, R. P. 1972. Vegetation pattern within northern desert shrub communities. Ph.D. diss., Univ. Wyoming, Laramie.

Gifford, G. F., and F. E. Busby, coords. 1975. The pinyon-juniper ecosystem: A symposium. College of Natural Resources, Utah State Univ., Logan.

Gifford, G. F., and R. H. Hawkins. 1978. Hydrologic impact of grazing on infiltration: A critical review. Water Res. Res. 14:305–13.

Girard, M. M., H. Goetz, and A. J. Bjugstad. 1989. Native woodland habitat types of southwestern North Dakota. U.S. For. Ser. Res. Paper RM-281.

Goldblum, D., and T. T. Veblen. 1992. Fire history of a ponderosa pine/Douglas-fir forest in the Colorado front range. Phys. Geogr. 13:133–48.

Goodin, J. R., and A. Mozafar. 1972. Physiology of salinity stress. In Wildland Shrubs: Their biology and utilization. U.S. For. Ser. Gen. Tech. Rep. INT-1, pp. 255–59.

Goodwin, D. L. 1956. Autecological studies of *Artemisia tridentata* Nutt. Ph.D. diss., Washington State Univ., Pullman.

Gorham, E., P. M. Vitousek, and W. A. Reiners. 1979. The regulation of chemical budgets over the course of terrestrial ecosystem succession. Annu. Rev. Ecol. Syst. 10:53–84.

Gosz, J. R. 1980. Biomass distribution and production budget for a nonaggrading forest ecosystem. Ecol. 61:507–14.

Grant, W. E. 1971. Comparisons of aboveground plant biomass on ungrazed pastures vs. pastures grazed by large herbivores, 1970 season. Technical Report 131, US-IBP Grassland Biome, Natural Resources Ecology Laboratory, Colorado State Univ., Fort Collins.

Grasso, D. 1990. Recognition and paleogeography of Quaternary Lake Wamsutter (a proposed lake in Wyoming's Great Divide Basin) combining Landsat remote sensing and digital elevation modeling. Ph.D. diss., Univ. Wyoming, Laramie.

Graves, H. L. 1899. Black Hills forest reserve. Annual Report 1897–98, U.S. Geol. Surv. 19:67–164.

Greater Yellowstone Coordinating Committee. 1987. The Greater Yellowstone Area: An aggregation of national park and national forest management plans. Greater Yellowstone Coordinating Committee, map supplement 22.

Green, A. W. 1978. Timber resources of western South Dakota. U.S. For. Ser. Res. Bull. INT-12.

Green, A. W., and R. C. Conner. 1989. Forest in Wyoming. U.S. For. Ser. Res. Bull. INT-61.

Greenland, D. 1989. The climate of Niwot Ridge, Front Range, Colorado, U.S.A. Arct. Alp. Res. 21:380–91.

Greenland, D. E., J. Burbank, J. Key, L. Klinger, J. Moorhouse, S. Oaks, and D. Shankman. 1985. The bioclimates of the Colorado Front Range. Mountain Res. Dev. 5:251–62.

Gresswell, R. E. 1984. The ecological profile as a monitoring tool for lakes in Yellowstone National Park. Yellowstone National Park Research Office.

Gribb, W. J., and D. J. Brosz. 1990. Irrigated acreage. In L. M. Ostresh, Jr., R. A. Marston, and W. M. Hudson, eds., Wyoming water atlas, pp. 38–39. Wyoming Water Development Commission, Cheyenne.

Grier, C. C., and S. W. Running. 1977. Leaf area of mature northwestern coniferous forests: Relation to site water balance. Ecol. 58:893–99.

Griggs, R. F. 1938. Timberlines in the northern Rocky Mountains. Ecol. 19:548–64.

———. 1946. The timberlines of North America and their interpretation. Ecol. 27:275–89.

———. 1956. Competition and succession on a Rocky Mountain fellfield. Ecol. 37:8–20.

Groves, C. R., and J. E. Anderson. 1981. Allelopathic effects of *Artemisia tridentata* leaves on germination and growth of two grass species. Am. Midl. Nat. 106:73–79.

Gruell, G. E. 1979. Wildlife habitat investigations and management implications on the Bridger-Teton National Forest. In M. S. Boyce and L. D. Hayden-Wing, eds., North American elk: Ecology, behavior and management, pp. 63–74. Univ. Wyoming, Laramie.

———. 1980. Fire's influence on wildlife habitat on the Bridger-Teton National Forest, Wyoming. 2 vols. U.S. For. Ser. Res. Paper INT-235 and INT-252.

———. 1983. Fire and vegetative trends in the northern

Rockies: Interpretations from 1871–1982 photographs. U.S. For. Ser. Gen. Tech. Rep. INT-158.

———. 1985. Fire on the early western landscape: An annotated record of wildland fires 1776–1900. Northw. Sci. 59:97–107.

Gruell, G. E., S. Bunting, and L. Neuenschwander. 1985. Influence of fire on curlleaf mountain-mahogany in the intermountain west. In J. K. Brown and J. Lotan, eds., Fire's effects on wildlife habitat, pp. 58–72. U.S. For. Ser. Gen. Tech. Rep. INT-186.

Gruell, G. E., and L. L. Loope. 1974. Relationships among aspen, fire, and ungulate browsing in Jackson Hole, Wyoming. U.S. For. Ser. (Intermountain Region) and U.S. National Park Service.

Gunderson, D. R. 1968. Floodplain use related to stream morphology and fish populations. J. Wildl. Mgmt. 32:153–96.

Haag, M. 1949. The range lands of Wyoming: A summary of the record of fifty years' study by the scientists of the Wyoming Agricultural Experiment Station. Wyo. Agric. Exp. Sta. Bull. 289.

Habeck, J. R. 1987. Present-day vegetation in the northern Rocky Mountains. Ann. Missouri Bot. Gard. 74:804–40.

Habeck, J. R., and R. W. Mutch. 1973. Fire-dependent forests in the northern Rocky Mountains. Quaternary Res. 3:408–24.

Hadley, J. L., and W. K. Smith. 1983. Influence of wind exposure on needle desiccation and mortality for timberline conifers in Wyoming, U.S.A. Arct. Alp. Res. 15:127–35.

———. 1986. Wind effects on needles of timberline conifers: Seasonal influence on mortality. Ecol. 67:12–19.

———. 1987. Influence of krummholz mat structure on microclimate and needle physiology. Oecologia 73:82–90.

———. 1989. Wind erosion of leaf surface wax in alpine timberline conifers. Arct. Alp. Res. 21:392–98.

———. 1990. Influence of leaf surface wax and the leaf area to water content ratio on cuticular transpiration in western conifers, U.S.A. Can. J. For. Res. 20:1306–11.

Haines, A. L. 1977. The Yellowstone story. 2 vols. Colorado Associated Univ. Press, Boulder.

———, ed. 1965. Osborne Russell's journal of a trapper. Univ. Nebraska Press, Lincoln.

Hallsten, G. P., Q. D. Skinner, and A. A. Beetle. Grasses of Wyoming. Wyo. Agric. Exp. Sta. Res. J. 202.

Hamerlynck, E. 1992. Subnivian and emergent growth, morphology, and physiology of Erythronium grandiflorum in relation to snow melt dynamics in the subalpine forest, Medicine Bow Mountains, Wyoming, U.S.A. M.S. thesis, Univ. Wyoming, Laramie.

Hamner, R. W. 1964. An ecological study of Sarcobatus vermiculatus communities of the Big Horn Basin, Wyoming. M.S. thesis, Univ. Wyoming, Laramie.

Hanna, L. A. 1934. The major plant communities of the headwater area of the Little Laramie River, Wyoming. Univ. Wyo. Publ. Sci. (Bot.) 1:243–66.

Hansen, D. J. 1977. Interrelations of valley vegetation, stream regimen, soils, and solar irradiation along the Rock Creek in the Uinta Mountains of Utah. Ph.D. diss., Univ. Michigan, Ann Arbor.

Hansen, H. P. 1938. Ring growth and reproduction cycle in Picea engelmannii near timberline. Univ. Wyoming Publ. 5:1–9.

———. 1940. Ring growth and dominance in a spruce-fir association in southern Wyoming. Am. Midl. Nat. 23:442–47.

Hansen, P. L., and G. R. Hoffman. 1988. The vegetation of the Grand River/Cedar River, Sioux, and Ashland Districts of the Custer National Forest: A habitat type classification. U.S. For. Ser. Gen. Tech. Rep. RM-157.

Hansen, R. M. 1962. Movements and survival of Thomomys talpoides in a mima-mound habitat. Ecol. 43:151–54.

Hansen, R. M., and I. K. Gold. 1977. Blacktail prairie dogs, desert cottontails and cattle trophic relations on shortgrass range. J. Range Mgmt. 30:210–14.

Hanson, C. E. 1974. An evaluation of sagebrush and juniper types as mule deer winter range at three locations in Wyoming. M.S. thesis, Univ. Wyoming, Laramie.

Hanson, C. L., C. W. Johnson, and J. R. Wight. 1982. Foliage mortality of mountain big sagebrush (Artemisia tridentata ssp. vaseyana) in southwestern Idaho during the winter of 1976–77. J. Range Mgmt. 35:142–45.

Hanson, H. C. 1955. Characteristics of the Stipa comata–Bouteloua gracilis–Bouteloua curtipendula association of northern Colorado. Ecol. 36:269–80.

Harder, L. D. 1979. Winter feeding by porcupines in southwestern Alberta: Preferences, density effects, and temporal changes. Can. J. Zool. 58:13–19.

Hardin, G. J. 1985. Filters against folly. Penguin, New York.

Harmon, M. E., and C. Hua. 1991. Coarse woody debris dynamics in two old-growth ecosystems. BioScience 41:604–10.

Harmon, M. E., J. F. Franklin, F. J. Swanson, P. Sollins, S. V. Gregory, J. D. Lattin, N. H. Anderson, S. P. Cline, N. G. Aumen, J. R. Sedell, G. W. Lienkaemper, K. Cromack, and K. W. Cummins. 1986. Ecology of coarse woody debris in temperate ecosystems. Adv. Ecol. Res. 15:133–302.

Harmon, M. E., and C. Hua. 1991. Coarse woody debris

dynamics in two old-growth ecosystems. BioScience 41:604–10.

Harniss, R. O., and R. B. Murray. 1973. Thirty years of vegetal change following burning of sagebrush-grass range. J. Range Mgmt. 26:322–25.

Harper, J. L. 1977. Population biology of plants. Academic Press, London.

Harper, K. T. 1959. Vegetational changes in a shadscale-winterfat plant association during twenty-three years of controlled grazing. M.S. thesis, Brigham Young Univ., Provo.

Harper, K. T., J. D. Brotherson, and F. R. Stradling. 1977. Vegetation data base for the environmental impact statement on proposed coal strip mining areas in Carbon and Sweetwater Counties, Wyoming. Dept. of Botany, Brigham Young Univ., Provo.

Harper, K. T., F. J. Wagstaff, and W. P. Clary. 1990. Shrub mortality over a fifty-four-year period in shadscale desert, west-central Utah. In E. D. McArthur, E. M. Romney, S. D. Smith, and P. T. Tueller, compilers, Symposium on cheatgrass invasion, shrub die-off, and other aspects of shrub biology and management, pp. 119–26. U.S. Forest Service Gen. Tech. Rep. INT-276.

Harper, K. T., F. J. Wagstaff, and L. M. Kunzler. 1985. Biology and management of the Gambel oak vegetative type: A literature review. U.S. For. Ser. Gen. Tech. Rep. INT-179.

Harper, K. T., R. A. Woodward, and K. B. McKnight. 1980. Interrelationships among precipitation, vegetation, and streamflow in the Unita Mountains, Utah. Encyclia 57:58–86.

Harr, R. D., and K. R. Price. 1972. Evapotranspiration from a greasewood-cheatgrass community. Water Res. Res. 8:1199–1203.

Harris, T. 1992. High Country News, 10 Feb. 1992, vol. 24(2).

Hart, G. E., N. V. DeByle, and R. W. Hennes. 1981. Slash treatment after clearcutting lodgepole pine affects nutrients in soil water. J. For. 79:446–50.

Hart, G. E., and D. A. Lomas. 1979. Effects of clearcutting on soil water depletion in an Engelmann spruce stand. Water Resour. Res. 15:1598–1602.

Hart, R. H., and M. J. Samuel. 1985. Precipitation, soils and herbage production on southeast Wyoming range sites. J. Range Mgmt. 38:522–25.

Harvey, A. E., M. F. Jurgensen, M. J. Larsen, and R. T. Graham. 1987. Relationships among soil microsite, ectomycorrhizae, and natural conifer regeneration of old-growth forests in western Montana. Can. J. For. Res. 17:58–62.

Hassan, M. A., and N. E. West. 1986. Dynamics of soil seed pools in burned and unburned sagebrush semideserts. Ecol. 67:269–72.

Hawksworth, F. G. 1975. Dwarf mistletoe and its role in lodgepole pine ecosystems. In D. M. Baumgartner, ed., Management of lodgepole pine ecosystems, vol. 1, pp. 342–58. Washington State Coop. Ext. Ser., Pullman.

Hawksworth, F. G., and D. W. Johnson. 1989. Biology and management of dwarf mistletoe in lodgepole pine in the Rocky Mountains. U.S. For. Ser. Gen. Tech. Rep. RM-169.

Hawksworth, F. G., and R. F. Scharpf, tech. coords. 1984. Biology of dwarf mistletoes. U.S. For. Ser. Gen. Tech. Rep. RM-111.

Hayden, F. V. 1879. Eleventh annual report of the United States geological and geographical survey of the territories, embracing Idaho and Wyoming, being a report of progress in the exploration for the year 1877. GPO, Washington, D.C.

Hayward, C. L. 1945. Biotic communities of the southern Wasatch and Uinta Mountains, Utah. Great Basin Nat. 6:1–124.

———. 1952. Alpine biotic communities of the Uinta Mountains, Utah. Ecol. Monogr. 22:93–120.

Hayward, H. E. 1928. Studies of plants in the Black Hills of South Dakota. Bot. Gaz. 85:353–412.

Hazlett, D. L., and G. R. Hoffman. 1975. Plant species distributional pattern in *Artemisia tridentata*- and *Artemisia cana*-dominated vegetation in western North Dakota. Bot. Gaz. 136:72–77.

Heady, H. F. 1950. Studies on bluebunch wheatgrass in Montana and height-weight relationships of certain range grasses. Ecol. Monogr. 20:55–81.

Heitschmidt, R. K. 1990. The role of livestock and other herbivores in improving rangeland vegetation. Rangelands 12:112–15.

Hennessy, J. T., R. P. Gibbens, J. M. Tromble, and M. Cardenas. 1983. Water properties of caliche. J. Range Mgmt. 36:723–26.

Henszey, R. J., Q. D. Skinner, and T. A. Wesche. 1991. Response of montane meadow vegetation after two years of streamflow augmentation. Regulated Rivers: Res. Mgmt. 6:29–38.

Henry, J. H. 1961. Biology of the sagebrush defoliator *Aroga websterii* Clarks in Idaho. M.S. thesis, Univ. Idaho, Moscow.

Hermes, K. 1955. Die Lage der oberen Waldgrenze in den Gebirgen der Erde und ihr Abstand zur Schneegrenze. Kölner Geographische Arbeiten 5:1–277 (kartenmappe). Selbstverlag der Geographischen Instituts, Univ. Cologne, Germany.

Hewitt, G. B. 1977. Review of forage losses caused by rangeland grasshoppers. U.S. Agric. Res. Ser. Misc. Publ. 1348.

Hewitt, G. B., W. H. Burleson, and J. A. Onsager. 1976. Forage losses caused by the grasshopper *Aulocara elliotti* on shortgrass rangeland. J. Range Mgmt. 29:376–80.

Hewitt, G. B., and J. A. Onsager. 1983. Control of grasshoppers on rangeland in the United States: A perspective. J. Range Mgmt. 36:202–7.

Higgins, K. F. 1984. Lightning fires in North Dakota grasslands and in pine-savanna lands of South Dakota and Montana. J. Range Mgmt. 37:100–103.

Hinds, T. E. 1976. Aspen mortality in Rocky Mountain campgrounds. U.S. For. Ser. Res. Paper RM-164.

Hironaka, M. 1963. Plant environment relations of major species in sagebrush-grass vegetation of southern Idaho. Ph.D. diss., Univ. Wisconsin, Madison.

Hobbs, N. T., and D. S. Schimel. 1984. Fire effects on nitrogen mineralization and fixation in mountain shrub and grassland communities. J. Range Mgmt. 37:402–5.

Hodgkinson, H. S. 1983. Relationship between Cutler mormon-tea *(Ephedra cutleri)* and coppice dunes in determining range trend in northeastern Arizona. J. Range Mgmt. 36:375–77.

———. 1987. Relationship of saltbush species to soil chemical properties. J. Range Mgmt. 40:23–26.

Hoff, C. 1957. A comparison of soil, climate, and biota of conifer and aspen communities in the central Rocky Mountains. Am. Midl. Nat. 58:115–40.

Hoffman, G. O., A. H. Walder, and R. A. Darrow. 1955. Pricklypear good and bad. Texas Agric. Ext. Serv. Bull. 806.

Hoffman, G. R., and R. R. Alexander. 1976. Forest vegetation of the Bighorn Mountains, Wyoming: A habitat classification. U.S. For. Ser. Res. Paper RM-170.

———. 1987. Forest vegetation of the Black Hills National Forest of South Dakota and Wyoming: A habitat type classification. U.S. For. Ser. Res. Paper RM-276.

Hoffman, G. R., and D. L. Hazlett. 1977. Effects of aqueous *Artemisia* extracts and volatile substances on germination of selected species. J. Range Mgmt. 30:134–37.

Holechek, J. L. 1983. Considerations concerning grazing systems. Rangelands 5:208–11.

Holland, D. G. 1986. The role of forest insects and diseases in the Yellowstone ecosystem. W. Wildlands (Fall): 19–23.

Holland, E. A., and J. K. Detling. 1990. Plant response to herbivory and belowground nitrogen cycling. Ecol. 71:1040–49.

Holland, E. A., W. J. Parton, J. K. Detling, and D. L. Coppock. 1992. Physiological responses of plant populations to herbivory and their consequences for ecosystem nutrient flow. Am. Nat. 140:685–706.

Hollaway, J. G., and R. T. Ward. 1963. Snow and melt-water effects in an area of Colorado alpine. Am. Midl. Nat. 69:189–97.

Holm, T. H. 1927. The vegetation of the alpine region of the Rocky Mountains in Colorado. Natl. Acad. Sci. Mem. 19:1–45.

Holmes, N. D., D. S. Smith, and A. Johnston. 1979. Effect of grazing by cattle on the abundance of grasshoppers on fescue grassland. J. Range Mgmt. 32:310–11.

Holpp, F. A. 1977. Vegetative composition and soil analysis of selected playas of Campbell County, Wyoming. M.S. thesis, Univ. Wyoming, Laramie.

Homsher, L. M., ed. 1960. South Pass, 1868: James Chisholm's journal of the Wyoming gold rush. Univ. Nebraska Press, Lincoln.

Houston, D. B. 1967. The Shiras moose in Jackson Hole, Wyoming. Ph.D. diss., Univ. Wyoming, Laramie.

———. 1971. Ecosystems of national parks. Science 172:648–51.

———. 1973. Wildfires in northern Yellowstone National Park. Ecol. 54:1111–17.

———. 1982. The northern Yellowstone Elk: Ecology and management. MacMillan, New York.

Houston, W. R. 1961. Some interrelations of sagebrush, soils, and grazing intensity in the northern Great Plains. Ecol. 42:31–38.

———. 1963. Plains pricklypear, weather, and grazing in the northern Great Plains. Ecol. 44:569–74.

Houston, W. R., and D. N. Hyder. 1975. Ecological effect and fate of N following massive N fertilization of mixed grass plains. J. Range Mgmt. 28:56–60.

Houston, W. R., L. D. Sabatka, and D. N. Hyder. 1973. Nitrate-nitrogen accumulation in range plants after massive N fertilization on shortgrass plains. J. Range Mgmt. 26:54–57.

Hudson, H. J. 1986. Fungal biology. Edward Arnold, London.

Hull, A. C., Jr., and M. K. Hull. 1974. Presettlement vegetation of Cache Valley, Utah and Idaho. J. Range Mgmt. 27:27–29.

Hull, A. C., Jr., and J. R. Killough. 1951. Ants are consuming Big Horn Basin ranges. W. Farm Life 53:70.

Hungerford, R. D., M. G. Harrington, W. H. Fraudsen, K. C. Ryan, and G. J. Niehoff. 1991. Influence of fire on factors that affect site productivity. In A. E. Harvey and L. F. Neuenschwander, compilers, Management and productivity of western-montane forest soils, pp. 32–50. U.S. For. Ser. Gen. Tech. Rep. INT-280.

Hunt, H. W., D. C. Coleman, E. R. Ingham, R. E. Ingham, E. T. Elliott, J. C. Moore, S. L. Rose, C. P. P. Reid, and C. R. Morley. 1987. The detrital food web in a shortgrass prairie. Biol. Fertil. Soils 3:57–68.

Huntly, N. J. 1987. Influence of refuging consumers (pikas: *Ochotona princeps*) on subalpine meadow vegetation. Ecol. 68:274–83.

———. 1991. Herbivores and the dynamics of communities and ecosystems. Annual Rev. Ecol. Syst. 22: 477–503

Hurd, R. M. 1961. Grassland vegetation in the Big Horn Mountains, Wyoming. Ecol. 42:459–67.

Hutchins, H. E., and R. M. Lanner. 1982. The central role of Clark's nutcracker in the dispersal and establishment of whitebark pine. Oecologia 55:192–201.

Hutchison, B. A. 1965. Snow accumulation and disappearance influenced by big sagebrush. U.S. For. Ser. Res. Note RM-46.

Hyde, R. M., and A. A. Beetle. 1964. Range survey in Sunlight Basin, Park County, Wyoming. Wyo. Agric. Exp. Sta. Bull. 423.

Hyder, D. N. 1971. Morphogenesis and management of perennial grasses in the United States. U.S. Dept. Agric. Misc. Publ. 1271.

Hyder, D. N., R. E. Bement, E. E. Remmenga, and D. F. Hervey. 1975. Ecological response of native plants and guidelines for management of shortgrass range. U.S. Agric. Res. Ser. Tech. Bull. 1503.

Ingham, R. E., and J. K. Detling. 1984. Plant-herbivore interactions in a North American mixed-grass prairie. 3, Soil nematode populations and root biomass on *Cynomys ludovicianus* colonies and adjacent uncolonized areas. Oecologia 63:307–13.

———. 1990. Effects of root-feeding nematodes on aboveground net primary production in a North American grassland. Plant and Soil 121:279–81.

———. 1991. Effects of the root-feeding nematode *Tylenchorhynchus claytoni* on growth and leaf gas exchange of *Bouteloua gracilis*. Pedobiol. 35:219–44.

Ingham, R. E., J. A. Trofymow, E. R. Ingham, and D. C. Coleman. 1985. Interactions of bacteria, fungi, and their nematode grazers: Effects on nutrient cycling and plant growth. Ecol. Monogr. 55:119–40.

Inouye, R. S., N. J. Huntly, D. Tilman, and J. R. Tester. 1987. Pocket gophers *(Geomys bursarius)*, vegetation, and soil nitrogen along a successional sere in east central Minnesota. Oecologia 72:178–84.

Irvine, J. R., and N. E. West. 1979. Riparian tree species distribution and succession along the lower Escalante River, Utah. Southw. Nat. 24:331–46.

Ives, R. L. 1942. The beaver-meadow complex. J. Geomorph. 5:191–203.

Jackson, W. N. 1957. Some soil characteristics of several grassland-timber transitions in the Big Horn Mountains and the Laramie Plains. M.S. thesis, Univ. Wyoming, Laramie.

Jacobi, W. R., B. W. Geils, J. E. Taylor, and W. R. Zentz.

1993. Predicting the incidence of comandra blister rust on lodgepole pine: Site, stand, and alternate-host influences. Phytopath. 83:630–637.

Jacoby, P. W. 1971. Interrelationships of vegetation and environmental factors on a mountain watershed in southeastern Wyoming. Ph.D. diss., Univ. Wyoming, Laramie.

Jameson, D. A. 1966. Pinyon-juniper litter reduces growth of blue grama. J. Range Mgmt. 19:214–17.

———. 1970. Degradation and accumulation of inhibitory substances from *Juniperus osteosperma* (Torr.) Little. Plant and Soil 33:213–24.

Jaramillo, V. J., and J. K. Detling. 1988. Grazing history, defoliation, and competition: Effects on shortgrass production and nitrogen accumulation. Ecol. 69: 1599–1608.

———. 1992. Small scale heterogeneity in a semi-arid American grassland. 2, Cattle grazing of simulated urine patches. J. Appl. Ecol. 29:9–13.

Jaynes, R. A. 1978. A hydrologic model of aspen-conifer succession in the western United States. U.S. For. Ser. Res. Paper INT-213.

Jefferies, R. L. 1981. Osmotic adjustment and the response of halophytic plants to salinity. BioScience 31:42–46.

Johnson, D. A., and M. D. Rumbaugh. 1981. Nodulation and acetylene by certain rangeland legume species under field conditions. J. Range Mgmt. 34:78–181.

Johnson, D. F. 1950. Plant succession on the Missouri River floodplain near Vermillion, South Dakota. M.A. thesis, Univ. South Dakota, Vermillion.

Johnson, D. W., D. C. West, D. E. Todd, and L. K. Mann. 1982. Effects of sawlog vs whole-tree harvesting removal on the nitrogen, phosphorus, potassium, and calcium budgets of an upland mixed oak forest. Soil Sci. Soc. Am. Proc. 46:1304–09.

Johnson, E. A. 1987. The relative importance of snow avalanche disturbance and thinning on canopy plant populations. Ecol. 68:43–53.

Johnson, E. A., and G. I. Fryer. 1989. Population dynamics in lodgepole pine-Engelmann spruce forests. Ecol. 70:1335–45.

Johnson, E. A., and C. P. S. Larsen. 1991. Climatically induced change in fire frequency in the southern Canadian Rockies. Ecol. 72:194–201.

Johnson, H. N. 1949. A climatological review of the Black Hills. South Dakota School of Mines and Technology. Black Hills Engineer 29:3–15.

Johnson, K. L. 1987. Rangeland through time: A photographic study of vegetation change in Wyoming 1870–1986. Misc. Publ. 50, Wyo. Agric. Exp. Sta., Laramie.

Johnson, P. L., and W. D. Billings. 1962. The alpine vegetation of the Beartooth Plateau in relation to cryopedogenic processes and patterns. Ecol. Monogr. 32:105–35.

Johnson, R. E. 1950. An ecological analysis of a mountain mahogany community. M.S. thesis, Univ. Wyoming, Laramie.

Johnson, R. R., C. D. Ziebell, D. R. Patton, P. F. Folliott, and R. H. Hamre, coords. 1985. Riparian ecosystems and their management: Reconciling conflicting uses. First North American Riparian Conference. U.S. For. Ser. Gen. Tech. Rep. RM-120.

Johnson, W. C., R. L. Burgess, and W. R. Keammerer. 1976. Forest overstory vegetation and environment on the Missouri River floodplain in North Dakota. Ecol. Monogr. 46:59–84.

Johnson, W. M. 1962. Vegetation of high-altitude ranges in Wyoming as related to use by game and domestic sheep. Wyo. Agric. Exp. Sta. Bull. 387.

———. 1969. Life expectancy of a sagebrush control in central Wyoming. J. Range Mgmt. 22:177–82.

Jones, G. 1979. Vegetation productivity. Longman, London.

Jones, R. G. 1971. Ecology of gall midges on sagebrush in Idaho. Ph.D. diss. Univ. Idaho, Moscow.

Joyce, L. A. 1981. Climate/vegetation relationships in the northern Great Plains and the Wyoming north-central basins. Ph.D. diss., Colorado State Univ., Fort Collins.

Kamm, J. A., F. A. Sneva, and L. M. Rittenhouse. 1978. Insect grazers on the cold desert biome. In D. N. Hyder, ed., First International Rangeland Congress, pp. 479–83. Soc. Range Mgmt. Denver.

Kastner, W. M., D. E. Schild, and D. S. Spahr. 1989. Water-level changes in the high plains aquifer underlying parts of South Dakota, Wyoming, Nebraska, Colorado, Kansas, New Mexico, Oklahoma, and Texas: Predevelopment through nonirrigation season 1987–88. U.S. Geol. Surv. Water-Resources Invest. Rep. 89–4073. Denver.

Kauffman, J. B., and W. C. Krueger. 1984. Livestock impacts on riparian ecosystems and streamside management implications: A review. J. Range Mgmt. 37:430–38.

Kauffman, J. B., W. C. Krueger, and M. Vavra. 1983. Impacts of cattle on streambanks in northeastern Oregon. J. Range Mgmt. 36:683–85.

Kaufmann, M. R. 1985. Annual transpiration in subalpine forests: Large differences among four tree species. For. Ecol. Mgmt. 13:235–46.

Kaufmann, M. R., C. A. Troendle, M. G. Ryan, and H. Todd Mowrer. 1987. Trees: The link between silviculture and hydrology. In C. A. Troendle, M. R. Kaufmann, R. H. Hamre, and R. P. Winokur, tech. coords., Management of subalpine forests: Building on fifty years of research, pp. 54–60. U.S. For. Ser. Gen. Tech. Rep. RM-149.

Kaufmann, M. R., and R. K. Watkins. 1990. Characteristics of high- and low-vigor lodgepole pine trees in old-growth stands. Tree Phys. 7:239–46.

Kay, C. E. 1990. Yellowstone's northern elk herd: A critical evaluation of the "natural regulation" paradigm. Ph.D. diss., Utah State Univ., Logan.

———. 1993. Aspen seedlings in recently burned areas in Grand Teton and Yellowstone National Parks. Northw. Sci. 67:94–104.

———. 1993. Historical conditions of woody vegetation on Yellowstone's Northern Range: A critical evaluation of the "natural regulation" paradigm. In D. G. Despain and R. H. Hamre, eds., Plants and their environments: First biennial scientific conference on the Greater Yellowstone Ecosystem. National Park Serv. Tech. Rep., U.S. Gov. Printing Office, Denver.

Keefer, W. R. 1971. The geologic story of Yellowstone National Park. U.S. Geol. Surv. Bull. 1347. (Reprinted by Yellowstone Library and Museum Association, Mammoth, Wyo.)

Keiter, R. B. 1988. Natural ecosystem management in park and wilderness areas: Looking into the law. In J. K. Agee and D. R. Johnson, eds., Ecosystem management for parks and wilderness, pp. 15–40. Univ. Washington Press, Seattle.

———. 1989. Taking account of the ecosystem on the public domain: Law and ecology in the Greater Yellowstone region. Univ. Colorado Law Rev. 60:923–1007.

Keiter, R. B., and M. S. Boyce, eds. 1991a. The Greater Yellowstone ecosystem: Redefining America's wilderness heritage. Yale Univ. Press, New Haven.

———. 1991b. Greater Yellowstone's future: Ecosystem management in a wilderness environment. In Keiter and Boyce, The Greater Yellowstone ecosystem, pp. 379–413. Yale Univ. Press, New Haven.

Keiter, R. B., and P. H. Froelicher. 1993. Bison, brucellosis, and law in the Greater Yellowstone ecosystem. Land Water Law Rev. 28:1–75.

Kemp, P. R., and G. J. Williams. 1980. A physiological basis for niche separation between *Agropyron smithii* (C_3) and *Bouteloua gracilis* (C_4). Ecol. 61:846–58.

Kiener, W. 1967. Sociological studies of the alpine vegetation on Long's Peak. New Series 34, Univ. Nebraska Studies, Lincoln.

Kimball, S. L., B. D. Bennett, and F. B. Salisbury. 1973. The growth and development of montane species at near-freezing temperatures. Ecol. 54:166–73.

Kirkham, D. R., and H. G. Fisser. 1972. Rangeland relations and harvester ants in northcentral Wyoming. J. Range Mgmt. 25:55–60.

Klatt, L. E., and D. Hein. 1978. Vegetative differences among active and abandoned towns of black-tailed prairie dogs (Cynomys ludovicianus). J. Range Mgmt. 31:315–17.

Kleiner, E. F., and K. T. Harper. 1972. Environment and community organization in grasslands of Canyonlands National Park. Ecol. 53:299–309.

———. 1977. Soil properties in relation to cryptogamic groundcover in Canyonlands National Park. J. Range Mgmt. 30:202–5.

Klipple, G. E., and D. F. Costello. 1960. Vegetation and cattle responses to different intensities of grazing on short-grass ranges on the Central Great Plains. U.S. Dept. Agric. Tech. Bull. 1216, Washington, D.C.

Klock, G. O. 1969. Some autecological characteristics of elk sedge. U.S. For. Ser. Res. Note PNW-106.

Knapp, A. K., and T. R. Seastedt. 1986. Detritus accumulation limits productivity of tallgrass prairie. BioScience 36:662–68.

Knapp, A. K., and W. K. Smith. 1981. Water relations and succession in subalpine conifers in southeastern Wyoming. Bot. Gaz. 142:502–11.

———. 1982. Factors influencing understory seedling establishment of Engelmann spruce (Picea engelmannii) and subalpine fir (Abies lasiocarpa) in southeastern Wyoming. Can. J. Bot. 60:2753–61.

———. 1989. Influence of growth form on the ecophysiology of subalpine plants during variable sunlight. Ecol. 70:1069–82.

Knight, D. H. 1973. Leaf area dynamics of a shortgrass prairie in Colorado. Ecol. 54:891–95.

———. 1987. Parasites, lightning, and the vegetation mosaic in wilderness landscapes. In M. G. Turner, ed., Landscape heterogeneity and disturbance, pp. 59–83. Springer-Verlag, New York.

———. 1989. Vegetation map of Wyoming. In S. Roberts, ed., Wyoming geomaps, pp. 33–34. Geol. Surv. Wyo., Laramie.

———. 1991. Pine forests: A comparative overview of ecosystem structure and function. In N. Nakagoshi and F. B. Golley, eds., Coniferous forest ecology from an international perspective, pp. 121–35. SPB Academic Publishing, The Hague.

Knight, D. H., R. J. Hill, and A. T. Harrison. 1976. Potential natural landmarks in the Wyoming Basin. Final report for National Park Service, Denver Service Center. (Contract 9900X20047)

Knight, D. H., G. P. Jones, Y. Akashi, and R. W. Myers. 1987. Vegetation ecology in the Bighorn Canyon National Recreation Area. Final report submitted to the University of Wyoming–National Park Service Research Center, Laramie.

Knight, D. H., B. S. Rogers, and C. R. Kyte. 1977. Understory plant growth in relation to snow duration in Wyoming subalpine forest. Bull. Torrey Bot. Club 104:314–19.

Knight, D. H., S. W. Running, and T. J. Fahey. 1985. Water and nutrient outflow from contrasting lodgepole pine forests in Wyoming. Ecol. Monogr. 55:29–48.

Knight, D. H., and L. L. Wallace. 1989. The Yellowstone fires: Issues in landscape ecology. BioScience 39: 700–706.

Knight, D. H., J. B. Yavitt, and G. D. Joyce. 1991. Water and nitrogen outflow from lodgepole pine forest after two levels of tree mortality. Forest Ecol. Mgmt. 46:215–25.

Knight, S. H. 1974. Geologic history of Wyoming landscapes. In M. Wilson, ed., Applied geology and archaeology: The Holocene history of Wyoming, pp. 1–7. Geol. Surv. Wyo. Rep. Invest. 10.

———. 1990. Illustrated geologic history of the Medicine Bow Mountains and adjacent areas, Wyoming. Memoir 4. Wyo. Geol. Surv., Laramie.

Knopf, F. L., and M. L. Scott. 1990. Altered flows and created landscapes in the Platte River headwaters 1840–1990. In J. M. Sweeney, ed., Management of dynamic ecosystems, pp. 47–70. North Central Section, the Wildlife Society, West Lafayette, Indiana.

Knowles, P., and M. C. Grant. 1983. Age and size structure analyses of Engelmann spruce, ponderosa pine, lodgepole pine, and limber pine in Colorado. Ecol. 64:1–9.

Knutson, R. M. 1981. Flowers that make heat while the sun shines. Natural History (Oct.) 90:75–81.

Koford, C. B. 1958. Prairie dogs, whitefaces, and blue grama. Wildl. Monogr. 3.

Kolm, K. E. 1974. ERTS MSS imagery applied to mapping and economic evaluation of sand dunes in Wyoming. Special report prepared for NASA/Goddard Space Flight Center, Greenbelt, Maryland.

———. 1975. Selenium in soils of the lower Wasatch formation, Campbell County, Wyoming: Geochemistry, distribution, and environmental hazards. M.S. thesis, Univ. Wyoming, Laramie.

Komarek, E. V., Sr. 1964. The natural history of lightning. Third Annual Tall Timber Fire Ecol. Conf., pp. 139–83.

Komarkova, V. 1976. Alpine vegetation of the Indian Peaks Area, Front Range, Colorado Rocky Mountains. Ph.D. diss., Univ. Colorado, Boulder.

Komarkova, V., and P. J. Webber. 1978. An alpine vegetation map of Niwot Ridge, Colorado. Arct. Alp. Res. 10:1–29.

Koske, R. E., and W. R. Polson. 1984. Are VA mycorrhizae required for sand dune stabilization? BioScience 34:420–24.

Krebill, R. G. 1972. Mortality of aspen on the Gros Ventre elk winter range. U.S. For. Ser. Res. Paper INT-129.

———. 1975. Lodgepole pine's fungus-caused diseases and decays. In D. M. Baumgartner, ed., Management of lodgepole pine ecosystems, pp. 377–405. Washington State Univ. Coop. Ext. Ser., Pullman.

Krueger, K. 1986. Feeding relationships among bison, pronghorn, and prairie dogs: An experimental analysis. Ecol. 67:760–70.

———. 1988. Prairie dog overpopulation: Value judgement or ecological reality. In D. W. Uresk, G. L. Schenbeck, and R. Cefkin, tech. coords., Eighth Great Plains wildlife damage control workshop, pp. 39–45. U.S. For. Ser. Gen. Tech. Rep. RM-154.

Küchler, A. W. 1966. Potential natural vegetation of the conterminous United States. Am. Geog. Soc. Spec. Publ. 36. New York.

Kyte, C. R. 1975. An autecological study of *Vaccinium scoparium* in the Medicine Bow Mountains, Wyoming. M.S. thesis, Univ. Wyoming, Laramie.

Lacey, J. R., and H. W. Van Poollen. 1981. Comparison of herbage production on moderately grazed and ungrazed western ranges. J. Range Mgmt. 34:210–12.

Lageson, D. R., and D. R. Spearing. 1988. Roadside geology of Wyoming. Mountain Press Publishing, Missoula, Montana.

Landis, T. D., and E. W. Mogren. 1975. Tree strata biomass of subalpine spruce-fir stands in southwestern Colorado. For. Sci. 21:9–12.

Lang, R. L. 1973. Vegetation changes between 1943 and 1965 on the shortgrass plains of Wyoming. J. Range Mgmt. 26:407–9.

Langenheim, J. H. 1962. Vegetation and environmental patterns in the Crested Butte area, Gunnison County, Colorado. Ecol. Monogr. 32:249–85.

Langer, R. H. M. 1979. How grasses grow. 2d ed. Edward Arnold, London.

Lanier, J. E. 1978. Population dynamics of Rocky Mountain Douglas fir *(Psuedotsuga menziesii* var. *glauca)* along an elevational gradient in the front range of Colorado. M.A. thesis, Univ. Colorado, Boulder.

Larsen, F., and W. Whitman. 1942. A comparison of used and unused grassland mesas in the badlands of South Dakota. Ecol. 23:438–45.

Larson, F. 1940. The role of bison in maintaining the short grass plains. Ecol. 21:113–21.

Larson, T. A. 1977. Wyoming: A bicentennial history. Norton, New York.

———. 1978. History of Wyoming. 2d ed. Univ. Nebraska Press, Lincoln.

Lauenroth, W. K., and J. L. Dodd. 1978. The effects of water- and nitrogen-induced stesses on plant community structure in a semiarid grassland. Oecologia 36:211–22.

———. 1979. Response of native grassland legumes to water and nitrogen treatments. J. Range Mgmt. 32:292–94.

Lauenroth, W. K., J. L. Dodd, and P. L. Sims. 1978. The effects of water- and nitrogen-induced stresses on plant community structure in a semiarid grassland. Oecologia 36:211–22.

Lauenroth, W. K., D. G. Milchunas, J. L. Dodd, R. H. Hart, R. K. Heitschmidt, and L. R. Rittenhouse. 1993. Grazing in the Great Plains of the United States. In M. Vavra, W. A. Laycock, and R. D. Pieper, eds., Ecological implications of livestock herbivory in the west. Soc. Range Mgmt., Denver.

Lawrence, D. B., and E. G. Lawrence. 1984. Historic lower landslide of the Gros Ventre Valley. Naturalist 35 (Winter): 24–28.

Laycock, W. A. 1958. The initial pattern of revegetation of pocket gopher mounds. Ecol. 39:346–51.

———. 1991. Stable states and thresholds of range condition on North American rangelands: A viewpoint. J. Range Mgmt. 44:427–33.

Laycock, W. A., and B. S. Mihlbachler. 1987. Ecology and management of pricklypear cactus on the Great Plains. In J. L. Capinera, ed., Integrated pest management on rangeland, pp. 81–99. Westview Press, Boulder.

Laycock, W. A., and B. Z. Richardson. 1975. Long-term effects of pocket gopher control on vegetation and soils of a subalpine grassland. J. Range Mgmt. 28:458–62.

Leaf, C. F. 1975. Watershed management in the Rocky Mountain subalpine zone: The status of our knowledge. U.S. For. Serv. Res. Paper RM-137.

Leopold, A. S., S. A. Cain, C. M. Cottam, I. N. Gabrielson, and T. L. Kimball. 1963. Wildlife management in the national parks. Trans. N. Am. Wildl. Nat. Res. Conf. 28:28–45.

Leopold, E. B., and H. D. MacGinitie. 1972. Development and affinities of Tertiary floras in the Rocky Mountains. In A. Graham, ed., Floristics and paleofloristics of Asia and eastern North America, pp. 147–200. Elsevier, Amsterdam.

Lepper, M. G. 1974. *Pinus flexilis* and its environmental relationships. Ph.D. diss., Univ. California, Davis.

Lepper, M. G., and M. Fleschner. 1977. Nitrogen fixation by *Cercocarpus ledifolius* (Rosaceae) in pioneer habitats. Oecologia 27:333–38.

Lewis, M. E. 1970. Alpine rangelands of the Uinta Mountains, Ashley and Wasatch National Forests. U.S. Forest Service, Region 4, Ogden, Utah.

Ligon, J. D. 1978. Reproductive interdependence of piñon jays and piñon pines. Ecol. Monogr. 48:111–26.

Lillegraven, J. A., M. J. Kraus, and T. M. Brown. 1979. Paleogeography of the world of the Mesozoic. In J. A. Lillegraven, Z. Kielan-Jaworowska, and W. A. Clemens, eds., Mesozoic mammals: The first two-thirds of mammalian history, pp. 277–308. Univ. California Press, Berkeley.

Lillegraven, J. A., and L. M. Ostresh, Jr. 1988. Evolution of Wyoming's early Cenozoic topography and drainage patterns. Natl. Geogr. Res. 4:303–27.

Limbach, W. E. 1974. Comparative physiological ecology of *Elymus canadensis* L., a C_3 grass and *Andropogon hallii* Hack., a C_4 grass in eastern Wyoming. M.S. thesis, Univ. Wyoming, Laramie.

Lindeberg-Johnson, M. 1981. Ecological physiology of populations of *Caltha leptosepala* DC. along altitudinal and latitudinal gradients in the central Rocky Mountains. Ph.D. diss., Univ. Colorado, Boulder.

Lindsay, J. H. 1971. Annual cycle of leaf water potential in *Picea engelmannii* and *Abies lasiocarpa* at timberline in Wyoming. Arct. Alp. Res. 3:131–38.

Linhart, Y. B., J. B. Mitton, K. C. Sturgeon, and M. L. Davis. 1981. Genetic variation in space and time in a population of ponderosa pine. Heredity 46:402–26.

Lisenbee, A., F. Karner, E. Kashbaugh, D. Halvorson, F. O'Toole, S. White, M. Wilkinson, and J. Kirchener. 1981. Geology of the Tertiary intrusive province of the Northern Black Hills, South Dakota and Wyoming. In F. J. Rich, ed., Geology of the Black Hills, South Dakota and Wyoming, pp. 33–105. American Geological Institute, Falls Church, Virginia.

Little, S. N., and J. L. Ohmann. 1988. Estimating nitrogen lost from forest floor during prescribed fires in Douglas-fir/western hemlock clearcuts. For. Sci. 34:152–64.

Livingston, R. B. 1952. Relict true prairie communities in central Colorado. Ecol. 33:72–86.

Lockwood, J. A., T. J. McNary, J. C. Larsen, and J. Cole. 1993. Distribution atlas for grasshoppers and the mormon cricket in Wyoming, 1988–1992. Univ. Wyo. Agric. Exp. Sta. Bull. 976.

Lommasson, T. 1948. Succession in sagebrush. J. Range Mgmt. 1:19–21.

Loope, L. L., and G. E. Gruell. 1973. The ecological role of fire in Jackson Hole, northwestern Wyoming. Quaternary Res. 3:425–43.

Lopushinsky, W. 1969. Stomatal closure in conifer seedlings in response to leaf moisture stress. Bot. Gaz. 130:258–63.

Lotan, J. E. 1975. The role of cone serotiny in lodgepole pine forests. In D. M. Baumgartner, ed., Management of lodgepole pine ecosystems, 471–95. Washington State Univ. Coop. Ext. Serv., Pullman.

———. 1976. Cone serotiny-fire relationships in lodgepole pine. Tall Timber Fire Ecol. Conf. Proc. 14:267–78.

Lotan, J. E., and D. A. Perry. 1983. Ecology and regeneration of lodgepole pine. U.S. Dept. Agric. Handb. 606.

Lovaas, A. L. 1976. Introduction of prescribed burning to Wind Cave National Park. Wildl. Soc. Bull. 4:69–73.

Love, J. D. 1960. Cenozoic sedimentation and crustal movement in Wyoming. Am. J. Sci. 258-A:201–14.

———. 1984. Geology and man in the Teton region. Naturalist 34 (Winter): 1–13.

———. 1989. Yellowstone and Grand Teton National Parks and the Middle Rocky Mountains. Field Trip Guidebook T328. American Geophysical Union, Washington, D.C.

Love, J. D., and A. C. Christensen. 1985. Geologic map of Wyoming. U.S. Geol. Surv., Denver.

Love, J. D., and J. M. Love. 1988. Geologic road log of part of the Gros Ventre River valley including the Lower Gros Ventre slide. Reprint 46. Wyo. Geol. Surv. Wyo, Laramie.

Love, J. D., and J. C. Reed, Jr. 1971. Creation of the Teton landscape. 2d ed. Grand Teton Natural History Association, Moose, Wyoming.

Love, J. D., J. C. Reed, Jr., and A. C. Christensen. 1992. Geologic map of Grand Teton National Park, Teton County, Wyoming. U.S. Geol. Surv. Misc. Invest. Series Map 1–2031.

Lowdermilk, W. C. 1925. Factors affecting reproduction of Engelmann spruce. J. Agric. Res. 30:995–1009.

Ludlow, W. 1875. Report of a reconnaissance of the Black Hills of Dakota made in the summer of 1874. U.S. Army, Dept. of Engineers, Washington, D.C.

Ludwig, J. A. 1969. Environmental interpretation of foothill grassland communities of northern Utah. Ph.D. diss., Univ. Utah, Salt Lake City.

Lundberg, C. E. 1977. The local composition and distribution of vegetation in the soda ash area of Sweetwater County, Wyoming. M.S. thesis, Univ. Wyoming, Laramie.

———. 1981. A moisture gradient in desert shrub communities in southwestern Wyoming as revealed by ordination. Ph.D. diss., Univ. Wyoming, Laramie.

Lunt, O. R., J. Letey, and S. B. Clark. 1973. Oxygen requirements for root growth in three species of desert shrubs. Ecol. 54:1356–62.

McArthur, E. D., E. M. Romney, S. D. Smith, and P. T. Tueller, eds. 1990. Symposium on cheatgrass invasion, shrub die-off, and other aspects of shrub biology and management. U.S. For. Ser. Gen. Tech. Rep. INT-276.

McBride, J. R., and J. Strahan. 1984. Establishment and survival of woody riparian species on gravel bars of an intermittent stream. Am. Midl. Nat. 112:235–45.

McCambridge, W. F., and F. B. Knight. 1972. Factors affecting spruce beetles during a small outbreak. Ecol. 53:830–39.

MacCracken, J. G., L. E. Alexander, and D. W. Uresk. 1983. An important lichen of southeastern Montana rangelands. J. Range Mgmt. 36:35–37.

MacCracken, J. G., D. W. Uresk, and R. M. Hansen. 1983. Plant community variability on a small area in southeastern Montana. Great Basin Nat. 43:660–68.

McCullough, H. A. 1948. Plant succession on fallen logs in a virgin spruce-fir forest. Ecol. 20:508–13.

McCune, B. 1983. Fire frequency reduced two orders of magnitude in the Bitterroot Canyons, Montana. Can. J. For. Res. 13:212–18.

McDonough, W. T. 1985. Sexual reproduction, seeds, and seedlings. See DeByle and Winokur, eds.

McDonough, W. T., and R. O. Harniss. 1974. Seed dormancy in Artemisia tridentata Nutt. subspecies vaseyana Rydb. Northw. Sci. 48:17–20.

McFaul, M. 1979. A geomorphic and pedological interpretation of the mima-mound prairies, south Puget Lowland, Washington State. M.A. thesis, Univ. Wyoming, Laramie.

MacGinitie, H. D., E. B. Leopold, and W. J. Hail, Jr. 1974. An early middle Eocene flora from the Yellowstone-Absaroka volcanic province, northwestern Wind River basin, Wyoming. Univ. California Publ. Geol. Sci. 108:1–66.

McGinnies, W. J. 1960. Effect of mima-type microrelief on herbage production of five seeded grasses in western Colorado. J. Range Mgmt. 13:231–34.

McGinnies, W. J., L. W. Osburn, and W. A. Berg. 1976. Plant-soil-microsite relationships on a saltgrass meadow. J. Range Mgmt. 29:395–400.

McGinnies, W. J., H. L. Shantz, and W. G. McGinnies. 1991. Changes in vegetation and land use in eastern Colorado: A photographic study. U.S. Agric. Res. Ser. ARS-85.

McGregor, M. D., and D. M. Cole. 1985. Integrating management strategies for the mountain pine beetle with multiple-resource management of lodgepole pine forests. U. S. For. Serv. Gen. Tech. Rep. INT-174.

McGrew, P. O., T. M. Brown, M. W. Hager, and B. Mears. 1974. Inventory of significant geological areas in the Wyoming Basin natural region. Final Report for National Park Service, Denver Service Center, Denver. (Contract 9900X20047)

McIntire, P. W. 1984. Fungus consumption by the Siskiyou chipmunk within a variously treated forest. Ecol. 65:137–46.

McIntosh, A. C. 1931. A botanical survey of the Black Hills of South Dakota. Black Hills Engr. 19:159–276.

Mack, R. N. 1971. Mineral cycling in Artemisia tridentata. Ph.D. diss., Washington State Univ, Pullman.

———. 1981. Invasion of Bromus tectorum L. into western North America: An ecological chronicle. Agro-Ecosyst. 7:145–65.

———. 1984. Invaders at home on the range. Natural History (Feb.): 40–46.

Mack, R. N., and D. A. Pyke. 1983. The demography of Bromus tectorum: Variation in time and space. J. Ecol. 71:69–93.

Mack, R. N., and J. N. Thompson. 1982. Evolution in steppe with few large, hoofed mammals. Am. Nat. 119:757–73.

McLean, A. 1953. The autecology of Atriplex nuttallii S. Wats. in southwestern Saskatchewan. M.S. thesis, Utah State Univ., Logan.

———. 1969. Fire resistance of forest species as influenced by root systems. J. Range Mgmt. 22:120–22.

MacMahon, J. A., and D. C. Anderson. 1982. Subalpine forests: A world perspective with emphasis on western North America. Prog. Phys. Geogr. 6:368–425.

McNaughton, G. M. 1984. Comparative water relations of Pinus flexilis at high and low elevations in the central Rocky Mountains. M.S. thesis, Univ. Wyoming, Laramie.

McNaughton, S. J. 1976. Serengeti migratory wildebeest: Facilitation of energy flow by grazing. Science 191:92–94.

———. 1979. Grazing as an optimization process: Grass-ungulate relationships in the Serengeti. Am. Nat. 113:691–703.

———. 1985. Ecology of a grazing ecosystem: The Serengeti. Ecol. Monogr. 55:259–94.

McNaughton, S. J., and F. S. Chapin. 1985. Effects of phosphorus nutrition and defoliation on C_4 graminoids from the Serengeti plains. Ecol. 66:1617–29.

McNaughton, S. J., L. L. Wallace, and M. B. Coughenour. 1983. Plant adaptation in an ecosystem context: Effects of defoliation, nitrogen, and water on growth of an African C_4 sedge. Ecology 64:307–18.

McNulty, I. B. 1947. The ecology of bitterbrush in Utah. M.S. thesis, Univ. Utah, Salt Lake City.

McNulty, M. E. 1969. Edaphic nitrogen transformations in aspen and conifer communities. M.S. thesis, Utah State Univ., Logan.

MacVean, C. M. 1987. Ecology and management of mormon cricket, *Anabrus simplex* Haldeman. In J. L. Capinera, ed., Integrated pest management on rangeland, pp. 116–36. Westview Press, Boulder.

McVeigh, B. 1989. The Yellowstone fires of 1988: An analysis based on sales tax collections and various park-related data. Wyoming Quarterly Update 8: 26–32.

McVeigh, B., J. Sarles, and S. Lamb. 1992. 1991 Wyoming data handbook. Wyo. Dept. Adm. Inf., Div. Econ. Anal., Cheyenne.

Madany, M. H., and N. E. West. 1983. Livestock grazing-fire regime interactions within montane forests of Zion National Park, Utah. Ecol. 64:661–67.

Maddock, T., Jr. 1976. A primer on floodplain dynamics. J. Soil Water Conserv. 31:44–47.

Major, J., and M. Rejmanek. 1992. *Amelanchier alnifolia* vegetation in eastern Idaho, U.S.A. and its environmental relationships. Vegetatio 98:141–56.

Malanson, G. P., and D. R. Butler. 1984. Transverse pattern of vegetation on avalanche paths in the northern Rocky Mountains, Montana. Great Basin Nat. 44:453–58.

Malcolm, W. M. 1966. Root parasitism of *Castilleja coccinea*. Ecol. 47:179–86.

Markgraf, V., and T. Lennon. 1993. Paleoenvironmental history of the last thirteen thousand years of the eastern Powder River Basin, Wyoming, and its implication for prehistoric cultural patterns. Plains Anthropologist.

Marquis, R. J., and E. G. Voss. 1981. Distributions of some western North American plants disjunct in the Great Lakes region. Michigan Botanist 20:53–82.

Marquiss, R., and R. Lang. 1959. Vegetational composition and ground cover of two natural relict areas and their associated grazed areas in the Red Desert of Wyoming. J. Range Mgmt. 12:104–9.

Marr, J. W. 1961. Ecosystems of the east slope of the Front Range in Colorado. Univ. Colorado Stud. Ser. Biol. 8.

———. 1977. The development and movement of tree islands near the upper limit of tree growth in the southern Rocky Mountains. Ecol. 58:1159–64.

Marriott, H. 1985. Flora of the northwestern Black Hills, Crook and Weston Counties, Wyoming. M.S. thesis, Univ. Wyoming, Laramie.

Marston, R. A. 1990. Changes in geomorphic processes in the Snake River following impoundment of Jackson Lake and potential changes due to 1988 fires in the watershed. Annual Report 14:85–90. Univ. of Wyoming–National Park Service Research Center, Laramie.

Marston, R. A., and D. J. Brosz. 1990. Drainage basins. In L. M. Ostresh, Jr., R. A. Marston, and W. M. Hudson, eds., Wyoming water atlas, pp. 46–47. Wyoming Water Development Commission, Cheyenne.

Martin, P. R. 1972. Ecology of skunkbush sumac *(Rhus trilobata* Nutt.) in Montana with special reference to use by mule deer. M.S. thesis, Montana State Univ., Bozeman.

Martin, P. S., and R. G. Klein, eds. 1989. Quaternary extinctions: A prehistoric revolution. Univ. Ariz. Press, Tucson.

Martin, R. E., and C. H. Driver. 1983. Factors affecting antelope bitterbrush reestablishment following fire. U.S. For. Ser. Intermountain For. Range Exp. Sta., Salt Lake City. U.S. Forest Service Gen. Tech. Rep. INT-152:266–79.

Martner, B. E. 1986. Wyoming climate atlas. Univ. Nebraska Press, Lincoln.

Maser, C., R. F. Tarrant, J. M. Trappe, and J. F. Franklin, eds. 1988. From the forest to the sea: A story of fallen trees. U.S. For. Ser. Gen. Tech. Rep. PNW-GTR-229.

Maser, C., J. M. Trappe, and R. A. Nussbaum. 1978. Fungus-small mammal interrelationships with emphasis on Oregon coniferous forests. Ecol. 59:799–809.

Matson, P. A., and R. D. Boone. 1984. Natural disturbance and nitrogen mineralization: Wave-form dieback of mountain hemlock in Oregon. Ecol. 65: 1511–16.

Matson, P. A., C. Volkmann, K. Coppinger, and W. A. Reiners. 1991. Annual nitrous oxide flux and soil nitrogen characteristics in sagebrush steppe ecosystems. Biogeochemistry 14:1–2.

Mattson, D. J. 1984. Classification and environmental relationships of wetland vegetation in central Yellowstone National Park. M.S. Thesis, Univ. Idaho, Moscow.

Mattson, W. J., and N. D. Addy. 1975. Phytophagous insects as regulators of forest primary productivity. Science 190:515–22.

Maun, M. A. 1981. Seed germination and seedling establishment of *Calamovilfa longifolia* on Lake Huron sand dunes. Can. J. Bot. 59:460–69.

May, D. E. 1976. The response of alpine tundra vegetation in Colorado to environmental variation. Ph.D. diss., Univ. Colorado, Boulder.

Mead, K. E. 1982. Nitrogen accretion in a Rocky Mountain subalpine succession. M.S. thesis, Utah State Univ., Logan.

Mears, A. I. 1975. Dynamics of dense-snow avalanches interpreted from broken trees. Geol. 3:521–23.

Mears, B., Jr. 1962. Stone nets on Medicine Bow Peak. Contributions to Geology 1:48.

———. 1974. The evolution of the Rocky Mountain glacial model. In D. R. Coates, ed., Glacial geomorphology, pp. 11–40. Publications in Geomorphology, State Univ. New York, Binghamton.

———. 1981. Periglacial wedges and the late Pleistocene environment of Wyoming's intermountain basins. Quaternary Res. 15:171–98.

———. 1987. Late Pleistocene periglacial wedge sites in Wyoming: An illustrated compendium. Memoir 3. Wyo. Geol. Surv., Laramie.

Medin, D. E. 1960. Physical site factors influencing annual production of true mountain mahogany, *Cercocarpus montanus*. Ecol. 41:454–60.

Megahan, W. F. 1983. Hydrologic effects of clearcutting and wildfire on steep granitic slopes in Idaho. Water Resources Res. 19:811–19.

Mehringer, P. J., Jr., and P. E. Wigand. 1990. Comparison of late Holocene environments from woodrat middens and pollen: Diamond Craters, Oregon. In J. L. Betancourt, T. R. Van Devender, and P. S. Martin, eds., Packrat middens: The last forty thousand years of biotic change, pp. 294–325. Univ. Arizona Press, Tucson.

Merkel, D. L., and H. H. Hopkins. 1957. Life history of saltcedar *(Tamarix gallica* L.). Kansas Acad. Sci. Trans. 60:366–69.

Merrill, E. H. 1978. Nutrient cycling in an ungulate-vegetation complex, Selway River, Idaho. Ph.D. diss., Univ. Idaho, Moscow.

Meyer, S. E., and S. B. Monsen. 1991. Habitat-correlated variation in mountain big sagebrush (*Artemisia tridentata* ssp. *vaseyana*) seed germination patterns. Ecology 72:739–42.

Mielke, H. W. 1977. Mound building by pocket gophers (Geomyidae): Their impact on soils and vegetation in North America. J. Biogeogr. 4:171–80.

Mihlbachler, B. S. 1986. Effects of prescribed burning on perennial grass vigor and production, plant water stress, and soil erosion on cactus rangelands in east-central Wyoming. M.S. thesis, Univ. Wyoming, Laramie.

Milchunas, D. G., W. K. Lauenroth, P. L. Chapman, and M. K. Kazempour. 1989. Effects of grazing, topography, and precipitation on the structure of a semiarid grassland. Vegetatio 80:11–23.

Milchunas, D. G., O. E. Sala, and W. K. Lauenroth. 1988. A generalized model of the effect of grazing by large herbivores on grassland community structure. Am. Nat. 132:87–106.

Miles, S. R., and P. C. Singleton. 1975. Vegetational history of Cinnabar Park in Medicine Bow National Forest, Wyoming. Soil Sci. Soc. Am. Proc. 39:1204–08.

Miller, J. R., T. T. Schulz, N. T. Hobbs, K. R. Wilson, D. L. Schrupp, and W. L. Baker. 1993. Flood regimes and landscape structure: Changes in a riparian zone following shifts in stream dynamics. Landscape Ecol.

Miller, R. C. 1979. Relationships between the vegetation and environmental characteristics associated with an alluvial valley floor. M.S. thesis, Univ. Wyoming, Laramie.

Miller, R. F., and L. I. Shultz. 1987. Development and longevity of ephemeral and perennial leaves on *Artemisia tridentata* ssp. *wyomingensis*. Great Basin Nat. 47:227–30.

Miller, R. F., T. Svejcar, and N. West. 1993. Implications of livestock grazing in the intermountain sagebrush region: Plant composition. In M. Vavra, W. A. Laycock, and R. D. Pieper, eds., Ecological implications of livestock herbivory in the west. Soc. Range Mgmt., Denver.

Miller, S. L., and E. B. Allen. 1992. Mycorrhizae, nutrient translocation and interaction between plants. In M. F. Allen, ed., Mycorrhizal functioning: An integrative plant-fungal process, pp. 301–32. Chapman and Hall, New York.

Miller, W. B. 1964. An ecological study of the mountain mahogany community. M.S. thesis, Univ. Wyoming, Laramie.

Mills, J. D. 1991. Wyoming's Jackson Lake dam, horizontal channel stability, and floodplain vegetation dynamics. M.S. thesis, Univ. Wyoming, Laramie.

Minshall, G. W., J. T. Brock, and J. D. Varley. 1989. Wildfires and Yellowstone's stream ecosystems. BioScience 39:707–15.

Mitchell, J. E., and R. H. Hart. 1987. Winter of 1886–87: The death knell of open range. Rangelands 9:3–8.

Mitchell, V. L. 1976. The regionalization of climate in the western United States. J. Appl. Meteorol. 15: 920–27.

Mitton, J. B. 1985. So grows the tree. Natural History (Jan.): 58–64.

Mock, C. J. 1991. Drought and precipitation fluctuations in the Great Plains during the late nineteenth century. Great Plains Res. 1:26–57.

Moir, W. H. 1969. The lodgepole pine zone in Colorado. Am. Midl. Nat. 81:87–98.

Moline, B. R., R. R. Fletcher, and D. T. Taylor. 1991. Impact of agriculture on Wyoming's economy. Wyo. Coop. Ext. Ser. Bull. 954, Dept. Agric. Econ., Univ. Wyoming, Laramie.

Mooney, H. A., and W. D. Billings. 1960. The annual carbohydrate cycle of alpine plants as related to growth. Am. J. Bot. 47:594–98.

———. 1961. Comparative physiological ecology of arctic and alpine populations of *Oxyria digyna*. Ecological Monogr. 31:1–29.

Moore, C. T. 1972. Man and fire in the central North American grassland 1535–1890: A documentary historical geography. Ph.D. diss., Univ. California, Los Angeles.

Moore, R. T. 1977. Gas exchange in photosynthetic pathways in range plants. In R. E. Sosebee, ed., Rangeland plant physiology, pp. 1–46. Range Science Series 4. Soc. Range Mgmt., Denver.

Moss, J. H. 1951. Early man in the Eden Valley. Univ. Pennsylvania Mus. Monogr., Philadelphia.

Mueggler, W. F. 1956. Is sagebrush seed residual in the soil of burns or is it wind-borne? U.S. For. Ser. Res. Note 35.

———. 1967. Response of mountain grassland vegetation to clipping in southwestern Montana. Ecol. 48:942–49.

———. 1975. Rate and pattern of vigor recovery in Idaho fescue and bluebunch wheatgrass. J. Range Mgmt. 28:198–204.

———. 1988. Aspen community types of the Intermountain region. U.S. For. Ser. Gen. Tech. Rep. INT-250.

Muir, P. S., and J. E. Lotan. 1985a. Disturbance history and serotiny of *Pinus contorta* in western Montana. Ecol. 66:1658–68.

———. 1985b. Serotiny and life history of *Pinus contorta* var *latifolia*. Can. J. Bot. 63:938–45.

Munn, L. C. 1977. Relationships of soils to mountain and foothill range habitat types and production in western Montana. Ph.D. diss., Montana State Univ., Bozeman.

Murdock, J. R. 1951. Alpine plant succession near Mount Emmons, Uinta Mountains, Utah. M.S. thesis, Brigham Young Univ., Provo.

Murray, R. B. 1975. Effect of *Artemisia tridentata* removal on mineral cycling. Ph.D. diss., Washinton State Univ. Pullman.

Mutel, C. D. F. 1973. An ecological study of the plant communities of certain montane meadows in the front range of Colorado. M.A. thesis, Univ. Colorado, Boulder.

Myers, W. E. 1969. Relationship of vegetation to geologic formations along Sheep Mountain, Albany County, Wyoming. M.S. thesis, Univ. Wyoming, Laramie.

Naeth, M. A., R. L. Rothwell, D. S. Chanasyk, and A. W. Bailey. 1990. Grazing impacts on infiltration in mixed prairie and fescue grassland ecosystems of Alberta. Can. J. Soil Sci. 70:593–605.

Naiman, R. J., J. M. Melillo, and J. E. Hobbie. 1986. Ecosystem alteration of boreal forest streams by beaver *(Castor canadensis)*. Ecol. 67:1254–69.

Neff, D. J. 1957. Ecological effects of beaver habitat abandonment in the Colorado Rockies. J. Wildl. Mgmt. 21:80–84.

Neilson, R. P., and L. H. Wullstein. 1983. Biogeography of two southwestern American oaks in relation to seedling drought response and atmospheric flow structure. Biogeography 10:275–97.

Nelson, D. L. 1983. Occurrence and nature of actinorhizae on *Cowania stansburiana* and other Rosaceae. In A. R. Tiedemann and K. L. Johnson, eds., Research and management of bitterbrush an cliffrose in western North America, pp. 225–39. U.S. For. Ser. Gen. Tech. Rep. INT-152

Nelson, D. L., and D. L. Sturges. 1986. A snowmold disease of mountain big sagebrush. Phytopathology 76:946–51.

Nelson, D. L., and C. F. Tiernan. 1983. Winter injury of sagebrush and other wildland shrubs in the western United States. U.S. For. Ser. Res. Paper INT-314.

Nelson, R. A., and R. L. Williams. 1992. Handbook of Rocky Mountain plants. Roberts Rinehart Publishers, Niwot, Colorado.

Newbauer, J. J., III, L. M. White, R. M. Moy, and D. A. Perry. 1980. Effects of increased rainfall on native forage production in eastern Montana. J. Range Mgmt. 33:246–50.

Newcomb, R. C. 1952. Origin of the mima mounds, Thurston County region, Washington. J. Geol. 60:461–72.

Newton, H., and W. P. Jenney. 1880. Report on the geology and resources of the Black Hills of Dakota, with atlas. GPO, Washington, D.C.

Nichols, J. T. 1964. Soil-vegetation relationships of the fifteen-mile drainage, Washakie County, Wyoming. Ph.D. diss., Univ. Wyoming, Laramie.

Nord, E. C., D. R. Christensen, and A. P. Plummer. 1969. *Atriplex* species (or taxa) that spread by root sprouts, stem layers, and by seed. Ecol. 50:324–26.

Noss, R. F. 1983. A regional landscape approach to maintain diversity. BioScience 33:700–706.

Nowak, R. S., and M. M. Caldwell. 1984. A test of compensatory photosynthesis in the field: Implications for herbivory tolerance. Oecologia 61:311–18.

Noy-Meir, I. 1973. Desert ecosystems: Environment and producers. Annu. Rev. Ecol. Syst. 4:25–51.

———. 1985. Desert ecosystem structure and function. In M. Evenari, I. Noy-Meir, and D. W. Goodall, eds., Ecosystems of the world: Hot deserts and arid shrublands, pp. 93–104. Elsevier, New York.

Nyren, P. E. 1979. Fertilization of northern Great Plains rangelands: A review. Rangelands 1:110–12, 154–56.

Oberbauer, S. F., and W. D. Billings. 1981. Drought tolerance and water use by plants along an alpine topographic gradient. Oecologia 50:325–31.

Ode, D. J., and L. Tieszen. 1980. The seasonal contribution of C_3 and C_4 plant species to primary production in a mixed prairie. Ecol. 61:1304–11.

Oliver, C. D. 1981. Forest development in North America following major disturbances. Forest Ecol. Mgmt. 3:169–82.

Olmsted, C. E. 1979. The ecology of aspen with reference to utilization by large herbivores in Rocky Mountain National Park. In M. S. Boyce and L. D. Hayden-Wing, eds., North American elk: Ecology, behavior and management, pp. 89–97. Univ. Wyoming, Laramie.

Olson, K. D., R. S. White, and B. W. Sindelar. 1985. Response of vegetation of the Northern Great Plains to precipitation amount and grazing intensity. J. Range Mgmt. 38:357–61.

Olson, R. A., and W. A. Gerhart. 1982. A physical and biological characterization of riparian habitat and its importance in Wyoming. Wyoming Game and Fish Department, Cheyenne.

Olson, T. E., and F. L. Knopf. 1986. Naturalization of Russian olive in the western United States. West. J. Appl. For. 1:65–69.

Oosting, H. J., and J. F. Reed. 1952. Virgin spruce-fir of the Medicine Bow Mountains, Wyoming. Ecol. Monogr. 22:69–91.

Orr, H. K. 1959. Precipitation and streamflow in the Black Hills. U.S. For. Ser. Res. Paper 44.

———. 1968. Soil-moisture trends after thinning and clearcutting in a second-growth ponderosa pine stand in the Black Hills. U.S. For. Ser. Res. Note RM-99. Fort Collins, Colorado.

———. 1975. Watershed management in the Black Hills: The status of our knowledge. U.S. For. Ser. Res. Paper RM-141.

Osmund, C. B., L. F. Pitelka, and G. M. Hidy. 1990. Plant biology of the basin and range. Springer-Verlag, New York.

Ostresh, L. M., R. A. Marston, and W. M. Hudson, eds. Wyoming water atlas. 1990. Wyoming Water Development Commission, Cheyenne.

Oswald, E. T. 1966. A synecological study of the forested moraines of the valley floor of Grand Teton National Park, Wyoming. Ph.D. diss., Montana State Univ., Bozeman.

Owen, D. F., and R. G. Wiegert. 1976. Do consumers maximize plant fitness? Oikos 27:488–92.

———. 1981. Mutualism between grasses and grazers: An evolutionary hypothesis. Oikos 36:376–78.

———. 1982a. Grasses and grazers: Is there a mutualism? Oikos 38:258–59.

———. 1982b. Beating the walnut tree: More on grass/grazer mutualism. Oikos 39:115–16.

Painter, E. L., and J. Belsky. 1993. Application of herbivore optimization theory to rangelands of the western United States. Ecol. Applications 3:2–9.

Painter, E. L., and J. K. Detling. 1981. Effects of defoliation on net photosynthesis and regrowth of western wheatgrass. J. Range Mgmt. 34:68–71.

Palmer, T. 1991. The Snake River: Window to the west. Island Press, Covelo, California.

Parker, A. J. 1986. Persistence of lodgepole pine forests in the central Sierra Nevada. Ecol. 67:1560–67.

Parker, A. J., and K. C. Parker. 1983. Comparative successional roles of trembling aspen and lodgepole pine in the southern Rocky Mountains. Great Basin Nat. 43:447–55.

Parker, J. 1953. Photosynthesis in *Picea excelsa* in winter. Ecol. 34:605–9.

Parker, M. 1986. Beaver, water quality, and riparian systems. In D. J. Brosz and J. D. Rogers, coords., Wyoming Water 1986 and Streamside Zone Conference, pp. 88–94. Wyoming Water Research Center, Univ. Wyoming, Laramie.

Parsons, W. F. J., D. H. Knight, and S. L. Miller. 1994. Root gap dynamics in lodgepole pine forests: Nitrogen transformations in gaps of different size. Ecol. Applications.

Parmenter, R. R., M. R. Mesch, and J. A. MacMahon. 1987. Shrub litter production in a sagebrush-steppe ecosystem: Rodent population cycles as a regulating factor. J. Range Mgmt. 40:50–54.

Pase, C. P. 1958. Herbage production and composition under immature ponderosa pine stands in the Black Hills. J. Range Mgmt. 11:238–42.

Pase, C. P., and R. M. Hurd. 1957. Understory vegetation as related to basal area, crown cover, and litter produced by immature ponderosa pine stands in the Black Hills. Proc. Soc. Am. Foresters, Syracuse, New York.

Pase, C. P., and J. F. Thilenius. 1968. Composition, productivity, and site factors of some grasslands in the Black Hills of South Dakota. U.S. For. Ser. Res. Note RM-103.

Passey, H. B., and V. K. Hugie. 1963. Variation in bluebunch wheatgrass in relation to environment and geographic location. Ecol. 44:158–61.

Patten, D. T. 1959. An ecological study of the "parks" of the Rocky Mountains in the Gallatin area of Montana. M.S. thesis, Univ. Massachusetts, Amherst.

———. 1963. Vegetational patterns in relation to environments in the Madison Range, Montana. Ecol. Monogr. 33:375–406.

———. 1968. Dynamics of the shrub continuum along the Gallatin River in Yellowstone National Park. Ecol. 49:1107–12.

———. 1969. Succession from sagebrush to mixed conifer forest in the Northern Rocky Mountains. Am. Midl. Nat. 82:229–40.

Patten, R. S. 1987. Snow avalanches and vegetation pattern in Cascade Canyon, Grand Teton National Park. M.S. thesis, Univ. Wyoming, Laramie.

Paul, E. A., F. E. Clark, and V. O. Biederbeck. 1979. Micro-organisms. In R. T. Coupland. ed., Grassland ecosystems of the world: Analysis of grasslands and their uses, pp. 87–96. Cambridge University Press, Cambridge.

Paulsen, H. A., Jr. 1975. Range management in the central and southern Rocky Mountains: A summary of the status of our knowledge by range ecosystems. U.S. For. Ser. Res. Paper RM-154.

Pearcy, R. W., and R. T. Ward. 1972. Phenology and growth of Rocky Mountain populations of *Deschampsia casespitosa* at three elevations in Colorado. Ecol. 53:1171–78.

Pearson, J. A., T. J. Fahey, and D. H. Knight. 1984. Biomass and leaf area in contrasting lodgepole pine forests. Can. J. For. Res. 14:259–65.

Pearson, J. A., D. H. Knight, and T. J. Fahey. 1987. Biomass and nutrient accumulation during stand development in Wyoming lodgepole pine forests. Ecol. 68:1966–73.

Pearson, L. C. 1965a. Primary production in grazed and ungrazed desert communities of eastern Colorado. Ecol. 46:278–85.

———. 1965b. Primary productivity in a northern desert area. Oikos 15:211–28.

Pearson, L. C., and D. B. Lawrence. 1957. Photosynthesis in aspen bark during winter months. Proc. Minnesota Acad. Sci. 25:101–7.

———. 1958. Photosynthesis in aspen bark. Am. J. Bot. 45:383–87.

Peden, D. G. 1976. Botanical composition of bison diets on shortgrass plains. Am. Midl. Nat. 96:225–29.

Peden, D. G., G. M. Van Dyne, R. W. Rice, and R. M. Hansen. 1974. The trophic ecology of *Bison bison* L. on shortgrass plains. J. Appl. Ecol. 11:489–98.

Peet, R. K. 1988. Forests of the Rocky Mountains. In M. G. Barbour and W. D. Billings, eds., North American terrestrial vegetation, pp. 63–101. Cambridge Univ. Press, New York.

Perry, D. A., and J. E. Lotan. 1977. Opening temperatures in serotinous cones of lodgepole pine. U.S. For. Ser. Res. Note INT-228.

Perry, T. O. 1971. Winter-season photosynthesis and respiration by twigs and seedlings of deciduous and evergreen trees. For. Sci. 17:41–43.

Peterson, D. L., and M. J. Arbaugh. 1986. Postfire survival in Douglas-fir and lodgepole pine: Comparing the effects of crown and bole damage. Can. J. For. Res. 16:1175–79.

Peterson, R. A. 1962. Factors affecting resistance to heavy grazing in needle-and-thread grass. J. Range Mgmt. 15:183–89.

Péwé, T. L. 1983. Alpine permafrost in the contiguous United States: A review. Arct. Alp. Res. 15:145–56.

Pfadt, R. E., and D. M. Hardy. 1987. A historical look at rangeland grasshoppers and the value of grasshopper control programs. In J. Capinera, ed., Integrated pest management on rangeland: A shortgrass prairie perspective, pp. 183–95. Westview Press, Boulder.

Phillips, C. M. 1977. Willow carrs of the upper Laramie River Valley, Colorado. M.S. thesis, Colorado State Univ., Fort Collins.

Phillips, V. D. 1982. Responses by alpine plants and soils to micro-topography within sorted polygons. Ph.D. diss., Univ. Colorado, Boulder.

Phillips, W. S. 1963. Photographic documentation: Vegetational changes in northern Great Plains. Report 214. Ariz. Agric. Exp. Sta., Tucson, Arizona.

Pierce, K. L., and J. M. Good. 1990. Quaternary geology of Jackson Hole, Wyoming. In S. Roberts, ed., Geologic field tours of western Wyoming and parts of adjacent Idaho, Montana, and Utah, pp. 79–88. Public Information Circular 29, Wyo. Geol. Surv., Laramie.

Pierce, W. G. 1961. Permafrost and thaw depressions in a peat deposit in the Beartooth Mountains, northwestern Wyoming. U.S. Geol. Surv. Prof. Paper 424B:B154–56.

Platt, P., and N. Slater. 1852. Travelers' guide across the plains upon the Overland Route to California. Daily Journal Office, Chicago. (Reprinted 1963 by John Howell Books, San Francisco.)

Platt, W. J. 1975. The colonization and formation of equilibrium plant species associations on badger disturbances in a tall-grass prairie. Ecol. Monogr. 45:285–305.

Platts, W. S. 1981. Effects of sheep grazing on a riparian-stream environment. U.S. For. Ser. Res. Note INT-307.

———. 1982. Livestock and riparian-fishery interactions: What are the facts? In K. Sabol, ed., Trans. No.

Am. Wildl. Nat. Res. Conf., no. 47, pp. 507–15. Wildlife Management Institute, Washington, D.C.

Platts, W. S., W. F. Megahan, and G. W. Minshall. 1983. Methods for evaluating stream, riparian, and biotic conditions. U.S. For. Ser. Gen. Tech. Rep. INT-138.

Plummer, A. P., D. R. Christensen, and S. B. Monsen. 1968. Restoring big game range in Utah. Publ. 68–3. Utah Div. Fish and Game, Salt Lake City, Utah.

Polley, H. W., and J. K. Detling. 1988. Herbivory tolerance of Agropyron smithii populations with different grazing histories. Oecologia 77:261–67.

Pond, F. W., and D. R. Smith. 1971. Ecology and management of subalpine ranges on the Big Horn Mountains of Wyoming. Wyo. Agric. Exp. Sta. Res. J. 53.

Porter, L. K. 1969. Nitrogen in grassland ecosystems. In R. L. Dix and R. G. Beidleman, eds., The grassland ecosystem: A preliminary synthesis, pp. 377–402. Colorado State Univ. Range Sci. Ser. 2, Fort Collins.

Potkin, M. A. 1991. Soil-vegetation relationships of subalpine and alpine environments, Wind River Range, Wyoming. M.S. thesis, Univ. Wyoming. Laramie.

Potter, L. D., and D. L. Green. 1964. Ecology of ponderosa pine in western North Dakota. Ecol. 45:10–23.

Potter, N., Jr. 1969. Tree-ring dating of snow avalanche tracks and the geomorphic activity of avalanches, northern Absaroka Mountains, Wyoming. Spec. Paper 123, Geological Society of America.

Potvin, M. A., and A. T. Harrison. 1984. Vegetation and litter changes of a Nebraska sandhills prairie protected from grazing. J. Range Mgmt. 37:55–58.

Pound, R., and F. E. Clements. 1897. The phytogeography of Nebraska. Jacob North, Lincoln, Nebraska.

Power, J. F. 1972. Fate of fertilizer nitrogen applied to a Northern Great Plains ecosystem. J. Range Mgmt. 25:367–71.

Prescott, C. E., J. P. Corbin, and D. Parkinson. 1989a. Biomass, productivity and nutrient-use efficiency of aboveground vegetation in four Rocky Mountain coniferous forests. Can. J. For. Res. 19:309–17.

———. 1989b. Input, accumulation, and residence times of carbon, nitrogen, and phosphorus in four Rocky Mountain coniferous forests. Can. J. For. Res. 19:489–98.

Pringle, W. L. 1960. The effect of a leaf feeding beetle on big sagebrush in British Columbia. J. Range Mgmt. 13:139–42.

Progulske, D. R. 1974. Yellow ore, yellow hair, yellow pine: A photographic study of a century of forest ecology. S. Dak. Agric. Exp. Sta. Bull. 616:1–169.

Pyne, S. J. 1989. The summer we let wild fire loose. Natural History (Aug.): 45–49.

Quinn, M. A., and D. D. Walgenbach. 1990. Influence of grazing history on the community structure of grasshoppers of a mixed-grass prairie. Environ. Ent. 19:1756–66.

Quinnild, C. L., and H. E. Cosby. 1958. Relics of climax vegetation on two mesas in western North Dakota. Ecol. 39:29–32.

Raffa, K. F., and A. A. Berryman. 1982. Gustatory clues in the orientation of Dendroctonus ponderosae (Coleoptera: Scolytidae) to host trees. Can. Ent. 114:97–104.

———. 1983. Physiological aspects of lodgepole pine wound responses to a fungal symbiont of the mountain pine beetle, Dendroctonus ponderosae (Coleoptera: Scolytidae). Can. Ent. 115:723–34.

Ralston, R. D. 1960. The structure and ecology of the north slope juniper stands of the Little Missouri badlands. M.S. thesis, Univ. Utah, Salt Lake City.

Ramaley, F. 1939. Sand-hill vegetation of northeastern Colorado. Ecol. Monogr. 9:1–51.

Randall, D. 1983. A new role for beavers. Defenders 58:29–32.

Raunkiaer, C. 1934. The life forms of plants and statistical plant geography. Clarendon Press, Oxford.

Raup, H. M. 1951. Vegetation and cryoplanation. Ohio J. Sci. 51: 105–116.

Rauzi, F. 1963. Water intake and plant composition as affected by differential grazing on rangeland. J. Soil Water Conserv. 18:114–16.

Rauzi, F., and M. L. Fairbourn. 1983. Effects of annual applications of low N fertilizer rates on a mixed grass prairie. J. Range Mgmt. 36:359–62.

Rauzi, F., and F. M. Smith. 1973. Infiltration rates: Three soils with three grazing levels in northeastern Colorado. J. Range Mgmt. 26:126–29.

Raven, P. H., and D. I. Axelrod. 1981. Angiosperm biogeography and past continental movements. Ann. Missouri Bot. Gard. 61:539–673.

Raynolds, W. F. 1868. Report of Brevet Colonel W. F. Raynolds, U.S.A., Corps of Engineers, on the exploration of the Yellowstone and Missouri rivers, in 1859–60. 40th Congress. Serial 1317. Senate Exec. Doc. 77.

Ream, R. R. 1964. The vegetation of the Wasatch Mountains, Utah and Idaho. Ph.D. diss., Univ. Wisconsin, Madison.

Reardon, P.O., C. L. Leinweber, and L. B. Merrill. 1974. Response of sideoats grama to animal saliva and thiamine. J. Range Mgmt. 27:400–401.

Redman, R. E. 1978. Plant and soil water potentials following fire in a northern mixed grassland. J. Range Mgmt. 31:443–45.

Reed, J. F. 1952. The vegetation of the Jackson Hole

Wildlife Park, Wyoming. Am. Midl. Nat. 48:700–29.

Reed, M. J., and R. A. Peterson. 1961. Vegetation, soil, and cattle responses to grazing on northern Great Plains range. U.S. For. Ser. Tech. Bull. 1252.

Reed, R. M. 1971. Aspen forests of the Wind River Mountains, Wyoming. Am. Midl. Nat. 86:327–43.

———. 1976. Coniferous forest habitat types of the Wind River Mountains, Wyoming. Am. Midl. Nat. 95:159–73.

Reese, R. 1984. Greater Yellowstone: The national park and adjacent wildlands. Montana Geographic Series, no. 6. Montana Magazine, Helena.

Rehfeldt, G. E. 1985. Ecological genetics of *Pinus contorta* in the Wasatch and Uinta Mountains of Utah. Can. J. Forest Res. 15:524–30.

Reichhardt, K. L. 1982. Succession of abandoned fields on the shortgrass prairie, northeastern Colorado. Southw. Nat. 27:299–304.

Reider, R. G. 1983. A soil catena in the Medicine Bow Mountains, Wyoming, U.S.A., with reference to paleoenvironmental influences. Arct. Alp. Res. 15:181–92.

———. 1990. Late Pleistocene and Holocene pedogenic and environmental trends at archaeological sites in plains and mountain areas of Colorado and Wyoming. In N. P. Lasca and J. Donahue, eds., Archaeological geology of North America, pp. 335–60. Geol. Soc. Am., Centennial Special Volume 4.

Reider, R. G., G. A. Huckleberry, and G. C. Frison. 1988. Soil evidence for postglacial forest-grassland fluctuation in the Absaroka Mountains of northwestern Wyoming, U.S.A. Arct. Alp. Res. 20:188–98.

Reider, R. G., N. J. Kuniansky, D. M. Stiller, and P. J. Uhl. 1974. Preliminary investigation of comparative soil development on Pleistocene and Holocene geomorphic surfaces of the Laramie Basin, Wyoming. In M. Wilson, ed., Applied geology and archeology: The Holocene history of Wyoming, pp. 27–33. Wyo. Geol. Surv. Report of Investigations 10.

Reider, R. G., and P. J. Uhl. 1977. Soil differences within spruce-fir forested and century-old burned areas of Libby Flats, Medicine Bow Range, Wyoming. Arct. Alp. Res. 9:383–92.

Reiners, W. A., L. L. Strong, P. A. Matson, I. C. Burke, and D. S. Ojima. 1989. Estimating biogeochemical fluxes across sagebrush-steppe landscapes with thematic mapper imagery. Remote Sensing Environ. 28:121–29.

Rennick, R. B. 1981. Effects of prescribed burning on mixed prairie vegetation in southeastern Montana. M.S. thesis, Montana State Univ., Bozeman.

Reynolds, D. N. 1984a. Alpine annual plants: Phenol-ogy, germination, photosynthesis, and growth of three Rocky Mountain species. Ecol. 65:759–66.

———. 1984b. Populational dynamics of three annual species of alpine plants in the Rocky Mountains. Oecologia 62:250–55.

Rice, B., and M. Westoby. 1978. Vegetative responses of some Great Basin shrub communities protected against jackrabbits or domestic stock. J. Range Mgmt. 31:28–34.

Richards, J. H., and M. M. Caldwell. 1985. Soluble carbohydrates, concurrent photosynthesis and efficiency in regrowth following defoliation: A field study with *Agropyron* species. J. Appl. Ecol. 22:907–20.

Richmond, G. M. 1949. Stone nets, stone stripes, and soil stripes in the Wind River Mountains, Wyoming. J. Geol. 57:143–53.

Rickard, W. H. 1964. Demise of sagebrush through soil changes. BioScience 14:43–44.

———. 1967. Seasonal soil moisture patterns in adjacent greasewood and sagebrush stands. Ecol. 48:1034–38.

Riedl, G. 1959. Geology of the eastern portion of Shirley Basin, Albany and Carbon counties, Wyoming. M.S. thesis, Univ. Wyoming, Laramie.

Riedl, W. A., K. H. Asay, J. L. Nelson, and G. M. Telwar. 1964. Studies of *Eurotia lanata* (winterfat). Wyo. Agric. Exp. Sta. Bull. 425.

Riegel, A. 1941. Life history and habits of blue grama. Kansas Acad. Sci. Trans. 44:1–10.

Ries, R. E., and J. F. Power. 1981. Increased soil water storage and herbage production from snow catch in North Dakota. J. Range Mgmt. 34:485–88.

Righetti, T. L., C. H. Chard, and D. N. Munns. 1983. Opportunities and approaches for enhancing nitrogen fixation in *Purshia, Cowania*, and *Fallugia*. In A. R. Tiedemann and K. L. Johnson, eds., Research and management of bitterbrush and cliffrose in western North America, pp. 214–23. U.S. For. Ser. Gen. Tech. Rep. INT-152.

Righter, R. W. 1982. Crucible for conservation: The creation of Grand Teton National Park. Colorado Associated Univ. Press, Boulder.

Riley, S. P. 1986. The Black Hills aspen: Management implications from treatment results. M.S. thesis, South Dakota State Univ., Brookings.

Risser, P. G. 1985. Grasslands. In B. F. Chabot and H. A. Mooney, eds., Physiological ecology of North American plant communities, pp. 232–56. Chapman and Hall, New York.

Risser, P. G., and W. J. Parton. 1982. Ecosystem analysis of the tallgrass prairie: Nitrogen cycle. Ecol. 63:1342–51.

Roberts, S. 1989. Wyoming geomaps. Wyo. Geol. Surv. Edu. Ser. 1.

Robertson, D. R., J. L. Nielsen, and J. H. Bare. 1966. Vegetation and soils of alkali sagebrush and adjacent big sagebrush ranges in North Park, Colorado. J. Range Mgmt. 19:17–20.

Robertson, J. H. 1947. Response of range grasses to different intensities of competition with sagebrush. Ecol. 28:1–16.

Robertson, P. A., and R. T. Ward. 1970. Ecotypic differentiation in *Koeleria cristata* (L.) Pers. from Colorado and related areas. Ecol. 51:1083–87.

Robinson, J. L. 1982. Development and testing of a stream morphology evaluation method for measuring user impact on riparian zones. M.S. thesis, Univ. Wyoming, Laramie.

Robinson, L. D. 1966. The vegetation and small mammals of the juniper zone in north central Wyoming. M.S. thesis, Univ. Wyoming, Laramie.

Robinson, T. W. 1965. Introduction, spread, and areal extent of salt-cedar *(Tamarix)* in the western states. U.S. Geol. Surv. Prof. Paper 491-A.

Rochow, T. F. 1969. Growth, caloric content, and sugars in *Caltha leptosepala* in relation to alpine snowmelt. Bull. Torrey Bot. Club 96:689–98.

———. 1970. Ecological investigations of *Thlaspe alpestre* L. along an ecological gradient in the central Rocky Mountains. Ecol. 51:649–56.

Rodell, C. F. 1977. A grasshopper model for a grassland ecosystem. Ecol. 58:227–45.

Roe, F. E. 1951. The North American buffalo: A critical study of the species in its wild state. Univ. Toronto Press, Toronto.

Rogers, D. J. 1966. Woodlands in the Great Plains. Science 151:1483.

———. 1969. Isolated stands of lodgepole pine and limber pine in the the Black Hills. Proc. South Dakota Acad. Sci. 48:138–47.

Rogers, L. E. 1987. Ecology and management of harvester ants in the shortgrass plains. In J. L. Capinera, ed., Integrated pest management on rangeland, pp. 261–70. Westview Press, Boulder.

Rollins, P. A., ed. 1935. The discovery of the Oregon Trail. Scribner's, New York.

Rolston, L. K. 1961. The subalpine coniferous forest of the Big Horn Mountains, Wyoming. M.S. thesis, Univ. Wyoming, Laramie.

Romme, W. H. 1982. Fire and landscape diversity in subalpine forests of Yellowstone National Park. Ecol. Monogr. 52:199–221.

Romme, W. H., and D. G. Despain. 1989. Historical perspective on the Yellowstone fires of 1988. BioScience 39:695–99.

Romme, W. H., and D. H. Knight. 1981. Fire frequency and subalpine forest succession along a topographic gradient in Wyoming. Ecol. 62:319–26.

———. 1982. Landscape diversity: The concept applied to Yellowstone Park. BioScience 32:664–70.

Romme, W. H., D. H. Knight, and J. B. Yavitt. 1986. Mountain pine beetle outbreaks in the Rocky Mountains: Regulators of primary productivity? Am. Nat. 127:484–94.

Romme, W. H., and M. G. Turner. 1990. Implications of global climate change for biogeographic patterns in the Greater Yellowstone Ecosystem. Conserv. Biol. 5:373–86.

Romo, J. T. 1984. Water relations in *Artemisia tridentata* subsp. *wyomingensis, Sarcobatus vermiculatus,* and *Kochia prostrata.* Ph.D. diss., Oregon State Univ., Corvallis.

Romo, J. T., and L. E. Eddleman. 1985. Germination response of greasewood *(Sarcobatus vermiculatus)* to temperature, water potential and specific ions. J. Range Mgmt. 38:117–20.

Rood, S. B., and S. Heinze-Milne. 1989. Abrupt downstream forest decline following river damming in southern Alberta. Can. J. Bot. 67:1744–49.

Rood, S. B., and J. M. Mahoney. 1990. Collapse of riparian poplar forests downstream from dams in western prairies: Probable causes and prospects for mitigation. Environ. Mgmt. 14:451–64.

Roughton, R. G. 1972. Shrub age structure on a mule deer winter range in Colorado. Ecol. 53:615–25.

Ruess, R. W., and S. J. McNaughton. 1988. Ammonia volatilization and the effects of large grazing mammals on nutrient loss from East African grasslands. Oecologia 77:382–86.

Running, S. W. 1980. Environmental and physiological control of water flux through *Pinus contorta.* Can. J. Forest Res. 10:82–91.

Rushin, J. W., and J. E. Anderson. 1981. An examination of the leaf quaking adaptation and stomatal distribution in *Populus tremuloides* Mich. Plant Phys. 67:1264–66.

Russey, G. R. 1967. The effect of grazing intensities on the roots of Nuttall's saltbush *(Atriplex nuttallii).* M.S. thesis, Univ. Wyoming, Laramie.

Rutherford, W. H. 1954. Interrelationships of beavers and other wildlife on a high-altitude stream in Colorado. M.S. thesis, Colorado State Univ., Fort Collins.

Ryan, K. C. 1991. Vegetation and wildland fire: Implications of global climate change. Environmental Intern. 17:169–178.

Ryan, K. C., and E. D. Reinhardt. 1988. Predicting post-fire mortality of seven conifers. Can. J. For. Res. 18:1291–97.

Ryan, M. G., and R. H. Waring. 1992. Maintenance respiration and stand development in a subalpine lodgepole pine forest. Ecol. 73:2100–8.

Ryle, G. J. A., and C. E. Powell. 1975. Defoliation and regrowth in the graminaceous plant: The role of current assimilate. Ann. Bot. 39:297–310.

Sabinske, D. W., and D. H. Knight. 1978. Variation within the sagebrush vegetation of Grand Teton National Park. Northw. Sci. 52:195–204.

Sakai, A. K., and T. A. Burris. 1985. Growth in male and female aspen clones: A twenty-five-year longitudinal study. Ecol. 66:1921–27.

Sakai, A., and K. Otsuka. 1970. Freezing resistance of alpine plants. Ecol. 51:665–71.

Sakai, A., and C. J. Weiser. 1973. Freezing resistance of trees in North America with reference to tree regions. Ecol. 54:118–26.

Sala, O. E., W. K. Lauenroth, W. J. Parton, and M.J. Trlica. 1981. Water status of soil and vegetation in a shortgrass steppe. Oecologia 48:317–31.

Sala, O. E., W. K. Lauenroth, and C. P. P. Reid. 1982. Water relations: A new dimension for niche separation between *Bouteloua gracilis* and *Agropyron smithii* in North American semi-arid grasslands. J. Appl. Ecol. 19:647–57.

Sala, O. E., W. J. Parton, L. A. Joyce, and W. K. Lauenroth. 1988. Primary production of the central grassland region of the United States. Ecol. 69:40–45.

Salisbury, F. B. 1984. Light conditions and plant growth under the snow. In J. F. Merrit, ed., Winter ecology of small mammals, pp. 39–50. Carnegie Mus. Nat. Hist. (Pittsburgh) Spec. Publ. 10.

Sandford, S. 1983. Management of pastoral development in the third world. Wiley, New York.

Sanford, R. C. 1970. Skunkbush (*Rhus trilobata* Nutt.) in the North Dakota badlands: Ecology, phytosociology, browse production and utilization. Ph.D. diss., North Dakota State Univ., Fargo.

Sauer, R. H. 1978. Effects of removal of standing dead material on growth of *Agropyron spicatum*. J. Range Mgmt. 31:121–22.

Schacht, W., and J. Stubbendieck. 1985. Prescribed burning in the loess hills mixed prairie of southern Nebraska. J. Range Mgmt. 38:47–51.

Schier, G. A., J. R. Jones, and R. P. Winokur. 1985. Vegetative reproduction. In N. V. DeByle and R. P. Winokur, eds., Aspen: Ecology and management in the western United States, pp. 29–33. U.S. For. Ser. Gen. Tech. Report RM-119.

Schimel, D., W. J. Parton, F. J. Adamsen, R. G. Woodmansee, R. L. Senft, and M. A. Stillwell. 1986. The role of cattle in the volatile loss of nitrogen from a shortgrass prairie. Biogeochemistry 2:39–52.

Schimel, D., M. A. Stillwell, and R. G. Woodmansee. 1985. Biogeochemistry of C, N, and P in a soil catena of the shortgrass steppe. Ecol. 66:276–82.

Schimpf, D. J., J. A. Henderson, and J. A. MacMahon. 1980. Some aspects of succession in the spruce-fir forest zone of northern Utah. Great Basin Nat. 40:1–26.

Schlesinger, W. H., J. F. Reynolds, G. L. Cunningham, L. F. Huenneke, W. M. Jarrell, R. A. Virginia, and W. G. Whitford. 1990. Biological feedbacks in global desertification. Science 247:1043–48.

Schmid, J. M., and T. E. Hinds. 1974. Development of spruce-fir stands following spruce beetle outbreaks. U.S. For. Ser. Res. Paper RM-131.

Schmid, J. M., S. A. Mata, and A. M. Lynch. 1991. Red belt in lodgepole pine in the Front Range of Colorado. U.S. Forest Service Res. Note RM-503.

Schmidt, W. C., and K. J. McDonald, compilers. 1990. Symposium on whitebark pine ecosystems: Ecology and management of a high-mountain resource. U.S. For. Ser. Gen Tech. Report INT-270.

Schoettle, A. W. 1991. The interaction between leaf longevity and shoot growth and foliar biomass per shoot in *Pinus contorta* at two elevations. Tree Physiology 7:209–14.

Schreier, C. 1982. Grand Teton explorers guide. Homestead Publishing, Moose, Wyoming.

———. 1983. Yellowstone explorers guide. Homestead Publishing, Moose, Wyoming.

Schullery, P. 1989. The fires and fire policy. BioScience 39:686–94.

———. 1992. The bears of Yellowstone. High Plains Publishing, Worland, Wyoming.

Schullery, P., and Lee Whittlesey. 1992. The documentary record of wolves and related wildlife species in the Yellowstone National Park area prior to 1882. In J. D. Varley and W. G. Brewster, eds., Wolves in Yellowstone? A report to the United States Congress. Vol. 4, Research and Analysis, pp. 1-4 to 1-174. Natl. Park Service, Yellowstone National Park, Wyoming.

Schulz, T. T., and W. C. Leininger. 1990. Differences in riparian vegetation structures between grazed areas and exclosures. J. Range Mgmt. 43:295–99.

Schuster, W. S., D. L. Alles, and J. B. Mitton. 1989. Gene flow in limber pine: Evidence from pollination phenology and genetic differentiation along an elevational transect. Am. J. Bot. 76:1395–1403.

Schwartz, C. C., and J. Ellis. 1981. Feeding ecology and niche separation in some native and domestic ungulates on the shortgrass prairie. J. Appl. Ecol. 18:343–53.

Scott, D., and W. D. Billings. 1964. Effects of environmental factors on standing crop and productivity of an alpine tundra. Ecol. Monogr. 34:243–70.

Scott, H. W. 1951. The geological work of the mound-building ants in western United States. J. Geol. 59:173–75.

Scott, J. A., N. R. French, and J. W. Leetham. 1979. Patterns of consumption in grasslands. In N. R. French, ed., Perspectives in grassland ecology, pp. 89–105. Springer-Verlag, New York.

Scott, R. W. 1966. The alpine flora of northwestern Wyoming. M.S. thesis, Univ. Wyoming, Laramie.

Sears, P. B. 1961. A pollen profile from the grassland province. Science 134:2038–40.

Sedgwick, J. A., and F. L. Knopf. 1989. Demography, regeneration, and future projections for a bottom-land cottonwood community. In R. Sharitz and J. W. Gibbons, eds., Freshwater wetlands and wildlife, pp. 249–66. Conf. Series 61. U.S. Dept. of Energy, Office of Scientific and Technical Information, Oak Ridge, Tenn. (Conf. 8603101)

———. 1991. Prescribed grazing as a secondary impact in a western riparian floodplain. J. Range Mgmt. 44:369–73.

Sellers, R. E., and D. G. Despain. 1976. Fire management in Yellowstone National Park. Montana Tall Timbers Fire Ecol. Conf. 14.

Senft, R. L., L. R. Rittenhouse, and R. G. Woodmansee. 1985. Factors influencing patterns of cattle grazing behaviour on shortgrass steppe. J. Range Mgmt. 38:82–87.

Severson, K. E. 1963. A description and classification by composition of the aspen stands in the Sierra Madre Mountains, Wyoming. Univ. Wyoming, Laramie.

Severson, K. E., and J. J. Kranz. 1976. Understory production not predictable from aspen basal area or density. U.S. For. Ser. Res. Note RM-314.

Severson, K. E., and J. F. Thilenius. 1976. Classification of quaking aspen stands in the Black Hills and Bear Lodge Mountains. U.S. For. Ser. Res. Paper RM-166.

Sharma, M. L., and D. J. Tongway. 1973. Plant induced soil salinity patterns in two saltbush (Atriplex spp.) communities. J. Range Mgmt. 26:121–25.

Sharp, L. A., K. Sanders, and N. Rimbey. 1990. Forty years of change in a shadscale stand in Idaho. Rangelands 12:313–28.

Sharps, J. C., and D. W. Uresk. 1990. Ecological review of black-tailed prairie dogs and associated species in western South Dakota. Great Basin Nat. 50:339–45.

Shaw, C. G., III, and G. A. Kile. 1991. Armillaria root disease. U.S. Dep. Agric. Agricultural Handbook 691.

Shea, K. 1985. Demographic aspects of coexistence in Engelmann spruce and subalpine fir. Am. J. Bot. 72:1823–33.

Sheets, W. B. 1958. The effect of chemical control of big sagebrush on the water budget and vertical energy balance. Ph.D. diss., Univ. Wyoming, Laramie.

Sheppard, J. S. 1971. The influence of geothermal temperature gradients upon vegetation patterns in Yellowstone National Park. Ph.D. diss., Colorado State Univ., Fort Collins.

Sherman, R. J., and W. W. Chilcote. 1972. Spatial and chronological patterns of Purshia tridentata as influenced by Pinus ponderosa. Ecol. 53:294–98.

Shirley, H. L. 1931. Does light burning stimulate aspen suckering? J. Forestry 29:524–25.

Shumar, M. L., and J. E. Anderson. 1986. Gradient analysis of vegetation dominated by two subspecies of big sagebrush. J. Range Mgmt. 39:156–60.

Sims, P. L., and R. T. Coupland. 1979. Producers. In R. T. Coupland, ed., Grassland ecosystems of the world: Analysis of grasslands and their uses, pp. 49–72. Cambridge Univ. Press, London.

Singer, F. J., W. Schreier, J. Oppenheim, and E. O. Garton. 1989. Drought, fires, and large mammals. BioScience 39:716–22.

Singh, J. S., and D. E. Coleman. 1977. Evaluation of functional root biomass and translocation of photo-assimilated ^{14}C in shortgrass prairie. In J. K. Marshall, ed. The belowground ecosystem: A synthesis of plant-associated processes, pp. 123–31. Range Sci. Dept. Sci. Ser. 26. Colorado State Univ., Fort Collins.

Singh, J. S., W. K. Lauenroth, R. K. Heitschmidt, and J. L. Dodd. 1983. Structural and functional attributes of the vegetation of northern mixed prairie of North America. Bot. Rev. 49:117–49.

Singh, J. S., W. K. Lauenroth, and D. G. Milchunas. 1983. Geography of grassland ecosystems. Prog. Phys. Geogr. 7:46–80.

Skinner, Q. D. 1986. Riparian zones then and now. In D. J. Brosz and J. D. Rogers, coords., Wyoming water 1986 and streamside zone conference, pp. 8–22. Wyoming Water Research Center, Univ. Wyoming, Laramie. Wyo. Agric. Exp. Sta., Laramie.

Skinner, Q. D., J. E. Speck, M. A. Smith, and J. C. Adams. 1984. Stream water quality as influenced by beaver within grazing systems in Wyoming. J. Range Mgmt. 37:142–46.

Skougard, M. G., and J. D. Brotherson. 1979. Vegetational response to three environmental gradients in the salt playa near Goshen, Utah County, Utah. Great Basin Nat. 39:44–58.

Smith, C. C. 1970. The coevolution of pine squirrels (Tamiasciurus) and conifers. Ecol. Monogr. 40:349–71.

———. 1975. The coevolution of plants and seed predators. In L. E. Gilbert and P. H. Raven, eds., Coevolution of animals and plants, pp. 53–77. Univ. Texas Press, Austin.

Smith, D. R., and W. M. Johnson. 1965. Vegetation characteristics on a high altitude sheep range in Wyoming. Wyo. Agric. Exp. Sta. Bull. 430.

Smith, E. L. 1966. Soil-vegetation relationships of some *Artemisia* types in North Park, Colorado. Ph.D. diss., Colorado State Univ., Fort Collins.

Smith, F. W. 1987. Silvicultural research in coniferous subalpine forests of the Central Rocky Mountains. In C. A. Troendle, M. R. Kaufmann, R. H. Hamre, and R. P. Winokur, tech. coords., Management of subalpine forests: Building on fifty years of research, pp. 15–20. U.S. For. Ser. Gen. Tech. Rep. RM-149.

Smith, M. A., J. L. Dodd, and J. D. Rodgers. 1985. Prescribed burning on Wyoming rangeland. Bull. 810, Wyo. Agric. Ext. Ser. Bull. 810., Laramie.

Smith, S. D., and R. S. Nowak. 1990. Ecophysiology of plants in the intermountain lowlands. In C. B. Osmund, L. F. Pitelka, and G. M. Hidy, eds., Plant biology of the basin and range, pp. 179–241. Springer-Verlag, New York.

Smith, W. K. 1985. Western montane forests. In B. F. Chabot and H. A. Mooney, eds., Physiological ecology of North American plant communities, pp. 97–126. Chapman and Hall, New York.

Smith, W. K., and G. A. Carter. 1988. Shoot structural effects on needle temperature and photosynthesis in conifers. Am. J. Bot. 75:496–500.

Smith, W. K., and G. N. Geller. 1979. Plant transpiration at high elevation: Theory, field measurements, and comparisons of desert plants. Oecologia 41:109–22.

Smith, W. K., and A. K. Knapp. 1990. Ecophysiology of high elevation forests. In C. B. Osmund, L. F. Pitelka, and G. M. Hidy, eds., Plant biology of the basin and range, pp. 87–142. Springer-Verlag, New York.

Smolik, J. D., and J. L. Dodd. 1983. Effects of water and nitrogen and grazing on nematodes in a shortgrass prairie. J. Range Mgmt. 36:744–48.

Smolik, J. D., and J. K. Lewis. 1982. Effect of range condition on density and biomass of nematodes in a mixed prairie ecosystem. J. Range Mgmt. 35:657–63.

Smolik, J. D., and L. E. Rogers. 1976. Effects of cattle grazing and wildfire on soil dwelling nematodes of the shrub-steppe ecosystem. J. Range Mgmt. 29: 744–48.

Sneva, F. A. 1979. The western harvester ant: Their density and hill size in relation to herbaceous productivity and big sagebrush cover. J. Range Mgmt. 32:46–47.

Snoke, A. W., J. R. Steidtman, and S. M. Roberts, eds. 1993. Geology of Wyoming. Memoir 5. Wyo. Geol. Surv., Laramie.

Sollins, P., C. C. Grier, F. M. McCorison, K. Cromack, Jr., and R. Fogel. 1980. The internal element cycles of an old-growth Douglas fir stand in western Oregon. Ecol. Monogr. 50:261–85.

Sonder, L. W. 1959. Soil moisture retention and snow holding capacity as affected by chemical control of big sagebrush *(Artemisia tridentata* Nutt.) M.S. thesis, Univ. Wyoming, Laramie.

Sosebee, R. E., ed. 1977. Rangeland plant physiology. Range Science Series 4. Soc. Range Mgmt., Denver.

Spackman, L. K., and L. C. Munn. 1984. Genesis and morphology of soils associated with formation of Laramie Basin (mima-like) mounds in Wyoming. Soil Sci. Soc. Am. J. 48:1384–92.

Spaeth, K. E. 1981. Successional trends on revegetated *Juniperus osteosperma* communities in north central Wyoming. M.S. thesis, Univ. Wyoming, Laramie.

Spence, J. R. 1980. Vegetation of subalpine and alpine moraines in the Teton Range, Grand Teton National Park, Wyoming. M.S. thesis, Montana State Univ, Bozeman.

Spence, W. K. 1985. A floristic analysis of neoglacial deposits in the Teton Range, Wyoming, U.S.A. Arct. Alp. Res. 17:19–30.

Spomer, G. G. 1964. Physiological ecology studies of alpine cushion plants. Physiol. Plant. 17:717–24.

Stabler, D. F. 1985. Increasing summer flow in small streams through management of riparian areas and adjacent vegetation: A synthesis. In R. R. Johnson, C. D. Ziebell, D. R. Patton, P. F. Folliott, and R. H. Hamre, coords., Riparian ecosystems and their management: Reconciling conflicting uses, pp. 206–10. First North American Riparian Conference. U.S. For. Ser. Gen. Tech. Rep. RM-120.

Stahelin, R. 1943. Factors influencing the natural restocking of high altitude burns by coniferous trees in the central Rocky Mountains. Ecol. 24:19–30.

Stanton, N. L. 1983. The effect of clipping and phytophagous nematodes on net primary production of blue grama, *Bouteloua gracilis.* Oikos 40:249–57.

———. 1987. Nematodes in rangelands. In J. L. Capinera, ed., Integrated pest management on rangeland, pp. 291–311. Westview Press, Boulder.

———. 1988. The underground in grasslands. Ann. Rev. Ecol. Syst. 19:573–89.

Stanton, N. L., M. Allen, and M. Campion. 1981. The effect of the pesticide carbofuran on soil organisms and root and shoot production in shortgrass prairie. J. Appl. Ecol. 18:417–31.

Stanton, N. L., D. Morrison, and W. A. Laycock. 1984. The effect of phytophagous nematode grazing on blue grama die-off. J. Range Mgmt. 37:447–50.

Stark, N. M. 1977. Fire and nutrient cycling in a Douglas fir/larch forest. Ecol. 58:16–30.

Starr, C. R. 1974. Subalpine meadow vegetation in relation to environment at Headquarter's Park, Medicine Bow Mountains, Wyoming. M.S. thesis, Univ. Wyoming, Laramie.

States, J. B. 1968. Growth of ponderosa pine on three geologic formations in eastern Wyoming. M.S. thesis, Univ. Wyoming, Laramie.

Stauffer, J. M. 1976. Ecology and floristics of Ohio Slide and other avalanche tracks in Lost Horse Canyon, Bitterroot Mountains, Montana. M.S. thesis, Univ. Montana, Missoula.

Steele, R., S. V. Cooper, D. M. Ondov, D. W. Roberts, and R. D. Pfister. 1983. Forest habitat types of eastern Idaho-western Wyoming. U.S. For. Ser. Gen. Tech. Rep. INT-144.

Steger, R. E. 1970. Soil moisture and temperature relationships of six salt desert shrub communities in northcentral Wyoming. M.S. thesis, Univ. Wyoming, Laramie.

Steinauer, G. A. 1984. A classification of the *Cercocarpus montanus, Quercus macrocarpa, Populus tremuloides,* and *Picea glauca* habitat types of the Black Hills National Forest. M.A. thesis, Univ. South Dakota, Vermillion.

Stelfox, J. G., and D. Stelfox. 1979. Fairy rings and wildlife. J. Range Mgmt. 32:478–79.

Stevens, G. C., and J. F. Fox. 1991. The causes of treeline. Annual Rev. Ecol. Syst. 22:

Steward, D. G. 1977. Tree-island communities of Rocky Mountain National Park. M.A. thesis, Univ. Colorado, Boulder.

Stoddart, L. A. 1941. The Palouse grassland association in northern Utah. Ecol. 22:158–63.

Stottlemyer, R. 1987. Natural and anthropic factors as determinants of long-term streamwater chemistry. In C. A. Troendle, M. R. Kaufmann, R. H. Hamre, and R. P. Winokur, tech. coords., Management of subalpine forests: Building on fifty years of research, pp. 86–94. U.S. For. Ser. Gen. Tech. Rep. RM-149.

Strahler, A. N. 1969. Physical geography. 3d ed. Wiley, New York.

Strain, B. R., and P. L. Johnson. 1963. Corticular photosynthesis and growth in *Populus tremuloides.* Ecol. 44:581–84.

Strasia, C. A., M. Thorn, R. W. Rice, and D. R. Smith. 1970. Grazing habits, diet and performance of sheep on alpine ranges. J. Range Mgmt. 23:201–208.

Strickland, D. 1986. Building better grizzly country. Wyoming Wildlife (Aug.): 26–33.

Strong, W. E. 1968. A trip to the Yellowstone National Park in July, August, and September, 1875. Univ. Oklahoma Press, Norman.

Stroud, D. O., R. H. Hart, M. J. Samuel, and J. D. Rogers.

1985. Western wheatgrass responses to simulated grazing. J. Range Mgmt. 38:103–8.

Stubbendieck, J., and M. A. Foster. 1978. Herbage yield and quality of threadleaf sedge. J. Range Mgmt. 31:290–92.

Stuckey, R. K. 1990. Evolution of land mammal diversity in North America during the Cenozoic. Curr. Mammol. 2:375–432.

Sturgeon, K. B. 1980. Evolutionary interactions between the mountain pine beetle, *Dendroctonus ponderosae* Hopkins, and its host trees in the Colorado Rocky Mountains. Ph.D. diss., Univ. Colorado, Boulder.

Sturges, D. L. 1975. Hydrologic relations on undisturbed and converted big sagebrush lands: The status of our knowledge. U.S. For. Ser. Res. Paper RM-140.

———. 1977a. Soil moisture response to spraying big sagebrush: A seven-year study and literature interpretation. U.S. For. Ser. Res. Paper RM-188.

———. 1977b. Soil water withdrawal and root characteristics of big sagebrush. Am. Midl. Nat. 98:257–74.

———. 1986. Responses of vegetation and ground cover to spraying a high elevation, big sagebrush watershed with 2,4-D. J. Range Mgmt. 39:141–46.

Sturges, D. L., and M. J. Trlica. 1978. Root weights and carbohydrate reserves of big sagebrush. Ecol. 59: 1282–85.

Sun, K. R. 1986. Ranch management of streamside zones. In D. J. Brosz and J. D. Rogers, coords., Wyoming water 1986 and streamside zone conference. Wyoming Water Research Center, Univ. Wyoming, Laramie. Wyo. Agric. Exp. Sta., Laramie.

Summers, C. A., and R. L. Linder. 1978. Food habits of the black-tailed prairie dog in western South Dakota. J. Range Mgmt. 31:134–36.

Swetnam, T. W., and A. M. Lynch. 1989. A tree-ring reconstruction of western spruce budworm history in the southern Rocky Mountains. Forest Sci. 35: 962–86.

Swift, B. L. 1984. Status of riparian ecosystems in the United States. Water Resources Bull. 20:223–28.

Tabler, R. D. 1964. The root system of *Artemisia tridentata* at 9,500 feet in Wyoming. Ecol. 45:633–36.

———. 1968. Soil moisture response to spraying big sagebrush with 2,4-D. J. Range Mgmt. 21:12–15.

Tausch, R. J., and P. T. Tueller. 1977. Plant succession following chaining of pinyon-juniper woodlands in eastern Nevada. J. Range Mgmt. 30:44–49.

———. 1990. Foliage biomass and cover relationships between tree- and shrub-dominated communities in pinyon-juniper woodlands. Great Basin Nat. 50:121–34.

Taylor, D. L. 1973. Some ecological implications of forest fire control in Yellowstone National Park, Wyoming. Ecol. 54:1394–96.

———. 1974. Forest fires in Yellowstone National Park. Forest Hist. 18:69–77.

Taylor, D. L., and W. J. Barmore, Jr. 1980. Post-fire succession of avifauna in coniferous forests of Yellowstone and Grand Teton National Parks, Wyoming. In R. M. DeGraff, ed., Workshop on management of western forests and grasslands for nongame birds, pp. 130–45. U.S. For. Ser. Gen. Tech. Rep. INT-86.

Terjung, W. H., R. N. Kickert, G. L. Potter, and S. W. Swarts. 1969. Energy and moisture balances of an alpine tundra in mid July. Arct. Alp. Res. 1:247–66.

Thatcher, A. P. 1959. Distribution of sagebrush as related to site differences in Albany County, Wyoming. J. Range Mgmt. 12:55–61.

Thilenius, J. F. 1970. An isolated occurrence of limber pine (Pinus flexilis James) in the Black Hills of South Dakota. Am. Midl. Nat. 84:411–17.

———. 1975. Alpine range management in the western United States: Principles, practice, and problems: The status of our knowledge. U.S. For. Ser. Res. Paper RM-157.

———. 1979. Range management in the alpine zone: Practices and problems. In D. A. Johnson, ed., Special management needs of alpine ecosystems, pp. 43–64. Range Science Series 5. Soc. Range Mgmt., Denver.

Thilenius, J. F., and D. R. Smith. 1985. Vegetation and soils of an alpine range in the Absaroka Mountains, Wyoming. U.S. For. Ser. Gen. Tech. Rep. RM-121.

Thompson, L. S. 1983. Shoreline vegetation in relation to water level at Toston Reservoir, Broadwater Co., Montana. Proc. Mont. Acad. Sci. 42:7–16.

Thompson, W. W., and F. R. Gartner. 1971. Native forage response to clearing low quality ponderosa pine. J. Range Mgmt. 24:272–77.

Thorpe, J. 1931. The effects of vegetation and climate upon soil profiles in northern and northwestern Wyoming. Soil Sci. 32:283–301.

Thwaites, R. G. 1905. Original journals of the Lewis and Clark Expedition. Vol. 7. Dodd, Mead, New York.

Tidwell, W. D. 1975. Common fossil plants of western North America. Brigham Young Univ. Press, Provo, Utah.

Tiedemann, A. R., E. D. McArthur, H. C. Stutz, R. Stevens, and K. L. Johnson, eds. 1984. Symposium on the biology of Atriplex and related chenopods. U.S. For. Ser. Gen. Tech. Rep. INT-172.

Tieszen, L. L., and J. K. Detling. 1983. Productivity of grassland and tundra. In O. L. Lange, P. S. Nobel, C. B. Osmund, and H. Ziegler, eds., Physiological plant ecoplogy, chap. 4, pp. 173–203. Springer-Verlag, New York.

Tiku, B. L.. 1976a. Effect of salinity on the photosynthesis of Salicornia rubra and Distichlis stricta. Physiol. Plant. 37:23–28.

———. 1976b. Ecophysiological aspects of halophyte zonation in saline sloughs. Plant Soil 42:355–69.

Tileston, J. W., and R. R. Lechleitner. 1966. Some comparisons of the black-tailed and white-tailed prairie dogs in north-central Colorado. Am. Midl. Nat. 75:292–316.

Tilman, D. 1988. Plant strategies and the dynamics and structure of plant communities. Princeton Univ. Press, Princeton, New Jersey.

Tisdale, E. W., and M. Hironaka. 1981. The sagebrush-grass region: A review of the ecological literature. Forest, Wildlife and Range Exp. Sta. Bull. 33, Univ. Idaho, Moscow.

Tisdale, E. W., M. Hironaka, and F. A. Fosberg. 1965. An area of pristine vegetation in Craters of the Moon National Monument, Idaho. Ecol. 46:349–52.

Tkacz, B. M, and R. F. Schmidt. 1985. Association of an endemic mountain pine beetle population with lodgepole pine infected by Armillaria root disease in Utah. U.S. For. Ser. Res. Note INT-353.

Tolstead, W. L. 1947. Woodlands in northwestern Nebraska. Ecol. 28:180–88.

Tomback, D. F. 1983. Nutcrackers and pines: Coevolution or coadaptation? In M. H. Nitecki, ed., Coevolution, pp. 179–223. Univ. Chicago Press, Chicago.

Tranquillini, W. 1979. Physiological ecology of the alpine timberline. Springer-Verlag, New York.

Trelease, S. F., and O. A. Beath. 1949. Selenium: Its geological occurrence and its biological effects in relation to botany, chemistry, agriculture, nutrition, and medicine. Wyo. Agric. Exp. Sta., Univ. Wyo., Laramie.

Troendle, C. A. 1983. The potential for water yield augmentation from forest management in the Rocky Mountain region. Water Resour. Bull. 19:359–73.

———. 1987. The potential effect of partial cutting and thinning on streamflow from the subalpine forest. U.S. For. Ser. Res. Paper RM-274.

Troendle, C. A., and M. R. Kaufmann. 1987. Influence of forests on the hydrology of the subalpine forest. In C. A. Troendle, M. R. Kaufmann, R. H. Hamre, and R. P. Winokur, tech. coords., Management of subalpine forests: Building on fifty years of research, pp. 68–78. U.S. For. Ser. Gen. Tech. Rep. RM-149.

Troendle, C. A., and R. M. King. 1987. The effect of partial and clear-cutting on streamflow at Deadhorse Creek, Colorado. J. Hydrol. 90:145–57.

Troendle, C. A., and M. A. Nilles. 1987. The effect of

clearcutting on chemical exports in lateral flow from differing soil depths on a subalpine forested slope. Vancouver symposium on forest hydrology and watershed management. IAHS-AISH Publ. 167.

Turlo, J. B. 1963. An ecological investigation of *Populus tremuloides* at two elevational levels in the Medicine Bow National Forest of Wyoming. M.S. thesis, Univ. Wyoming, Laramie.

Turner, G. T., and D. F. Costello. 1942. Ecological aspects of the pricklypear problem in eastern Colorado and Wyoming. Ecol. 23:419–26.

Turner, G. T., and G. E. Klipple. 1952. Growth characteristics of blue grama in northeastern Colorado. J. Range Mgmt. 5:22–28.

Turner, G. T., and H. A. Paulsen, Jr. 1976. Management of mountain grasslands in the central Rocky Mountains: The status of our knowledge. U.S. For. Ser. Res. Paper RM-161.

Turner, M. G., ed. 1987. Landscape heterogeneity and disturbance. Springer-Verlag, New York.

Uhlich, J. W. 1982. Vegetation dynamics of non-grazed and grazed sagebrush grassland communities in the Big Horn Basin of Wyoming. M.S. thesis, Univ. Wyoming, Laramie.

Ungar, I. A. 1966. Salt tolerance of plants growing in saline areas of Kansas and Oklahoma. Ecol. 47:154–55.

Uresk, D. W. 1984. Black-tailed prairie dog food habits and forage relationships in western South Dakota. J. Range Mgmt. 37:325–29.

———. 1985. Effects of controlling black-tailed prairie dogs on plant production. J. Range Mgmt. 38:466–68.

———. 1987. Relation of black-tailed prairie dogs and control programs to vegetation, livestock, and wildlife. In J. L. Capinera, ed., Integrated pest management on rangeland: A shortgrass perspective, pp. 312–23. Westview Press, Boulder.

Uresk, D. W., and A. J. Bjugstad. 1983. Prairie dogs as ecosystem regulators on the northern high plains. In C. L. Kucera, ed., Seventh No. Am. Prairie Conf., 4–6 Aug. 1980, pp. 91–94. Southwest Missouri State Univ., Springfield.

Uresk, D. W., and C. E. Boldt. 1986. Effect of cultural treatment on regeneration of native woodlands on the northern Great Plains. Prairie Nat. 18:193–202.

Uresk, D. W., and W. W. Paintner. 1985. Cattle diets in a ponderosa pine forest in the northern Black Hills. J. Range Mgmt. 38:440–42.

Uresk, D. W., P. L. Sims, and D. A. Jameson. 1975. Dynamics of blue grama within a shortgrass ecosystem. J. Range Mgmt. 28:205–8.

Urness, P. 1989. Why did bison fail west of the Rockies? Utah Sci. 50:175–79.

Vale, T. R. 1975. Presettlement vegetation in the sagebrush-grass area of the intermountain west. J. Range Mgmt. 28: 32–36.

———. 1978. Tree invasion of Cinnabar Park in Wyoming. Am. Midl. Na. 100:277–84.

Van Haveren, B. P. 1974. Soil water phenomena of a shortgrass prairie site. M.S. thesis, Colorado State Univ., Fort Collins.

Van Haveren, B. P., and W. D. Striffler. 1976. Snowmelt recharge on a shortgrass prairie site. Western Snow Conf. Proc. 44:56–62.

Van Vuren, D. 1987. Bison west of the Rocky Mountains: A review of the theories and an explanation. Northw. Sci. 61:65–69.

Vander Wall, S. B. 1990. Food hoarding in animals. Univ. Chicago Press, Chicago.

Vander Wall, S. B., and R. P. Balda. 1977. Coadaptation of the Clark's nutcracker and the pinyon pine for efficient seed harvest and dispersal. Ecol. Monogr. 47: 89–111.

———. 1983. Remembrance of seeds stashed. Natural History (Sept.): 61–64.

Vass, A. F., and R. Lang. 1938. Vegetative composition, density, grazing capacity and grazing land values in the Red Desert area. Wyo. Agric. Exp. Sta. Bull. 229.

Vavra, M., W. A. Laycock, and R. D. Pieper, eds. Ecological implications of livestock herbivory in the west. Soc. Range Mgmt., Denver.

Vavra, M., R. W. Rice, R. M Hansen, and P. L. Sims. 1977. Food habits of cattle on shortgrass range in northeastern Colorado. J. Range Mgmt. 30: 261–63.

Veblen, T. T. 1986a. Treefalls and the coexistence of conifers in subalpine forests of the central Rockies. Ecol. 67:644–49.

———. 1986b. Age and size structure of subalpine forests in the Colorado Front Range. Bull. Torrey Bot. Club 113:225–40.

Veblen, T. T., K. S. Hadley, and M. S. Reid. 1991. Disturbance and stand development of a Colorado subalpine forest. J. Biogeogr. 18:707–16.

Veblen, T. T., K. S. Hadley, M. S. Reid, and A. J. Rebertus. 1991a. The response of subalpine forests to spruce beetle outbreaks in Colorado. Ecol. 72:213–31.

———. 1991b. Methods of detecting past spruce beetle outbreaks in Rocky Mountain subalpine forests. Can. J. For. Res. 21:242–54.

Veblen, T. T., and D. C. Lorenz. 1986. Anthropogenic disturbance and recovery patterns in montane forests, Colorado Front Range. Phys. Geog. 7:1–24.

———. 1991. The Colorado front range: A century of ecological change. Univ. Utah Press, Salt Lake City.

Vestal, A. G. 1914. Prairie vegetation of a mountain-front area in Colorado. Bot. Gaz. 58:377–400.

Vinton, M. A. 1987. Distribution and nitrogen-fixing

activity of *Lupinus argenteus* on sagebrush-steppe in Wyoming. Dept. of Botany, Univ. Wyo., Laramie.

Vitousek, P. M., and J. M. Melillo. 1979. Nitrate losses from disturbed forests: Patterns and mechanisms. For. Sci. 25:605–19.

Vlamis, J., A. M. Schultz, and H. H. Biswell. 1964. Nitrogen fixation by root nodules of western mountain mahogany. J. Range Mgmt. 17:73–74.

Vogl, R. J. 1974. Effects of fire on grassland. In T. T. Kozlowski and C. E. Ahlgren, eds., Fire and ecosystems, pp. 139–94. Academic Press, New York.

Vogt, K. A., C. C. Grier, C. E. Meier, and R. L. Edmonds. 1982. Mycorrhizal role in net primary production and nutrient cycling in *Abies amabilis* ecosystems in western Washington. Ecol. 63:370–80.

Voight, B. 1982. Lower Gros Ventre Slide, Wyoming, U.S.A. In B. Voight, ed., Natural phenomena: Rock slides and avalanches, pp. 113–66. Elsevier, Amsterdam.

Vosler, L. C. 1962. An ecological study of *Atriplex nuttalli* in the Big Horn Basin of Wyoming. M.S. thesis, Univ. Wyoming, Laramie.

Waddington, J. C. B., and H. E. Wright. 1974. Late Quaternary vegetational changes on the east side of Yellowstone National Park, Wyoming. Quaternary Res. 4:175–84.

Wagle, R. F., and J. Vlamis. 1961. Nutrient deficiencies in two bitterbrush soils. Ecol. 42:745–52.

Walker, D. A., J. C. Halfpenny, M. D. Walker, and C. A. Wessman. 1993. Long-term studies of snow-vegetation interactions. BioScience 43:287–301.

Walker, D. N. 1987. Late Pleistocene-Holocene environmental changes in Wyoming: The mammalian record. Illinois State Museum Scientific Papers 22: 334–93.

Walker, G. R., and J. D. Brotherson. 1982. Habitat relationships of basin wildrye in the high mountain valleys of central Utah. J. Range Mgmt. 35:628–33.

Wallace, A., and D. L. Nelson. 1990. Wildland shrub dieoffs following excessively wet periods: A synthesis. In E. Durant McArthur, E. M. Romney, S. D. Smith, and P. T. Tueller, compilers, Symposium on cheatgrass invasion, shrub die-off, and other aspects of shrub biology and management, pp. 81–83. U.S. For. Ser. Gen. Tech. Rep. INT-276.

Wallace, L. L. 1993. System stability and perturbation in Yellowstone's northern range: The effects of the 1988 drought on grassland communities. In D. G. Despain and R. H. Hamre, eds. Plants and their environments: First Biennial Scientific Conference on the Greater Yellowstone Ecosystem. National Park Serv. Tech. Rep., U.S. Gov. Printing Office, Denver.

Wallace, L. L., and A. T. Harrison. 1978. Carbohydrate

mobilization and movement in alpine plants. Am. J. Bot. 65:1035–40.

Walter, H. 1973. Vegetation of the earth. Springer-Verlag, New York.

Walter, H., E. Harnickell, and D. Mueller-Dombois. 1975. Climate diagram maps. Springer-Verlag, New York.

Walters, J. W., T. E. Hinds, D. W. Johnson, and J. Beatty. 1982. Effects of partial cutting on diseases, mortality, and regeneration of Rocky Mountain aspen stands. U.S. For. Ser. Res. Paper RM-240.

Walton, T. P. 1984. Reproductive mechanisms of plains silver sagebrush (*Artemisia cana cana*) in southeastern Montana. M.S. thesis, Montana State Univ., Bozeman.

Wamboldt, C. L. 1973. Conifer water potential as influenced by stand density and environmental factors. Can. J. Bot. 51:2333–37.

Ward, R. T. 1969. Ecotypic variation in *Deschampsia caespitosa* (L.) Beauv. from Colorado. Ecol. 50:519–22.

Wardle, P. 1965. A comparison of alpine timberlines in New Zealand and North America. New Zealand J. Bot. 3:113–35.

———. 1968. Engelmann spruce (*Picea engelmannii* Engel.) at its upper limits on the Front Range, Colorado. Ecol. 49:483–95.

———. 1971. An explanation for alpine timberline. New Zealand J. Bot. 9:371–402.

———. 1981. Winter desiccation of conifer needles simulated by artificial freezing. Arct. Alp. Res. 13: 419–23.

Waring, R. H., and G. B. Pitman. 1985. Modifying lodgepole pine stands to change susceptibility to mountain pine beetle attack. Ecol. 66:889–97.

Waring, R. H., and S. W. Running. 1978. Sapwood water storage: Its contribution to transpiration and effect upon water conductance through the stems of old-growth Douglas-fir. Plant Cell Environ. 1:131–40.

Waring, R. H., and W. H. Schlesinger. 1985. Forest ecosystems: Concepts and management. Academic Press, New York.

Waring, R. H., W. G. Thies, and D. Muscato. 1980. Stem growth per unit of leaf area: A measure of tree vigor. For. Sci. 26:112–17.

Washburn, A. L. 1988. Mima mounds, an evaluation of proposed origins with special reference to the Puget Lowland. Washington Div. Geol. Earth Res. Report of Investigations 29.

Watts, J. G., E. W. Huddleston, and J. C. Owens. 1982. Rangeland entomology. Ann. Rev. Ent. 27: 283–311.

Waugh, W. J. 1986. Verification, distribution, demography, and causality of *Juniperus osteosperma* encroachment at a Big Horn Basin, Wyoming, site. Ph.D. diss., Univ. Wyoming, Laramie.

Weakly, H. E. 1943. A tree-ring record of precipitation in western Nebraska. J. Forest. 41:816–19.

Weaver, J. E. 1943. Resurvey of grasses, forbs, and underground plant parts at the end of the great drought. Ecol. Monogr. 13:63–117.

———. 1954. North American prairie. Johnsen Publishing, Lincoln, Nebraska.

———. 1958. Summary and interpretation of underground development in natural grassland communities. Ecol. Monogr. 28:55–78.

———. 1960. Floodplain vegetation of the central Missouri valley and contacts of woodland with prairie. Ecol. Monogr. 30: 37–64.

———. 1965. Native vegetation of Nebraska. Univ. Nebraska Press, Lincoln.

———. 1968. Prairie plants and their environment. Univ. Nebraska Press, Lincoln.

Weaver, J. E., and F. W. Albertson. 1956. Grasslands of the Great Plains: Their nature and use. Johnson Publishing, Lincoln, Nebraska.

Weaver, J. E., and W. E. Brunner. 1954. Nature and place of transition from true prairie to mixed prairie. Ecol. 35:117–26.

Weaver, J. E., and F. E. Clements. 1938. Plant ecology. McGraw-Hill, New York.

Weaver, T. 1974. Ecological effects of weather modification: Effect of late snowmelt on *Festuca idahoensis* Elmer meadows. Am. Midl. Nat. 92:346–56.

———. 1978. Changes in soils along a vegetation-altitudinal gradient of the northern Rocky Mountains. In C. T. Youngberg, ed., Forest soils and land use, pp. 14–29. Fifth No. Am. Forest Soils Conf., Colorado State Univ., Fort Collins.

———. 1979. Climates of fescue grasslands of mountains in the western United States. Great Basin Nat. 39:283–88.

———. 1980. Climates of vegetation types of the northern Rocky Mountains and adjacent plains. Am. Midl. Nat. 103:392–98.

Weaver, T., and D. Dale. 1974. *Pinus albicaulis* in central Montana: Environment, vegetation and production. Am. Midl. Nat. 92:222–30.

Weaver, T., and F. Forcella. 1977. Biomass of fifty forests and nutrient exports associated with their harvest. Great Basin Nat. 37:395–401.

Weaver, T., and D. Klarich. 1977. Allelopathic effects of volatile substances from *Artemisia tridentata* Nutt. Am. Midl. Nat. 97:508–12.

Weaver, T., and D. Perry. 1978. Relationship of cover type to altitude, aspect, and substrate in the Bridger Range, Montana. Northw. Sci. 52:212–19.

Webb, J. J., Jr. 1941. The life history of buffalo grass. Kansas Acad. Sci. Trans. 44:58–75.

Webb., S. D. 1977. A history of savanna vertebrates in the new world. Pt. 1, North America. Ann. Rev. Ecol. Syst. 8:355–80.

Webb, W. L., W. K. Lauenroth, S. R. Szarek, and R. S. Kinerson. 1983. Primary production and abiotic controls in forests, grasslands, and desert ecosystem in the United States. Ecol. 64:134–51.

Webb, W. L., S. Szarek, W. Lauenroth, R. Kinerson, and M. Smith. 1978. Primary productivity and water use in native forest, grassland, and desert ecosystems. Ecol. 59:1239–47.

Webber, P. J., J. C. Emerick, D. C. Ebert May, and Vera Komarkova. 1976. The impact of increased snowfall on alpine vegetation. In H. W. Steinhoff and J. D. Ives, eds., Ecological impacts of snowpack augmentation on the San Juan Mountains, Colorado, pp. 201–64. Final Report, San Juan Ecology Project, Colorado State Univ. Publ., Fort Collins.

Weinstein, J. 1979. The condition and trend of aspen along Pacific Creek in Grand Teton National Park. In M. S. Boyce and L. D. Hayden-Wing, eds., North American elk: Ecology, behavior and management. Univ. Wyoming, Laramie.

Welch, B. L., F. J. Wagstaff, and J. A. Roberson. 1991. Preference of wintering sage grouse for big sagebrush. J. Range Mgmt. 44:462–65.

Weldon, L. W., D. W. Bohmont, and H. P. Alley. 1959. The interrelation of three environmental factors affecting germination of sagebrush seed. J. Range Mgmt. 12: 236–38.

Wells, P. V. 1965. Scarp woodlands, transported grassland soils, and concept of grassland climate in the Great Plains region. Science 148:246–49.

———. 1970a. Vegetational history of the Great Plains: A post-glacial record of coniferous woodland in southeastern Wyoming. In W. Dort, Jr., and J. K. Jones, Jr., eds., Pleistocene and recent environments of the central Great Plains, pp. 85–202. Dept. of Geology, Univ. Kansas, Special Publication 3. Univ. Press of Kansas, Lawrence.

———. 1970b. Postglacial vegetational history of the great plains. Science 167:1574–82.

Welsh, S. L. 1957. An ecological survey of the vegetation of the Dinosaur National Monument. M.S. thesis, Brigham Young Univ., Provo, Utah.

Wendtland, K. J., and J. L. Dodd. 1992. The fire history of Scotts Bluff National Monument. In D. D. Smith, and J. L. Dodd, eds., Recapturing a vanishing heritage, pp. 141–43. Twelfth No. Am. Prairie Conf.

Univ. Northern Iowa, Cedar Rapids, Iowa, 5–9 August 1990.

Wenger, L. E. 1943. Buffalo grass. Kansas Agric. Exp. Sta. Bull. 321:1–78.

Wentworth, E. N. 1948. America's sheep trails: History and personalities. Iowa State Univ. Press, Ames.

Wesche, T. A., C. M. Goertler, and C. B. Frye. 1985. Importance and evaluation of instream and riparian cover in smaller trout streams. In R. R. Johnson, C. D. Ziebell, D. R. Patton, P. F. Folliott, and R. H. Hamre, coords., Riparian ecosystems and their management: Reconciling conflicting uses. First North American Riparian Conference. U.S. For. Ser. Gen. Tech. Rep. RM-120, pp. 325–28.

West, N. E. 1983. Western intermountain sagebrush steppe. In N. E. West, ed., Ecosystems of the world. Vol. 5, Temperate deserts and semi-deserts. Elsevier, New York.

———. 1985. Aboveground litter production of three temperate semidesert shrubs. Am. Midl. Nat. 113: 158–69.

———. 1988. Intermountain deserts, shrub steppes, and woodlands. In M. B. Barbour and W. D. Billings, eds., North American terrestrial vegetation, pp. 209–30. Cambridge Univ. Press, Cambridge.

———. 1990. Structure and function of microphytic soil crusts in wildland ecosystems of arid and semi-arid regions. Adv. Ecol. Res. 20:179–223.

———. 1991. Nutrient cycling in soils of semiarid and arid regions. In J. Skujins, ed., Semiarid lands and deserts: Soil resource and reclamation, pp. 295–332. Marcel Dekker, New York.

———. 1993. Biodiversity of rangelands. J. Range Mgmt. 46:2–13.

West, N. E., and G. F. Gifford. 1976. Rainfall interception by cool-desert shrubs. J. Range Mgmt. 29:171–72.

West, N. E., and M. A. Hassan. 1985. Recovery of sagebrush-grass vegetation following wildfire. J. Range Mgmt. 38:131–34.

West, N. E., and J. O. Klemmedson. 1978. Structural distribution of nitrogen in desert ecosystems. In N. E. West and J. Skujins, eds., Nitrogen in desert ecosystems, pp. 1–16. Dowden, Hutchinson, and Ross, Stroudsburg, Pennsylvania.

West, N. E., K. H. Rea, and R. O. Harniss. 1979. Plant demographic studies in sagebrush-grass communities of southeastern Idaho. Ecology 60:376–88.

West, N. E., and J. Skujins. 1977. The nitrogen cycle in North American cold-winter semi-desert ecosystems. Oecologia Plant. 12:45–53.

———, eds. 1978. Nitrogen in desert ecosystems. Dowden, Hutchinson, and Ross, Stroudsburg, Pennsylvania.

Weynand, B., W. Behrends, D. Stout, and E. Raper. 1979. Bighorn River habitat management plan. Internal Report, Wyoming Game and Fish Dept., Cheyenne.

Wheeler, N. C., and R. P. Guries. 1982. Biogeography of lodgepole pine. Can. J. Bot. 60:1805–14.

Whicker, A. D., and J. K. Detling. 1988. Modification of vegetation structure and ecosystem processes by North American grassland mammals. In M. J. A. Werger, J. M. van der Aart, H. J. During, and J. T. A. Verhoeven, eds., Plant form and vegetation structure: Adaptation, plasticity and relation to herbivory, pp. 301–16. S.P.B. Academic Publishing, The Hague.

Whipple, S. A., and R. L. Dix. 1979. Age structure and successional dynamics of a Colorado subalpine forest. Am. Midl. Nat. 101:142–48.

Whisenant, S. G. 1990. Changing fire frequencies on Idaho's Snake River plains: Ecological and management implications. In E. D. McArthur, E. M. Romney, S. D. Smith, and P. T. Tueller, compilers, Symposium on cheatgrass invasion, shrub die-off, and other aspects of shrub biology and management, pp. 4–10. U.S. For. Ser. Gen. Tech. Rep. INT-276.

Whisenant, S. G., and D. W. Uresk. 1990. Spring burning Japanese brome in a western wheatgrass community. J. Range Mgmt. 43:205–8.

White, M. L. 1968. Ecology of *Amelanchier* in western Wyoming. Wyoming. M.S. thesis, Univ. Wyoming, Laramie.

White, R. S. 1976. Seasonal patterns of photosynthesis and respiration in *Atriplex confertifolia* and *Ceratoides lanata*. Ph.D. diss., Utah State Univ., Logan.

White, R. S., and P. O. Currie. 1983. The effects of prescribed burning on silver sagebrush. J. Range Mgmt. 36:611–13.

———. 1984. Phenological development and water relations in plains silver sagebrush. J. Range Mgmt. 37:503–7.

Whitford, W. G. 1988. Decomposition and nutrient cycling in disturbed arid ecosystems. In E. B. Allen, ed., The reconstruction of disturbed arid lands: An ecological approach, pp. 136–61. Am. Assn. Adv. Sci. Selected Symp 109. Westview Press, Boulder.

Whitlock, C. 1993. Postglacial vegetation and climate of Grand Teton and southern Yellowstone National Parks. Ecol. Monogr. 63:173–98.

Whitlock, C., S. C. Fritz, and D. R. Engstrom. 1991. A prehistoric perspective on the Northern Range. In R. B. Keiter and M. S. Boyce, eds., The Greater Yellowstone Ecosystem: Redefining America's wilderness heritage, pp. 289–305. Yale Univ. Press, New Haven.

Whitman, W. C., and H. C. Hanson. 1939. Vegetation on scoria and clay buttes in western North Dakota. Ecol. 20:455–57.

Whitman, W. C., H. T. Hanson, and G. Loder. 1943. Natural revegetation of abandoned fields in western North Dakota. North Dakota Agric. Exp. Sta. Bull. 321.

Wight, J. R. 1976. Range fertilization in the northern Great Plains. J. Range Mgmt. 29: 180–85.

Wight, J. R., and A. L. Black. 1979. Range fertilization: Plant response and water use. J. Range Mgmt. 32: 345–49.

Wight, J. R., and H. G. Fisser. 1968. *Juniperus osteosperma* in northwest Wyoming: Their distribution and ecology. Wyo. Agric. Exp. Sta. Sci. Monogr. 7. Univ. Wyoming, Laramie.

Wight, J. R., and E. B. Godfrey. 1985. Predicting yield response to nitrogen fertilization on Northern Great Plains rangeland. J. Range Mgmt. 38:238–41.

Wight, J. R., and J. T. Nichols. 1966. Effects of harvester ants on production of a saltbush community. J. Range Mgmt. 19:68–71.

Willard, B. E. 1979. Plant sociology of alpine tundra, Trail Ridge, Rocky Mountain National Park, Colorado. Colorado School of Mines Q. 74:vi, 119.

Willard, B. E., and J. W. Marr. 1971. Recovery of alpine tundra under protection after damage by human activities in the Rocky Mountains of Colorado. Biol. Conserv. 3:181–90.

Williams, B. D., and R. S. Johnston. 1984. Natural establishment of aspen from seed on a phosphate mine dump. J. Range Mgmt. 37:521–22.

Williams, C. S. 1961. Distribution of vegetation in the Wind River Canyon, Wyoming. M.S. thesis, Univ. Wyoming, Laramie.

———. 1963. Ecology of bluebunch wheatgrass in northwestern Wyoming. Ph.D. diss., Univ. Wyoming, Laramie.

Williams, D. E. 1976. Growth, production, and browse utilization characteristics of serviceberry *(Amelanchier alnifolia* Nutt.) in the badlands of southwestern North Dakota. M.S. thesis, North Dakota State Univ., Fargo.

Williamson, S. C., J. K. Detling, J. L. Dodd, and M. I. Dyer. 1989. Experimental evaluation of the grazing optimization hypothesis. J. Range Mgmt. 42:149–52.

Wilson, A. M. 1984. Leaf area, nonstructural carbohydrates, and root growth characteristics of blue grama seedlings. J. Range Mgmt. 37:514–17.

Wilson, H. C. 1969. Ecology and successional patterns of wet meadows, Rocky Mountain National Park, Colorado. Ph.D. diss., Univ. Utah, Salt Lake City.

Wilson, M. 1974. History of the bison in Wyoming with particular reference to early Holocene forms. In M. Wilson, ed., Applied geology and archaeology: The Holocene history of Wyoming, pp. 91–99. Report of Investigations 10, Wyo. Geol. Surv., Laramie.

Windell, J. T., B. E. Willard, D. J. Cooper, S. Q. Foster, C. F. Knud-Hansen, L. P. Rink, and G. N. Kiladis. 1986. An ecological characterization of Rocky Mountain montane and subalpine wetlands. U.S. Fish Wildl. Ser. Biol. Rep. 86:1–298.

Wing, S. L. 1981. A study of paleoecology and paleobotany in the Willwood Formation (Early Eocene, Wyoming). Ph.D. diss., Yale Univ., New Haven.

Winkel, V. 1986. Habitat differences between mountain big sagebrush and mountain silver sagebrush in the Strawberry Valley of central Utah. M.S. thesis, Brigham Young Univ., Provo.

Winward, A. H. 1970. Taxonomy and ecology of big sagebrush. Ph.D. diss., Univ. Idaho, Moscow.

Wolfe, C. W. 1973. Effects of fire on a sand hills grassland environment. Tall Timbers Fire Ecol. Conf. 12: 241–55.

Wolfe, J. A. 1978. A paleobotanical interpretation of Tertiary climates in the northern hemisphere. Am. Scientist 66: 694–703.

Wolff, S. W., T. A. Wesche, and W. A. Hubert. 1989. Stream channel and habitat changes due to flow augmentation. Regulated Rivers: Res. Mgmt. 4:225–33.

Wood, B. W. 1966. An ecological life history of budsage in western Utah. M.S. thesis, Brigham Young Univ, Provo, Utah.

Wood, G. W. 1988. Effects of prescribed fire on deer forage and nutrients. Wildl. Soc. Bull. 16:180–86.

Wood, M. K., W. H. Blackburn, R. E. Eckert, Jr., and F. F. Peterson. 1978. Interrelations of the physical properties of coppice dune and vesicular dune interspace soils with grass seedling emergence. J. Range Mgmt. 31:189–92.

Wood, M. K., R. W. Knight, and J. A. Young. 1976. Spiny hopsage germination. J. Range Mgmt. 29:53–56.

Woodmansee, R. G. 1978. Additions and losses of nitrogen in grassland ecosystems. BioScience 28:448–53.

Woodmansee, R. G., J. L. Dodd, R. A. Bowman, F. E. Clark, and C. E. Dickinson. 1978. Nitrogen budget of a shortgrass prairie ecosystem. Oecologia 34:363–76.

Woodmansee, R. G., I. Vallis, and J. J. Mott. 1981. Grassland nitrogen. In F. E. Clark and T. Rosswall, eds., Terrestrial nitrogen cycles, pp. 443–62. Ecological Bulletin 33, Stockholm.

Woods, R. F., D. R. Betters, and E. W. Mogren. 1982. Understory herbage production as a function of Rocky Mountain aspen stand density. J. Range Mgmt. 35: 380–81.

Workman, J. P., and N. E. West. 1967. Germination of *Eurotia lanata* in relation to temperature and salinity. Ecol. 48:659–61.

Wright, C. W. 1952. An ecological description of an isolated piñon pine grove. M.S. thesis, Univ. Colorado, Boulder.

Wright, H. A. 1971. Why squirreltail is more tolerant to burning than needle-and-thread. J. Range Mgmt. 24:277–84.

Wright, H. A., and A. W. Bailey. 1980. Fire ecology and prescribed burning in the Great Plains: A research review. U.S. For. Ser. Gen. Tech. Rep. INT-77.

———. 1982. Fire ecology. Wiley, New York.

Wright, H. E., Jr. 1970. Vegetational history of the Great Plains. In W. Dort, Jr., and J. K. Jones, Jr., eds., Pleistocene and recent environments of the central Great Plains, pp. 157–72. Univ. Kansas Press, Lawrence.

———. 1974. Landscape development, forest fires, and wilderness management. Science 186:487–95.

Wright, H., L. Neuenschwander, and C. Britton. 1979. The role and use of fire in sagebrush-grass and pinyon-juniper plant communities: A state-of-the-art review. U.S. For. Serv. Gen. Tech. Rep. INT-58.

Wright, J. C., and E. A. Wright. 1948. Grassland types of south central Montana. Ecol. 29:449–60.

Wu, Y. 1991. Fire history and potential fire behavior in a Rocky Mountain foothill landscape. Ph.D. diss., Univ. Wyoming, Laramie.

Wullstein, L. H. 1980. Nitrogen fixation (acetylene reduction) associated with rhizosheath of Indian ricegrass used in stabilization of the Slick Rock, Colorado, tailings pile. J. Range Mgmt. 33:204–6.

Wullstein, L. H., M. L. Bruening, and W. B. Bollen. 1979. Nitrogen fixation associated with rhizosheaths of certain perennial grasses. Physiol. Plantarum 46: 1–4.

Wyant, J. G. 1986. Fire induced tree mortality in a Colorado ponderosa pine/Douglas-fir stand. For. Sci. 32: 49–59.

Yake, S., and J. D. Brotherson. 1979. Differentiation of serviceberry habitats in the Wasatch Mountains of Utah. J. Range Mgmt. 32:379–83.

Yavitt, J. B., and T. J. Fahey. 1982. Loss of mass and nutrient changes of decaying woody roots in lodgepole pine forests, southeastern Wyoming. Can. J. For. Res. 12:745–52.

Yorks, T. P., N. E. West, and K. M. Capels. 1992. Vegetation differences in desert shrublands of western

Utah's Pine Valley between 1933 and 1989. J. Range Mgmt. 45:569–78.

Young, D. L., and J. A. Bailey. 1975. Effect of fire and mechanical treatment on *Cercocarpus montanus* and *Ribes cereum*. J. Range Mgmt. 28:495–97.

Young, J. A. 1978. Mormon crickets. Rangeman's J. 5:193–96.

Young, J. A., and R. A. Evans. 1973. Downy brome: Intruder in the plant succession of big sagebrush communities in the Great Basin. J. Range Mgmt. 26:410–15.

———. 1974. Population dynamics of green rabbitbrush in disturbed big sagebrush communities. J. Range Mgmt. 27:127–32.

———. 1975. Germinability of seed reserves in a big sagebrush community. Weed Sci. 23:358–64.

———. 1978. Population dynamics after wildfires in sagebrush grasslands. J. Range Mgmt. 31:283–89.

———. 1981. Demography and fire history of a western juniper stand. J. Range Mgmt. 34:501–5.

———. 1983. Seed physiology of antelope bitterbrush and related species. In A. R. Tiedemann and K. L. Johnson, eds., Research and management of bitterbrush and cliffrose in western North America, pp. 70–80. U.S. For. Ser. Gen. Tech. Rep. INT-152.

Young, J. A., R. A. Evans, and R. E. Eckert. 1969a. Emergence of medusahead and other grasses from four seedling depths. Weed Sci. 17:376–79.

———. 1969b. Population dynamics of downy brome. Weed Sci. 17:20–26.

Young, J. A., R. A. Evans, B. A. Roundy, and G. J. Cluff. 1984. Ecology of seed germination in representative Chenopodiaceae. In A. R. Tiedemann et al., eds., Symposium on the biology of *Atriplex* and related chenopods, pp. 159–65. U.S. For. Ser. Gen. Tech. Rep. INT-172.

Young, J. A., R. A. Evans, and P. T. Tueller. 1976. Great Basin plant communities, pristine and grazed. In R. Elston, ed., Holocene environmental change in the Great Basin, pp. 187–215. Nevada Archeol. Surv. Res. Paper 6.

Young, J. A., R. A. Evans, and R. A. Weaver. 1976. Estimating potential downy brome competition after wildfires. J. Range Mgmt. 29:322–25.

Young, R. P. 1983. Fire as a vegetation management tool in rangelands of the intermountain region. In S. B. Monsen and N. Shaw, compilers, Managing intermountain rangelands: Improvements of range and wildlife habitats, pp. 18–31. U.S. For. Ser. Gen. Tech. Rep. INT-15.

Youngberg, C. T., and L. Hu. 1972. Root nodules on mountain mahogany. For. Sci. 18:211–12.

Youngblood, A. P., and W. F. Mueggler. 1981. Aspen

community types on the Bridger-Teton National Forest in western Wyoming. U.S. For. Ser. Gen. Tech. Rep. INT-272.

Youngblood, A. P., W. G. Padgett, and A. H. Winward. 1985. Riparian community type classification of eastern Idaho–western Wyoming. U.S. For. Serv., Interm. Region R4-Ecol-85–01.

Younkin, W. E., Jr. 1970. A study of the vegetation of alpine rock outcrops in northern Colorado. M.S. thesis, Colorado State Univ., Fort Collins.

Zamora, B., and P. T. Tueller. 1973. *Artemisia arbuscula, A. longiloba* and *A. nova* habitat types in northern Nevada. Great Basin Nat. 33:225–41.

Zavitkovski, J., and M. Newton. 1968. Ecological importance of snowbrush *(Ceanothus velutinus)* in the Oregon Cascades. Ecol. 49:1132–44.

Zwingle, E. 1993. Ogallala aquifer: Wellspring of the high plains. National Geographic (March): 81–109.

Index

Adiabatic lapse rate, 30, 153
Agriculture, 3–4, 67, 257–58. *See also* Livestock
Alder, thinleaf, 47, 50
Alpine tundra, 8, 15, 203, 204–10, 216, 238
Altithermal, 18
Antelope. *See* Pronghorn antelope
Antelope bitterbrush, 102, 121
Ants. *See* Harvester ants
Ash, green, 50, 145, 248
Aspen: in Black Hills, 247, 248–49, 252; effects of browsing and fire on, 61, 147–50, 222–23, 226, 228, 229, 240; effects on streamflow of, 182; foothill woodlands with, 137, 143, 144, 145; forests of, 8, 156, 159, 171–73, 182; in GTNP, 236, 240; photosynthesis in bark of, 173; plants associated with, 136–37, 161–62; in relation to snow, 146, 198–99; in YNP, 218, 222–23, 226, 228, 229, 231
Avalanches, snow, 210

Badgers, 39, 68
Badlands, 34, 123–25
Bank storage, riparian, 56, 60
Bark beetles. *See* Douglas-fir beetle; Mountain pine beetle; Spruce beetle
Basins, map of, 4, 5

Bears: food preferences of, 223; grizzly, in Black Hills, 242; historic abundance of, 20; in YNP, 223–24
Bear Lodge Mountains, 242–53
Beaver, 19, 48, 53, 56–59, 253, 278n.9
Belle Fourche River, 3
Bentonite, 4, 34–35
Big sagebrush. *See* Sagebrush
Bighorn Canyon National Recreation Area, 51
Bighorn River, 50, 51, 52–55
Bighorn sheep, 19
Birch, paper, in Black Hills, 244, 247
Birdfoot sagewort, 92
Bison: association with prairie dogs, 81–82; brucellosis in, 223, 241; evolution of, 277n.1; food habits and grazing by, 73, 81, 86, 106–7; historic abundance of, 19, 20, 60
Bitterbrush, antelope, 102, 121, 142, 143, 144, 238
Black Hills: flora of the, 243–44, 245; geologic history, 242; geomorphic regions, 243, 244–46; human history, 242–43; management issues, 250, 253; vegetation of, 246–50
Blackfelt snowmolds, 198
Blister rust on whitebark pine, 224, 280n.2

Blue grama, 38, 67, 72, 113, 116
Blue spruce, 47, 50, 234, 278n.8
Bluebunch wheatgrass, 67, 103, 105, 116, 133, 142, 144, 193, 219
Bogs, 205, 205–6
Boxelder, 50, 145, 244, 248
Bristlecone pine, 159
Brucellosis. *See* Bison; Elk
Bud sagewort, 94, 110, 113, 115
Buffalo. *See* Bison
Buffalo grass, 67, 278n.1
Buffaloberry, silver, 50, 52
Bur oak, 145, 146, 147, 246, 247, 248
Burrowing animals, 39, 68, 87. *See also* Harvester ants; Pocket gophers; Prairie dogs

C_3 plants, 72–73, 83–84
C_4 plants, 72–73, 73
Cactus, pricklypear: adaptations of, 71; effect of burning on, 85; effect of grazing on, 85–86; protection of other plants by, 85
Caliche, 34
Cattail marshes, 52
Cattle: economic importance of, 3–5, 75; food preferences of, 73; in GTNP, 240, 241; introduction of, 20–22, 86; numbers in Wyoming in *1992*, 277n.7. *See also* Grazing

Chain-of-Lakes, 127

Cheatgrass: in desert shrubland, 117, 118; in foothills, 144; in sagebrush steppe, 103–5; seed in soil, 279n.8

Chokecherry, 50, 145, 247, 248

Cinnabar Park, 194, 197, 198

Cinquefoil, shrubby, 47, 146, 193, 219

Cirques, 15

Clark's nutcracker, pine seed dispersal by, 150–51, 203

Climate, Wyoming: change in, 11–19, 30; diagrams of, 28–29, 31; general description of, 23–33; limits to economic development, 258

Coal, 4

Cold air drainage, 30–31, 44

Colorado River, 44, 62

Comandra blister rust on lodgepole pine, 167, 175–76

Compensatory growth hypothesis, 79–82, 222

Continental divide, location of, 44

Continental drift, 11

Coppice dunes, 93, 115

Cottonwood: leaves of, as cattle forage, 278n.11; narrowleaf, 47, 50, 218, 234, 244, 125; plains, 50–55, 61, 248. See also Riparian woodlands

Cryoturbation, 202, 203, 206, 208–10

Cryptogamic crusts. See Microphytic crusts

Cushion plants: alpine, 205; lowland, 144–45

Dams. See Reservoirs

Debris jam, 43, 48, 57–59

Decomposition in forests, 179, 184–85, 186

Deer: food preferences of, 73; mule, 19, 241; whitetail, 19

Desert pavement, 94, 115

Desert shrubland: animals of, 117–18; in badlands, 125; hydrology of, 114, 117; NPP of, 117; plants of, 108, 111, 113, 116; shrub mortality in, 118; soils of, 108, 110, 112, 113–14, 116; vegetation mosaic in, 53, 108–13

Detritus in grasslands, 75, 77

Devil's Tower National Monument, 8, 144, 242, 244

Dinosaurs in Wyoming, 10–11

Dogwood, red-osier, 50, 248

Douglas-fir: in Fossil Butte National Monument, 125, 126; in GTNP, 52, 234, 236–37, 240; plants associated with, 161–62; prehistoric distribution of, 17–18, 216; stomatal closure of, 169; woodlands of, 8, 137, 140–41, 143, 156, 159, 163–65; in YNP, 217, 218, 225, 228, 229

Douglas-fir beetle, 179

Drainage basins in Wyoming, 44

Drought years: effect on grassland NPP, 84; frequency of, 277n.1; in grassland, 84–85; interaction with livestock grazing and grasshoppers, 84–86; recovery from, 85

Dunes. See Coppice dunes; Sand dunes

Ecology, science of, 5–9

Ecosystem concept, 6–9

Ecosystem management, 262–64

Elevation: effect on plant distribution, 23–26, 94, 135, 156, 238; effect on temperature, 30–33, 153; environmental effects of, 23–26, 31

Elk: brucellosis in, 241; food preferences of, 73; hunting of, in GTNP, 241; in lowlands, 19, 20, 277n.1; winter feeding of, 235; in YNP, 220–23, 224, 225, 226. See also Aspen

Elk Refuge, National, 235

Elm, American, 50, 145, 248

Energy flow in grasslands: NPP, 73–76; photosynthesis, 73, 76; solar radiation, 73–74, 76. See also Grasshoppers; Grazing

Engelmann spruce: in mountain forests, 159, 169–70; prehistoric abundance, 18, 216; in riparian woodlands, 47; at treeline, 201, 202, 203, 210. See also Spruce-fir forest

Erosion: effect on aquatic ecosystems, 189; after fire, 186–87,

189; after timber harvesting, 189

Escarpments, 25, 67, 133–45

Eutrophication, 61, 62

Evaporation, from reservoirs, 62

Evapotranspiration: from desert shrublands, 117; general, 30–31; from grasslands, 74; from irrigated croplands, 62; map of potential ET in Wyoming, 33; from mountain forests, 178, 181, 182; from sagebrush steppe, 99, 102. See also Transpiration

Fairy rings, 68, 69

Fellfield, 205, 206

Fertilizing rangelands, 78–79

Fir, subalpine, 170, 201, 202, 203, 210. See also Spruce-fir forest

Fir, white, 147, 281n.7

Fire: at alpine treeline, 203; in aspen woodlands, 148–49; in the Black Hills, 247, 250, 251–53; as a decomposition process, 187; in desert shrublands, 118; effects on animals, 187, 188; effects on aquatic ecosystems, 187; effects of grazing on, 38, 253; as an environmental factor, 13, 19–22, 37–39; in foothill shrublands, 141–44, 147; in grasslands, 20, 69, 75, 77, 83–84, 140, 144–45; in GTNP, 237, 240; interaction with insects, 186–88; recovery after, 191; return intervals of, 38; in riparian zone, 52, 53, 55; in sand dunes, 122; in YNP, 218, 225–31. See also Erosion; Fire suppression; Lightning; specific forest types

Fire suppression, effects of, 189–90; 247, 251–53

Fish habitat, 60–61

Flaming Gorge Reservoir, 61

Floods: control of, 53, 59–60; effects of, on riparian mosaic, 43, 53–55

Foothill shrublands, vegetation mosaic of, 133–34, 135, 250

Forest, coniferous: animal adaptations for, 159; animals found in, 163; biomass distribution in, 178–79; classification of, 160, 218, 236, 246–47; detritus as a food source in, 178–80, 188; effect of canopy gaps, 174; effects of insects on, 174–78, 179, 186–88; effects of parasites on, 175–78; evapotranspiration from, 178; fragmentation of, 5, 22, 191, 262; leaf area in, 181–82; location of, 8, 48, 49, 156; nutrient cycling in, 174, 179, 184–85, 186–91; old-growth, 191, 252, 261; plant adaptations for, 157–59; plants found in, 161–62; prehistoric distribution of, 16–18; regrowth after cutting, 20–21; soils of, 155–56; tree growth rates in, 179; understory density of, 174, 187; variation in, 156, 159–60; water outflow from, 180–84, 199. *See also* Aspen; Douglas-fir; Fire; Lodgepole pine; Ponderosa pine; Spruce-fir forests

Forest canopy, gaps in, 174

Forest floor, effect on nutrients, 184–85

Fossil Butte National Monument, 12, 13, 125, 126

Fossil fish, 12, 13

Fossil fuels, 4, 257, 259

Fossil ice wedges, 128

Fossil wood, 10–12, 216

Fragmentation of forests, 5, 22, 191, 262

Fungi: blackfelt snowmold, 198; bluestain, 177; in grasslands, 73; in forests, 179–80; role in nutrient cycling, 185. *See also* Mycorrhizae

Gambel oak, 27, 133, 145, 146–47

Gannett Peak, 3

Gardner saltbush, 38, 90, 92

Genetic variation in trees: aspen, 159, 172; conifers, 159

Geyser basins, 219, 220

Glaciation in Wyoming, 14, 15, 16, 17–18, 216, 233–35, 277*n.3*

Grand Teton National Park (GTNP): cattle in, 240, 241; elk

management in, 240–41; fire management in, 240; geology of, 233–35; history of, 235–36, 241; vegetation of, 236–39

Grasshoppers, 20, 39, 85–86, 102

Grasslands: animals of, 68, 71; in the Black Hills, 249; disturbances in, 82–89; fire in, 82–84; foothill, 67; herbivory in, 68–69, 70, 71–72, 75, 76, 77–78, 79–82; human food from, 75; map of, 8; nutrient cycling in, 75–79; origin of, 13; plant adaptations of, 68–73; rooting depth of plants in, 72; soils of, 67; water infiltration in, 77, 84, 86. *See also* Fire; Grasshoppers; Grazing; Mixed-grass prairie; Short-grass prairie

Grazing: in alpine tundra, 205; in desert shrublands, 118–19; effects on flammability, 77, 83–84, 104; effects on microphytic crusts, 106–7; effects on NPP, 77, 79–82; effects on nutrient cycling, 77–78; effects on plants, 75, 77, 79–82, 87; in GTNP, 240, 241; historic effects of, 20–22, 277*n.6;* in mountain meadows, 199–200; in riparian zones, 248, 250; in sagebrush steppe, 100, 103, 104–7

Greasewood: evapotranspiration from, 117; on hillsides, 125, 143; prehistoric occurrence of, 12, 18; in riparian shrublands, 55, 112; shrublands dominated by, 110–12; soils associated with, 35, 51, 58, 90, 110–11

Great Divide Basin, 25, 44

Great Plains, 25

Gros Ventre landslide, 239–40

Groundwater, 260

Ground squirrels, 68, 88

Growing season length, 28–31

Gullies: effect of beaver on, 58–59; historic, 20, 59

Habitat types, forest: in the Black Hills, 246–47; in GTNP, 236; in Wyoming, 160; in YNP, 218

Hackberry, 244, 248

Halogeton, 118

Halophytes, 35, 49, 51, 219

Hardpan, 34

Hardwood draws, 145, 248

Harvester ants, 39, 68, 88, 89, 118

Herbivory. *See* Grazing

Holocene epoch, 17–22

Hopsage. *See* Spiny hopsage

Horses, 17, 277*n.7*

Huckleberry, dwarf, 153

Hypsodont dentition, 72

Ice wedges, fossil, 18, 128

Idaho fescue, 48, 49, 193, 194, 196, 197, 219

Indians, Native American, 227, 257, 258

Infiltration. *See* Inverse texture effect; Soils; *names of specific vegetation types*

Insects: effects on forest ecosystems, 156, 171, 174–78, 179, 186–88; in grasslands, 39, 82. *See also* Douglas-fir beetle; Grasshoppers; Mormon cricket; Mountain pine beetle; Spruce beetle; Spruce budworm

Inverse texture effect, 25, 122

Irrigation, 3, 62–63, 258

Jackrabbits in desert shrublands, 117–18

Jackson Hole, 233–41

Jackson Lake Dam, 235, 241

Jenny Lake, 16

Juniper woodlands, 8, 26, 135–40, 250

Kentucky bluegrass, 51

Kettle moraines, 235

Keystone, 20, 21

Killpecker Sand Dunes, 120, 121, 122, 123, 277*n.1*

Krummholz, 201, 202

Lakes, prehistoric, 12, 16

Layering, 170, 203

Leaf area of forests, effects on forest hydrology, 181–82, 186, 190

Lichens. See *Xanthoparmelia chlorochroa*

Lightning-caused fires, 37–38, 83, 252

Limber pine: at alpine treeline, 201; association with bedrock fissures, 26, 35; in Black Hills, 248; in foothill woodlands, 133, 140–41, 142; in GTNP, 237; plants associated with, 161–62; seed dispersal by birds, 150–51, 203

Little Bighorn National Monument, 105

Livestock: economic effects of, 3–4, 257; effects on aspen, 148; historic effects on rangeland, 20–21, 86, 87; interactions with prairie dogs, 88; in mountain meadows, 199–200; numbers in Wyoming in *1992*, *277n.7;* in ponderosa pine, 160–61; recovery from heavy grazing, 86; in riparian zones, 60–61. *See also* Cattle; Grazing; Sheep

Lodgepole pine: leaf duration of, 18; predation by squirrels on seeds of, 167, 174, 175; rooting depth of, 182; serotinous cones of, 165–68; snowmolds on, 198; stomatal response to water stress, 168–69. *See also* Lodgepole pine forest

Lodgepole pine forest, 156, 159, 165–68, 169; in the Black Hills, 248; in GTNP, 52, 236, 237, 239; hydrology of, 180–84; map of, 8; NPP of, 178–80, 188; plants of, 161–62; recovery after harvesting, 20, 21; in YNP, 217, 218, 220, 224, 225;

Mammoth Hot Springs, 219
Maple, canyon, 147
Maple, Rocky Mountain, 210
Meadows, alpine, plants of, 207
Meadows, mountain: in Black Hills, 250, 253; characteristics of, 193–200; grazing in, 199–200; tree invasion of, 194–96, 198, 200, 253; in YNP, 219
Microphytic crusts, 101, 106, 107
Mima mounds, 125–30
Mineral resources, 4, 21, 258, 259

Mining, 4, 21
Mistletoe, dwarf, 167, 175, 247
Mites in grassland soils, 70, 73
Moose, 19, 241
Mormon crickets, 20, 39
Mountain-mahogany shrublands, 35, 134–35, 250
Mountain pine beetle, 39, 188, 237; in Black Hills, 247; in GTNP, 237; in YNP, 218
Mountains: characteristics of, 154–55; effect on climate, 153, 155; landscapes of, 25–26; map of, 4, 5; uplifting of, 11, 12, 14
Mud flats around reservoirs, 62
Mud springs, 125, 127
Mulch. *See* Detritus
Multiple-use on federal lands, 260–62
Mushrooms, 179
Mycorrhizae, 70, 123, 158, 179–80

National Elk Refuge, 235
National forests, 6, 260–64
National grasslands, 3, 6
National parks: map of, 6. *See also names of specific parks*
National recreation areas, 6
Native American Indians, 257, 258
Natural areas, 257, 263
Natural gas, 4
Natural regulation policy of YNP, 220–23, 231
Nematodes in grassland soils, 70, 73, 82, 278n.6
Net primary productivity (NPP). *See names of specific vegetation types*
Nitrogen: in forests, 184–85, 191; in grasslands, 77–79, 84. *See also* Nitrogen fixation
Nitrogen fixation: by antelope bitterbrush, 143; by ceanothus, 143; by microphytic crusts, 117; by mountain-mahogany, 135; in sand dunes, 123; by skunkbush sumac, 143
Nivation hollows, 30, 97
Northern Range of YNP, 217, 221, 228, 229
North Platte River, 50
Nutrients: effects on aquatic ecosystems, 186–87; effects of dis-

turbances on, 186–88, 189, 191; leaching from mountain forests, 31, 190; in mountain forests, 174, 179, 184–85, 186; reabsorbtion by leaves, 158, 185

Oak woodlands, 146–47; with ponderosa pine, 247, 248. *See also* Bur oak; Gambel oak
Ogallala aquifer, 260
Oil, 4
Old-growth forest, 191, 252, 261
Oregon Trail, 20–21, 258, 277n.6
Overthrust belt, 10

Packrats. *See* Woodrats
Pangaea, 11
Parks. *See* Meadows
Parmelia. See *Xanthoparmelia*
Permafrost, 17
Phreatophytes, 56
Physiographic regions, 4, 5, 25–26
Pine. *See* Bristlecone pine; Limber pine; Lodgepole pine; Pinyon pine; Ponderosa pine; Whitebark pine
Pinyon pine, 133, 140, 150–51
Playas, 110
Pleistocene epoch, 13–18, 216
Pocket gophers, 39, 81–82, 87, 198, 208
Point bars, 48, 52, 53
Pollen diagram, 17
Ponderosa pine, 19, 35, 67, 133, 159, 164, 169; in Black Hills, 244, 245, 246–47, 248, 249, 251–53; effects of fire suppression on, 190; invading grasslands, 251–53; location of, 8, 68, 156; woodlands of, 140–41, 160–63
Population, human, 257, 259
Porcupines, 281n.9
Potentilla, shrubby. *See* Cinquefoil
Prairie: animals of, 68; in Black Hills, 244, 249, 250; map of, 8; mixed-grass, 67; plants of, 67–68; shortgrass, 67; tallgrass, 67–68. *See also* Grasslands
Prairie dogs: effect on bison, 81, 243; effect on plants, 87–88, 279n.18; effect on soils, 87–88;

feeding behavior, 81; interaction with livestock, 88. *See also* Burrowing animals

Precipitation: effect of winter, 30; map of annual, 27; seasonality of, 28–31. *See also* Snowdrifting

Predator control in YNP, 220

Pronghorn antelope, 10, 17, 19, 73, 81, 277*n.1*

Quaternary period: mammals of, 17; vegetation of, 13–22, 216

Railroad ties, cutting of, 21–22

Rainfall. *See* Precipitation

Rainshadow, 13, 27

Reservoirs, 44, 53, 59–60, 61–62, 241, 260

Return flows from irrigation, 62–63

Ribbon forest, 159, 197, 198–99, 201

Riparian marshes, 52

Riparian meadows, 45–46, 47, 51, 216, 219

Riparian shrublands, 44–55, 219

Riparian woodlands: change in, 22, 278*n.8;* in GTNP, 236; in YNP, 218, 220. *See also* Bighorn River; Cottonwood

Riparian zone: animals found in, 47; dependence of animals on, 43; grazing in, 50, 60–61; plants found in, 45–46; succession in, 54–55; vegetation mosaic in, 43, 47–55. *See also* Bighorn River

Russian olive, 52, 278*n.5*

Russian thistle, 118

Sagebrush, alkali, 91, 95

Sagebrush, basin big, 50, 52, 90, 91, 94, 95, 114. *See also* Sagebrush; Wyoming big

Sagebrush, black, 91, 94, 95

Sagebrush, low, 91, 95, 237, 238

Sagebrush, mountain big, 48, 49, 52, 91, 94, 95, 102, 141, 142, 143, 144, 194. *See also* Sagebrush steppe; Sagebrush, Wyoming big

Sagebrush, sand, 67, 250

Sagebrush, silver: in Black Hills, 250; habitat of, 50, 67, 90, 94, 96, 145, 146, 193; in sand dunes, 121; subspecies of, 91, 95; susceptibility to drought and fire, 102, 103; in YNP, 219

Sagebrush, threetip, 94, 137, 144

Sagebrush, Wyoming big: adaptations of, 94, 96–99; allelopathic effect of, 279*n.5;* association with snow, 90, 92, 99; ephemeral leaves of, 97; habitat of, 50, 51, 67, 90, 91, 92, 93–96, 121; seedling establishment of, 90, 98–99; soil preferences of, 35, 38. *See also* Desert shrubland; Sagebrush steppe

Sagebrush steppe: animals of, 90; climate of, 90, 94, 100; effects of drought in, 102, 106; effects of fire on, 39, 100, 101–4; effects of grazing on, 105–7; effects of insects, 102; effects of spring frost, 102; in GTNP, 234, 236, 237, 238–39, 240; habitat of, 13, 18, 20, 22, 26, 35, 90–107, 114; hydrology of, 99–100, 102; invasion by lodgepole pine, 236; islands of fertility in, 93, 94, 98, 101; map of, 8; NPP of, 99–101, 102; nutrient cycling in, 101; succession after plowing, 103; in YNP, 217, 218, 221, 229. *See also* Sagebrush, black; Sagebrush, mountain big; Sagebrush, Wyoming big

Sagewort, birdfoot, 92, 94

Sagewort, bud, 94, 110, 113, 115

Salicornia, 112, 113

Salinity, plant adaptations to, 108–10

Salinization by irrigation, 62–63, 258

Saltbush desert shrublands, 115–16, 125. *See also* Gardner saltbush; Shadscale

Saltcedar, 52, 55, 62, 278*n.6*

Saltgrass, inland, 51, 110, 112, 113

Sand dunes, 8, 31–33, 34, 120–23

Sand wedge relics, 128

Scoria, 67

Sedimentation: behind beaver dams, 56–59; behind reservoirs, 61

Selenium: effects of land management on, 115; plant indicators of, 115

Serotinous cones. *See* Lodgepole pine

Serviceberry, 142, 143, 144, 210

Shadscale, 97, 102, 110, 113, 118, 119

Sheep, 3–5, 21, 73

Shoshone National Forest, 3

Shoshone River, 51, 54

Silt dunes. *See* Coppice dunes

Skunkbush sumac, 50, 52, 53, 103, 125

Smooth brome. *See* Cheatgrass

Snake River, 44, 52, 241, 278*n.8*

Snow: avalanches, 210; plant growth under, 153; water equivalent of, 180. *See also* Snowdrifting

Snowberry, western, 103

Snowdrifting: in alpine tundra, 133, 142, 180, 203, 204, 206; effect of forest openings on, 153, 155, 183–84, 198; effect on soil water, 30–31, 153, 155; effect on tree seedling establishment, 198; in grasslands, 74–75; in ribbon forests, 198–99

Snowglades, 30, 146, 159, 197, 198–99

Snowmolds: on sagebrush, 102; blackfelt, on lodgepole pine, 198, 203

Soil: aeration, 35; alkalinity, 34–35; development, 16, 18; effect of bedrock on, 34; in relation to elevation, 34; fertility, 34; infiltration rate, 34–35; salinity, 34–35; sodicity, 34–35; texture, effect on plant distribution, 34–35, 38; types in Wyoming, table of, 36–37; water-holding capacity, 34–35; variation in, 33–37

Solifluction, 208, 209

South Pass, 20

Spiny hopsage, 94, 110, 113, 121

Spruce. *See* Blue spruce; Engelmann spruce; White spruce

Spruce beetles, 39, 177, 179

Spruce budworm, western, 39, 164, 176

Spruce-fir forest: effect of snow-drifting on, 199; in GTNP, 236, 237, 239; habitat of, 156, 168–71; hydrology of, 182; NPP of, 178–80; plants of, 161–62; in YNP, 217, 218

Squirrel, red, 167, 174, 175

Stellar's jay, 203

Stone nets, alpine, 208, 209

Stream channel shifting, 48, 49, 53

Streambank: erosion of, 48, 49, 52, 60; water storage, 56, 60. *See also* Gullies

Streamflow: effect of forest disturbances on, 180, 183–84, 186–87, 190; effect of increased tree density on, 253; effect of irrigation on, 62–63; from mountain forests, 180–84, 198

Succession. *See names of specific vegetation types*

Sumac, skunkbush, 140, 142, 145, 250

Sustainable land management, 257–65

Sweetwater River, 50, 54

Temperature change: daily in tundra, 33; with elevation, 30–33, 153; with season, 28–33

Teton Range: geology of, 12, 233; glaciation in, 233–34; vegetation of, 16

Ties, railroad, 21–22, 261

Timber atolls, 199

Timber harvesting: in Black Hills, 247, 250, 252; clearcut, 186; compared to fire, 186, 188–89; contribution to economy, 4;

effect on streamflow, 182–84; historical trends in, 260–64; for railroad ties, 21–22, 261; selective, 190

Timothy, 51

Togwotee pass, 48

Topography: map of, 24; effects of, 24–26, 30–31, 156

Tourism, 258–59

Towns and cities, map of, 7

Transpiration: from riparian zone, 56; vernal, 181. *See also* Evapotranspiration

Treeline, alpine, 201–4, 218

Trona, 4

Tundra. *See* Alpine tundra

Union Pacific Railroad, 21, 258

Uranium, 259

Urban-industrial landscapes, 258

Vegetation: change during Quaternary, 10–19, 216; map of, 8. *See names of specific vegetation types*

Volcanic eruptions, 10, 12, 13

Wagon trains, 20, 258, 277n.6

Water development, 258, 259–60

Waterfalls Canyon fire, 240–41, 282n.2

Weeds: riparian zone, 51. *See also* Cheatgrass; Saltcedar

Wheatgrass, bluebunch. *See* Bluebunch wheatgrass

White fir, 147, 281n.7

White spruce, 244, 247, 249

Whitebark pine: blister rust on, 224; in YNP, 218, 223, 224; habitat of, 159, 201; plants associated with, 161–62; seed

dispersal by birds, 150–51, 203; in GTNP, 236, 237;

Wilderness, 3, 6, 261

Wildlands, value of, 257, 258, 261

Willow, peachleaf, 50, 52

Willow shrublands: effect of browsing on, 61, 222–23, 224, 225, 248, 250; in foothills, 50; in GTNP, 236; in mountains, 48, 49; plants associated with, 45–46, 207; in YNP, 219

Wind Cave National Park: interaction between prairie dogs and bison, 81–82

Wind disturbances in forest, 31, 39, 156–57

Winter range, 25, 133, 150

Winterfat, 92, 102, 103, 110, 113, 119

Wolves, 17, 20, 223

Wood and forest soils, 184–85

Woodrats, middens of, 18

Wyoming: geologic history of, 10–18; history of human settlement, 257; land management in, 257, 264; population of, 257

Xanthoparmelia chlorochroa, 78

Yellowstone National Park (YNP): climate of, 216–17; early history of, 3, 215, 220; fire management in, 225–31; fires of *1988,* 165, 166, 168; geologic history of, 215–16; glaciation in, 216–17; lakes of, 216–17; landscape mosaic in, 217–20; Northern Range of, 217, 221, 228, 229; postglacial vegetation development, 216. *See also* Elk; Fire

Yellowstone plateau, 13–14, 215–32